NATO ASI Series

Advanced Science Institutes Series

A series presenting the results of activities sponsored by the NATO Science Committee, which aims at the dissemination of advanced scientific and technological knowledge, with a view to strengthening links between scientific communities.

The Series is published by an international board of publishers in conjunction with the NATO Scientific Affairs Division

A Life Sciences B Physics	Plenum Publishing Corporation London and New York
C Mathematical and Physical Sciences D Behavioural and Social Sciences E Applied Sciences	Kluwer Academic Publishers Dordrecht, Boston and London
F Computer and Systems Sciences G Ecological Sciences H Cell Biology I Global Environmental Change	Springer-Verlag Berlin Heidelberg New York London Paris Tokyo Hong Kong Barcelona Budapest

PARTNERSHIP SUB-SERIES

1. Disarmament Technologies	Kluwer Academic Publishers
2. Environment	Springer-Verlag/Kluwer Academic Publishers
3. High Technology	Kluwer Academic Publishers
4. Science and Technology Policy	Kluwer Academic Publishers
5. Computer Networking	Kluwer Academic Publishers

The Partnership Sub-Series incorporates activities undertaken in collaboration with NATO's Cooperation Partners, the countries of the CIS and Central and Eastern Europe, in Priority Areas of concern to those countries.

NATO-PCO DATABASE

The electronic index to the NATO ASI Series provides full bibliographical references (with keywords and/or abstracts) to about 50000 contributions from international scientists published in all sections of the NATO ASI Series. Access to the NATO-PCO DATABASE compiled by the NATO Publication Coordination Office is possible in two ways:

- via online FILE 128 (NATO-PCO DATABASE) hosted by ESRIN, Via Galileo Galilei, I-00044 Frascati, Italy.
- via CD-ROM "NATO Science & Technology Disk" with user-friendly retrieval software in English, French and German (© WTV GmbH and DATAWARE Technologies Inc. 1992).

The CD-ROM can be ordered through any member of the Board of Publishers or through NATO-PCO, Overijse, Belgium.

Series H: Cell Biology, Vol. 98

Springer
Berlin
Heidelberg
New York
Barcelona
Budapest
Hong Kong
London
Milan
Paris
Santa Clara
Singapore
Tokyo

Lactic Acid Bacteria:

Current Advances in Metabolism, Genetics and Applications

Edited by

T. Faruk Bozoğlu

Department of Food Engineering
Middle East Technical University
06531 Ankara, Turkey

Bibek Ray

Department of Animal Science
University of Wyoming
Laramie, WY 82070, USA

Springer

Published in cooperation with NATO Scientific Affairs Division

Proceedings of the NATO Advanced Study Institute on "Current Advances in Genetics, Metabolism and Application of Lactic Acid Bacteria", held in Kuşadası (İzmir), Turkey, September 10–22, 1995

Library of Congress Cataloging-in-Publication Data applied for

Die Deutsche Bibliothek - CIP-Einheitsaufnahme

Lactic acid bacteria : current advances in metabolism, genetics and applications ; [proceedings of the NATO Advanced Study Institute on "Current Advances in Genetics, Metabolism and Application of Lactic Acid Bacteria", held in Kuşadası (Izmir), Turkey, September 10 - 22, 1995] / ed. by T. Faruk Bozoğlu ; Bibek Ray. Publ. in cooperation with NATO Scientific Affairs Division. - Berlin ; Heidelberg ; New York ; Barcelona ; Budapest ; Hong Kong ; London ; Milan ; Paris ; Santa Clara ; Singapore ; Tokyo : Springer, 1996
(NATO ASI series : Ser. H, Cell biology ; Vol. 98)
ISBN-13: 978-3-642-64850-2 e-ISBN-13: 978-3-642-61462-0
DOI: 10.1007/978-3-642-61462-0
NE: Bozoğlu, T. Faruk [Hrsg.]; Advanced Study Institute on Current Advances in Genetics, Metabolism and Application of Lactic Acid Bacteria <1995, Kuşadası>; NATO: NATO ASI series / H

ISBN-13: 978-3-642-64850-2

Softcover reprint of the hardcover 1st edition 1996

Typesetting: Camera ready by authors/editors

SPIN 10477110 31/3137 - 5 4 3 2 1 0

PREFACE

This volume represents the proceedings of a NATO Advanced Study Institute on 'Current Advances in Genetics , Metabolism and Application of Lactic Acid Bacteria' which was held in Kuşadası (İzmir) Turkey, September 10 to September 22 1995.

The Institute emphasized the recent research and developments and industrial applications on Lactic Acid Bacteria. Techniques that were developed to engineer proteins, expression of biologically active proteins from food grade Gram-positive lactic acid bacteria were discussed. Development and the use of non-pathogenic lactic acid bacteria was presented as a growing interest for vaccine delivery systems for mucosal immunization. The role of lactic acid bacteria in various food fermentation their important metabolic activities, their use for carbohydrate modification for commercial applications, key systems for proteolysis and lantibiotic production have been introduced at the genetic and biochemical level. Transformation of organic wastes by use of lactic acid bacteria to transform them into nutritionally beneficial products of feed, fertilizers and food and the processes were emphasized. Key for the identification of food-relevant genera of lactic acid bacteria and factors affecting their survival and growth and potential lactic acid bacteria for food spoilage were discussed. β-casein breakdown products and peptides liberated by the proteinases, peptide transport systems and proteolytic pathways were presented. Beneficial antibacterial metabolites and specific cell components of various desirable intestinal indigenous bacterial species and some non-intestinal lactic acid bacteria were presented.

Mechanism of nisin-induced inhibition, induction of β-lactamese activity, kinetic representation for pure and mixed cultures of lactic acid bacteria and stabilization of important enzymes were also the important subjects in this meeting.

The editors believe this book will prove useful both to engineers and scientists of both beginners and to those with a long term interest in genetics, metabolism and kinetics of lactic acid bacteria.We would like to thank to NATO Scientific Affairs Division and all participants that contributed and made the ASI successful both in academic and social ways.

F. Bozoğlu
B.Ray

TABLE OF CONTENTS

Genetic Engineering Strategies

Pierre Renault,
Laboratoire de Génétique Microbienne,
INRA, Biotechnologie,
78352 Jouy en Josas Cedex
France

Lactic acid bacteria (LAB) are widely used in food technology, and these last years an increasing number of laboratories studies their genetics in order to better know their metabolic capacities and improve them. It should be stressed out that most studies on the genetics of LAB have been done on *Lactococci*, mesophilic bacteria used as starters of many cheese and fresh dairy products. Most examples of the strategies that will be presented here are coming from work performed on *Lactococcus lactis*, even if in the last two years, some of these technologies have been successfully transfered to other members of LAB. I must also state that this course will cover most of the strategies used in the field, but with a limited number of examples for each, so that this report does not cover exhaustively the genetics of LAB, but is simply a guide for a first approach to this field.

STRATEGIES FOR WHAT?

Genetic engineering is a concept for the use of a wide variety of tools, often built for different purposes, the common point of which is that they were conceived to modify the information encoded in the genome. Techniques have been developed to read this information and then use it to establish phylogeny, detect the presence, even at very low level, of bacterial populations, and screen for the presence of genes relevant in academic or applied research. Once detected, a gene can be isolated, a step which is often limiting for further studies. The isolation of a gene, or its cloning, might be time consuming although several alternative strategies can be used to achieve it. The characterization of the gene(s), like its primary sequence, with all the features controlling its expression or the activity of the encoded products is an important step toward its engineering. Lastly, the modified gene(s) can be

NATO ASI Series, Vol. H 98
Lactic Acid Bacteria:
Current Advances in Metabolism, Genetics and Applications
Edited by T. Faruk Bozoğlu and Bibek Ray

reintroduced into the bacterium or in another host of interest, and the effect of the modification studied.

Although all this work could be done entirely in LAB, it is very frequent that isolation, partial characterization and modification of their gene(s) are easier to perform in an heterologous host as *Escherichia coli* and sometimes *Bacillus subtilis.*

WHAT CAN BE DONE OUTSIDE LAB:

A number of techniques necessitate only a small sample of DNA or RNA of the bacterium. For example, detecting certain RNA species as 16S RNA requires two PCR primers the sequences of which are conserved among all bacteria. Subsequent sequencing of the PCR product or its use for ribotyping allow the identification of strains at the subspecies level (Kohler *et al.*, 1991; Salama *et al.*, 1991).

Most LAB genes characterized up to now have been detected and isolated in *E. coli* or *B. subtilis* by classical techniques. This approach has been widely used to clone genes encoding for amino acid or base biosynthesis by complementation of mutants which are available in both species (Renault *et al.*, 1995a; Chopin, 1993). Genes were also retrieved by plate assay for the presence of an activity producing a colored compound (Chapot-Chartier *et al.*, 1993). Heterologous gene expression has also been identified by immunodetection provided that the encoded proteins have been purified or cross-react with antibodies raised against the homologs from another bacterium (Ansanay *et al.*, 1993). When the N-terminal extremity of a protein is sequenced, degenerate oligonucleotides can be designed and used directly to screen a library or to amplify by PCR a fragment which would subsequently be used as probe (Monnet *et al.*, 1994). The increasing knowledge of a number of homologous genes in different bacterial species has boosted PCR based strategies using degenerate oligonucleotides deduced from protein motifs conserved among organisms (Duwat *et al.*, 1992).

WHAT SHOULD BE DONE INSIDE LAB

Cloning specific genes

Most of the genes able to confer a phenotype have been cloned in *E. coli* or *B. subtilis* as described above. However, certain genes of interest have been directly characterized in LAB. Most of them are naturally carried on plasmids, and were detected due to the loss of a given property concomitant with that of the plasmid

(Gasson & Davies, 1984). The genes for lactose utilization, that for the cell-envelope associated proteinase and that of the citrate permease of *L. lactis* were thus the first genes to be localized. The location of these genes on plasmids has greatly facilitated their subcloning in *L. lactis* and their further characterization (Kok *et al.*, 1985). The development of a high efficiency transformation system in *L. lactis* also allowed the direct cloning of chromosomal genes by complementation of mutants (Nardi *et al.*, 1991). This led to the detection of genes which can not be selected in a heterologous host as specific regulators (Renault *et al.*, 1989) or participating in a mechanism of defence against phages (Cluzel *et al.*, 1991). However, most LAB can not be transformed with sufficient efficiency and this approach is limited to few lactococcal genes.

Mutagenesis

The isolation of mutants affected in a biochemical pathway, and their characterization is a prerequisite for metabolic studies when the pathway is not well defined. Random mutagenesis by chemicals like nitrosoguanidine is an easy procedure and does not require precise knowledge of the genetics of the bacterium. However, the rescue of a gene by complementation requires an efficient transformation procedure. Transposition mutagenesis can overcome this problem since the mutation is labelled by the transposon, and the mutated gene can be recovered in a heterologous host by selecting for the antibiotic marker carried by the transposon.

TOOLS AVAILABLE FOR LAB

Introducing genes in LAB

As it was mentioned previously, the goal of genetic engineering is to reintroduce, once characterized, a modified gene in the appropriate strains. A number of different strategies based on transformation, conjugation or transduction have been devised for this purpose.

The first transformation procedure described for LAB relies on the treatment of protoplasts in the presence of PEG (Kondo & Mac Kay *et al.*, 1984). Although good efficiency was obtained for several LAB, this technique is not used anymore because it is time consuming and can not be easily applied to many species. Electroporation offers a good alternative, and requires only a moderately expensive equipment. Moreover the preparation of electrocompetant cell is very simple and rapid, the cells being just collected, washed and frozen before use. The prepared cell culture are melted, mixed with DNA and submitted to an intense electrical field. Transformed cells are then resuspended for a few hours in a non selective medium before plating on media with the proper selective agent.

Although very simple, number of factors have to be optimized to achieve high yield of transformants (Holo & Nes, *et al.*, 1989). The state of the culture can be critical. Weakening the cell-wall by high D-L threonine or glycine concentrations which interfere with peptidoglycan synthesis may significantly improve cell competence. This may require an osmotic protection by addition of saccharose. The composition of the washing and storage buffers can differ significantly in function of the bacterial species. The final cell density has also some influence on the results. The electric parameters and the shape of the cuve may be critical for transformation efficiency and cell survival. Lastly a proper medium of expression and plating should be used to increase cell recovery after the electric shock.

Almost all species of LAB can be transformed with reasonable yield after optimization of the procedures with the exception of some *Lactobacilli* species for which, procedures involving conjugative or mobilizable plasmids, conjugative transposons and transducing phages have been proposed (Gasson & Fitzgerald, 1994; Langella *et al.*, 1993; Mercenier *et al.*, 1994).

Genetic markers and replicons suitable for LAB

Introduction of genes in bacteria necessitate suitable vectors which can be of two types, replicative or integrative. Both types require good markers ensuring selection of the transformants. Unfortunately, the wide series of vectors developed in gram negative bacteria as *E. coli* and most of those of *B. subtilis* or *Staphylococcus aureus* are not working in LAB. However most "gram positive antibiotic markers" are expressed in LAB and allow the isolation of suitable replicons.

Genetic markers

Most of the markers available for LAB have been isolated in *B. subtilis* or in pathogenic Gram positive bacteria and confer resistance to antibiotics (de Vos & Simons, 1994). The erythromycin gene from pAMβ1 or pE194 are the most widely used. *ery*-β1 is expressed more strongly than *ery*-194 which allows a better selection of integrants in the chromosome in a single copy. The use of *ery*-194 leads often to tandem integration in the same conditions. The advantage of this latter marker is the possibility it offers to amplify a chromosomal integration by selecting bacteria for resistance to clindamycin, an antibiotic for which several copies of *ery*-194 are required. Other markers commonly used are genes for resistance to chloramphenicol (*cat*-194 from *Staphylococcus aureus*, *cat*-86 from *Bacillus pumilus*), tetracycline (*tetM* from Tn916 of *Enterococcus faecalis*, *tetL* from *S. aureus*), kanamycin (*aphA3* from *E. faecalis*, *knt*-110 from *Bacillus*), Spectinomycin (from *Streptococcus ferrus*).

However, the use of these markers is absolutely prohibited in bacteria that will be used in food processing, and alternative "food grade" markers have been designed. The *lacF* gene carried on a plasmid can complement a mutation introduced in the *lac* operon of the lactose plasmid (De Vos & Simons, 1994). A gene conferring resistance to the lantibiotic nisin isolated from a natural plasmid of *Lactococci* was also proposed (Froseth & Mac Kay, 1991). Vectors carrying non sense suppressors were also devised (Johansen *et al.*, 1995). The alternative for the use of a marker is to introduce the gene of interest into the chromosome by double cross over as described below (Biswas *et al.*, 1993).

Mode of replication

The second important feature of a vector is its ability to replicate. Two distinct mechanisms of replication have been described, the sigma and the theta type, regarding the shape of their intermediate of replication visualized by electronic microscopy (Janière *et al.*, 1993). The mode of replication of a plasmid is an important factor to take into account as it may influence the structural stability of the contructs.

Replication by sigma mechanism is initiated by the binding of a plasmid encoded protein (Rep) to the origin, the introduction of a nick in one strand and the fixation of Rep at the 5'-end of the nicked strand. A new leading strand is synthesized through the extension of the free 3'OH at the nick site which results in the displacement of the original leading strand. This replication intermediate has the shape of a sigma. Displacement of the leading strand continues until the origin is reached, the Rep protein cleaves the leading strand at this site and the displaced parental strand is ligated (termination). This yields a single stranded circle and a duplex plasmid. Replication is then completed by the conversion of the single strand circle into a double strand plasmid. Several errors might occur during the process leading to aberrant forms of DNA and structural instability. First , if termination does not occur, the replication goes on for a second round or more, giving rise to linear tandem molecules of duplex plasmid also called HMW (high molecular weight DNA). These molecules are frequently found when heterologous inserts are cloned into the plasmid (Biswas *et al.*, 1995). A second kind of mistake might occur due to the generation of a single strand replication intermediate which enhances certain recombination processes as copy choice and produces deletions of parts of the plasmid.

The theta mechanism does not generate a single strand intermediate. Several models have been proposed for the initiation of replication of these plasmids, leading to the formation of replication forks similar to those of the chromosome. As a result, theta type replication allows the stable maintenance of large plasmids in contrast to the sigma type (Bruand *et al.*, 1993). Indeed, almost all plasmids larger than 12 kb are replicating by theta mechanism.

LAB cloning vectors

Although using theta replicative plasmids offers the advantage of an increased stability, especially when large fragment have to be cloned, a number of useful vectors have been built from small cryptic plasmids isolated from different LAB like pWV01 or pSH71, which replicate with a sigma mode (De Vos & Simons, 1994). Low and high copy number derivatives with different markers are available and have subsequently been used to design vectors for special purposes. Lastly, a temperature sensitive rep protein of pWV01 has been isolated allowing the construction of plasmids the replication of which can be controlled by temperature (Maguin *et al.*, 1992).

In the last years, a number of replicons from large natural plasmids have been characterized like those from the "lactose" or the "citrate" plasmids (Seegers *et al.*, 1994). They offer the advantages conferred by their theta mechanism of replication, but have a rather narrow host range. pAMβ1, isolated from *E. faecalis* has been extensively studied and has been used to built the backbone of many vectors which replicate in almost all Gram positive bacteria (Bruand *et al.*, 1993). Recently, derivatives of pAMβ1 which can be switched from high to low copy number per cell were designed (Renault *et al.*, 1996). The basis of this switch is the insertion of a linker of 22 bp in a restriction site present in a gene encoding a repressor which controls the production of the Rep protein (Lechatelier *et al.*, 1994). The linker inactivates the repressor gene by introducing a frameshift early in its coding region. Once the linker is removed, the repressor gene is reactivated and the copy number decreases about 20 fold. This allows to make a construct on a high copy vector, and test it at a copy number close to chromosomal gene dosage.

Integrative vectors have also been developed for LAB. They offer several advantages over those replicating. Once introduced in the chromosome, they are structurally and segregationaly stable (Biswas *et al.*, 1993; Leenhouts *et al.*, 1991; Leenhouts, 1995). The construction is present at only one or few copies per chromosome, and variation in the copy number occurs rarely in function of the growth conditions. Finally the state of supercoiling and the bending of the DNA which are known to affect the expression of many genes might differ when genes are present in the chromosome or in plasmids.

Integration of a construct in the chromosome can be achieved by two major mechanisms, homologous recombination and site specific recombination. Homologous recombination is mediated by the cell machinery and involves a set of proteins including *RecA* which plays a central role. The integration requires also the presence in the vector of a DNA fragment homologous to a chromosomal target where the cross over will occur (Fig. 1). The frequency of integration will be the result of the efficiency of DNA entry into the cell, and of the length and the percentage of homology between the plasmid and the chromosomal homologous fragments. Any plasmid, with a proper marker and a chromosomal DNA fragment could be considered as an integration vector. A plasmid with thermosensitive replication fuctions can also be used as backbone for an integration vector, offering the advantage of overcoming the requirement of high transformation efficiency transformation which is the bottle neck of direct plasmid integration (Maguin *et al.*, 1992).

A procedure has been developed with thermosensitive vectors to stably integrate DNA into the chromosome (Fig. 1; Biswas *et al.*, 1993). Indeed, once the plasmid is integrated by homologous recombination, replication can be reactivated by growing the cells again at the permissive temperature. Stimulating replication enhances the frequency of recombination between the two homologous regions flanking the plasmid and leads to the inverse reaction than that occurred for its integration. When a mutation, deletion or an additional DNA fragment is introduced in the middle of the fragment directing the homologous recombination, the second cross over can allow an exchange of alleles between the plasmid and the chromosome. We should mentione that the exchange might occur theoretically at a frequency of 50%. Depending on the length of homology present at each side of the introduced change, this frequency may vary dramatically.

Vectors which integrate by site specific recombination are built with fragments isolated from phages, transposon or IS. These systems direct insertion at preferred

EXAMPLE OF HOMOLOGOUS RECOMBINATION: FOOD GRADE CHROMOSOMAL INSERTION BY DOUBLE CROSS-OVER.

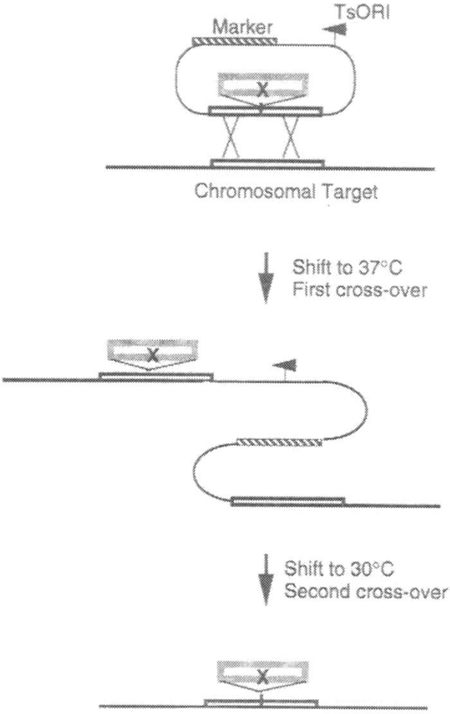

sites in the chromosome, for example tRNAser in the case of the *Lactobacillus delbrukii* phage pMC1 derivative (Dupont *et al.*, 1995), or almost randomly for ISS1 derivatives (Maguin *et al.*, 1996). Intermediate situations have been described in the case of Tn916 derivatives (Renault *et al.*, 1995b).

Vectors for special purposes

We described previously "cloning vectors" with the ability to carry DNA fragments and transform LAB. These vectors can also contain gene or signals which can be used for special purposes as screening for cloned inserts (Leloir *et al.*, 1994), or directing gene expression (de Vos & Simons, 1994).

Expression of genes allows the modification of the properties of bacteria and metabolic engineering. To achieve gene expression, proper signals of transcription and translation should be present upstream of the coding frames. Vectors containing a selectable marker without these signals can be used to isolate promotors and eventually RBS. These signals are the special features of expression vectors. The first generation of characterized promotors was isolated by screening for chloramphenicol resistant clones. These promotors were constitutive or their regulation was unknown, so that the expression they directed could not be controlled (van der Vossen *et al.*, 1987).

For this reason, promotors for regulated genes were searched by isolating the corresponding gene, or promotors which are expressed in different environmental conditions (Israelsen *et al.*, 1995). More sophisticated promotor probes were used like transposons or integrative plasmids containing *lacZ* fusions which allow screening by plating the clones in different media and selecting for the appearance of blue color (Israelsen *et al.*, 1993). Vectors directing gene expression in presence of lactose, nisin (see de Vos in this issue), high salt concentration, low pH, or in several other conditions are now available. A system based on the controlled expression of T7 polymerase expression have also been adapted from *E. coli* and allows very high expression level (Wells *et al.*, 1993).

Lastly, expression can be directed in time, but also in space. Vectors allowing the isolation of secretion signals have been constructed and many promotors followed by an open reading frame of an exported protein have been characterized. Heterologous proteins containing signals allowing their secretion and even their anchoring in the cell wall were produced in this way (de Vos & Simons, 1994).

Mutagenesis tools

Mutagenesis is a powerful method to detect genes conferring a phenotype. Gene inactivation by insertion, although restricted to non essential genes, is a convenient way to carry out genetic studies as it allows a rapid physical characterization of the targets. Transposons have been widely used for this purpose.

Tn916 is the best known example of broad host range conjugative transposon (Gawron & Clewell, 1984). It can be delivered in most LAB by conjugation from an *E. faecalis* donnor to a recipient which can be selected against the donor. Spontaneous streptomycin mutants are usually used for this purpose. The recipients containing the transposon are recovered at frequencies ranging from 10^{-4} to 10^{-7} by double selection on tetracycline (conferred by Tn916) and streptomycin. The chromosomal fragment carrying the transposon can be isolated with enzymes which do not cut in the transposon and then selected in *E. coli* by the resistance to tetracycline it confers. Lastly the transposon excises spontaneously from its target in *E. coli* to restore the original sequence. This system have been proved very useful in different species of *Streptococci* but it does not insert randomly in the chromosome of *L. lactis* (Renault *et al.*, 1995b).

Another kind of transposition system based on different IS elements have been developed for several LAB (Fig. 2, Maguin *et al.*, 1996). IS inserts quite randomly in the chromosome when introduced by transformation on a selectable plasmid. The bottle-neck of this approach is that an efficient transformation procedure is required. A variant of this system is to carry the IS on a vector the replication of which can be controlled by temperature. Once the plasmid with the IS is established in the cell, its replication can be inactivated. Clones containing the inserted IS on the chromosome will also retain the plasmid marker and are selectable.

Lastly, it has been shown that derivatives of Tn917 developed in *B. subtilis* can be used in *L. lactis*. This allowed to construct *lacZ* fusions in genes regulated by environmental conditions (Israelsen *et al.*, 1993 and 1995). Other insertion systems have been devised to produce mutants or *lacZ* fusions by homologous recombination. However, these systems are based on integrative vectors and require highly transformable strains (Leenhouts, 1995).

EXAMPLE OF SITE SPECIFIC RECOMBINATION: ISOLATION OF MUTANTS WITH IS*S1*/TS VECTOR, AND CLONING OF THE TARGET SEQUENCE

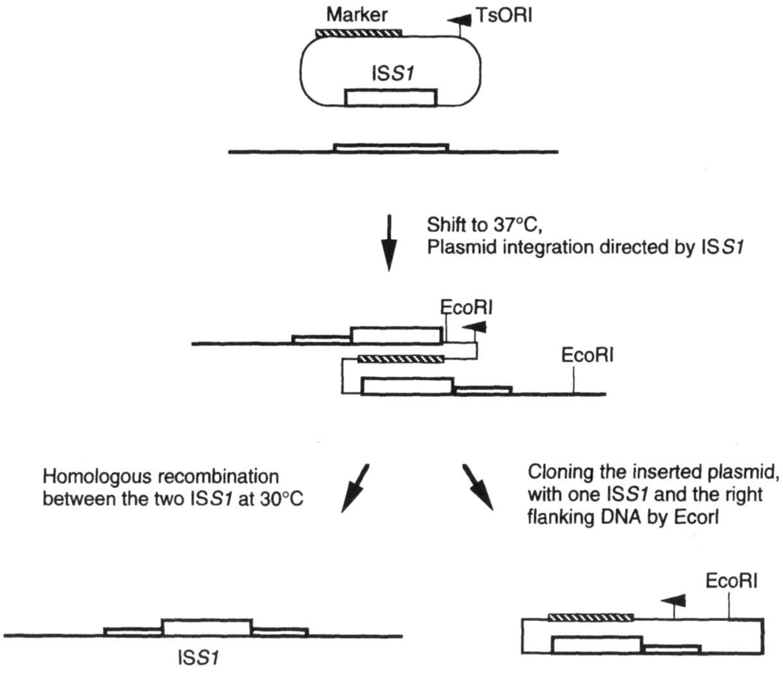

Concluding remarks

LAB are of major economic significance and may be of value in food technology and in promoting the health of humans. Improving strains with pertinent traits depends on efficient methods of genetic engineering. Taken together, the systems described in this chapter offer a large choice of tools which will allow further studies on the genetics of LAB. However, progress is still needed to facilitate the study of LAB other than *Lactococci* (Mercenier *et al.*, 1994). The knowledge that will be gained in the coming years has reasonable chances to produce potentially useful strains.

ACKNOWLEDGEMENTS

I Thank Costa Anagnostopoulos for critical reading of the manuscript

REFERENCES

Ansanay, V., S. Dequin, B. Blondin and P. Barre. 1993. Cloning, sequence and expression of the gene encoding the malolactic enzyme from Lactoccus lactis. FEBS 332:74-80.

Biswas, I., A. Gruss, S. D. Ehrlich and E. Maguin. 1993. High-Efficiency Gene Inactivation and Replacement System for Gram-Positive Bacteria. J. Bacteriol. 175:3628-3635.

Biswas, I., E. Maguin, S. D. Ehrlich and A. Gruss. 1995. A 7-base-pair sequence protects DNA from exonucleolytic degradation in Lactococcus lactis. PNAS USA 92:2244-2248.

Bruand, C., E. Lechatelier, S. D. Ehrlich and L. Janniere. 1993. A Fourth Class of Theta-Replicating Plasmids - The pAMbeta 1 Family from Gram-Positive Bacteria. PNAS USA 90:11668-11672.

Chapot-Chartier, M. P., M. Nardi, M. C. Chopin, A. Chopin and J. C. Gripon. 1993. Cloning and Sequencing of pepC, a Cysteine Aminopeptidase Gene from Lactococcus lactis subsp. cremoris AM2. Appl. Environm. Microbiol. 59:330-333.

Chopin, A. 1993. Organization and Regulation of Genes for Amino Acid Biosynthesis in Lactic Acid Bacteria. FEMS Microbiology Reviews 12:21-38.

Cluzel, P. J., A. Chopin, S. D. Ehrlich and M. C. Chopin. 1991. Phage abortive infection mechanism from lactococcus lactis subsp. lactis, expression of which is mediated by an Iso-ISS1 element. Appl. Environ. Microbiol. 57:3547-3551.

de Vos, W. M. and S. F. M. Simons. 1994. Gene cloning and expression systems in Lactococci. in Genetic and biotechnology of lactic acid bacteria, (Ed.) Gasson M.J. and de Vos W.M., Blackie Academic & Professional, London. pp. 52-105.

Dupont, L., B. Boizetbonhoure, M. Coddeville, F. Auvray and P. Ritzenthaler. 1995. Characterization of genetic elements required for site-specific integration of Lactobacillus delbrueckii subsp bulgaricus bacteriophage mv4 and construction of an integration-proficient vector for Lactobacillus plantarum. J. Bacteriol. 177:586-595.

Duwat, P., S. D. Ehrlich and A. Gruss. 1992. Use of Degenerate Primers for Polymerase Chain Reaction Cloning and Sequencing of the Lactococcus lactis subsp. lactis recA Gene. Appl. Environm. Microbiol. 58:2674-2678.

Froseth, B. R. and L. L. Mac Kay. 1991. Molecular Characterization of the Nisin Resistance Region of Lactococcus-Lactis Subsp Lactis Biovar Diacetylactis DRC3. Appl. Environm. Microbiol. 57:804-811.

Gasson, M. J. and F. L. Davis. 1984. The genetic of dairy lactic acid bacteria. In Advance in the Microbiology and Biochemistry of Cheese pp. 99-126.

Gasson, M. J. and G. F. Fitzgerald. 1994. Gene transfer systems and transposition. in Genetic and biotechnology of lactic acid bacteria, (Ed.) Gasson M.J. and de Vos W.M., Blackie Academic & Professional, London. :pp. 1-51.

Gawron-Burke, C. and D. B. Clewell. 1984. Regeneration of insertionally inactivated Streptococcal DNA fragments after excision of transposon Tn916 in Escherichia coli : Strategy for targeting and cloning of genes from Gram-positive bacteria. J. Bacteriol. 159:214-221.

Holo, H. and I. F. Nes. 1989. High-frequency transformation by electroporation of Lactococcus lactis subsp.cremoris grown with glycine in osmotically stabilized media. Appl. Environm. Microbiol. 55:3119-3123.

Israelsen, H. and E. B. Hansen. 1993. Insertion of Transposon Tn917 Derivatives into the Lactococcus lactis subsp. lactis Chromosome. Appl. Environm. Microbiol. 59:21-26.

Israelsen, H., S. M. Madsen, A. Vrang, E. B. Hansen and E. Johansen. 1995. Cloning and partial characterization of regulated promoters from Lactococcus lactis Tn917-lacZ integrants with the new promoter probe vector, pAK80. Appl. Environm. Microbiol. 61:2540-2547.

Janniere, L., A. Gruss and S. D. Ehrlich. 1993. Plasmids. in Bacillus subtilis and other gram-positive bacteria (Ed. Sonenshein A.L., Hoch J.A. and Losick R.) ASM, Washington, DC pp.625-644.

Johansen, E., F. Dickely, D. Nilson and E. B. Hansen. 1995. Nonsense suppression in Lactococcus lactis. Construction of food grade cloning vector. in Genetics of Streptococci, Enterococci and Lactococci. Ferretti JJ., Gilmore M.S. Klaenhammer T.R. and F. Brown (eds) Dev. Biol. Stand. Basel, Karger, SW 85:531-534.

Kohler, G., W. Ludwig and K. H. Scleifer. 1991. Differenciation of lactococci by rRNA gene restriction analysis. FEMS Microbiol. Letters 84:307-312.

Kok, J., J. Maarten, J. M. van der Vossen and G. Venema. 1985. Cloning and Expression of a Streptococcus cremoris Proteinase in Bacillus subtilis and Streptococcus lactis. Appl. Environm. Microbiol. 50:94-101.

Kondo, J. K. and L. L. McKay. 1984. Plasmid transformation of Streptococcus lactis protoplasts: optimisation and use in molecular cloning. Appl. Environm. Microbiol. 48:252-259.

Langella, P., Y. Leloir, S. D. Ehrlich and A. Gruss. 1993. Efficient Plasmid Mobilization by pIP501 in Lactococcus-lactis subsp. lactis. J. Bacteriol. 175(18):5806-5813.

Lechatelier, E., S. D. Ehrlich and L. Janniere. 1994. The pAMbeta1 CopF repressor regulates plasmid copy number by controlling transcription of the repE gene. Molecular Microbiology 14:463-471.

Leenhouts, K. J. 1995. Integration strategies and vectors. in Genetics of Streptococci, Enterococci and Lactococci. Ferretti JJ., Gilmore M.S. Klaenhammer T.R. and F. Brown (eds) Dev. Biol. Stand. Basel, Karger, SW 85:523-530.

Leenhouts, K. J., J. Kok and G. Venema. 1991. Lactococcal Plasmid pWVO1 As an Integration Vector for Lactococci. Appl. Environm. Microbiol. 57:2562-2567.

Leloir, Y., A. Gruss, S. D. Ehrlich and P. Langella. 1994. Direct screening of recombinants in gram-positive bacteria using the secreted staphylococcal nuclease as a reporter. Journal of Bacteriology 176(16):5135-5139.

Maguin, E., P. Duwat, T. Hege, S. D. Ehrlich and A. Gruss. 1992. New thermosensitive plasmid for gram-positive bacteria. J. Bacteriol. 17:5633-5638.

Maguin, E., H. Prevost, S. D. Ehrlich and A. Gruss. 1996. Efficient insertional mutagenesis in Lactococci and other Gram-posistive bacteria. J. Bacteriol. 178:in press.

Mercenier, A., P. H. Pouwels and B. M. Chassy. 1994. Genetic engineering of lactobacilli, leuconostocs and Sterptococcus thermophilus. in Genetic and biotechnology of lactic acid bacteria, (Ed.) Gasson M.J. and de Vos W.M., Blackie Academic & Professional, London. :pp. 252-294.

Monnet, V., M. Nardi, A. Chopin, M. C. Chopin and J. C. Gripon. 1994. Biochemical and genetic characterization of PepF, an oligopeptidase from Lactococcus lactis. Journal of Biological Chemistry 269(51):32070-32076.

Nardi, M., M. C. Chopin, A. Chopin, M. M. Cals and J. C. Gripon. 1991. Cloning and DNA Sequence Analysis of an X-Prolyl Dipeptidyl Aminopeptidase Gene from Lactococcus-Lactis Subsp Lactis NCDO-763. Applied and Environmental Microbiology 57(1):45-50.

Renault, P., G. Corthier, N. Goupil, C. Delorme and S. D. Ehrlich. 1996. A set of vectors for Gram positive bacteria which can be switched from high to low copy number. Submitted

Renault, P., C. Gaillardin and H. Heslot. 1989. Product of the Lactococcus lactis gene required for malolactic fermentation is homologous to a family of positive regulators. J. Bacteriol. 171:3108-3114.

Renault, P., J.-J. Godon, N. Goupil, C. Delorme, G. Corthier and S. D. Ehrlich. 1995a. Metabolic operons in Lactococci. in Genetics of Streptococci, Enterococci and Lactococci. Ferretti JJ., Gilmore M.S. Klaenhammer T.R. and F. Brown (eds) Dev. Biol. Stand. Basel, Karger, SW 85:431-441.

Renault, P., J. F. Nogrette, N. Galleron, J.-J. Godon and S. D. Ehrlich. 1995b. Specificity of insertion of Tn1545 transposon family in Lactococcus lactis. in Genetics of Streptococci, Enterococci and Lactococci. Ferretti JJ., Gilmore M.S. Klaenhammer T.R. and F. Brown (eds) Dev. Biol. Stand. Basel, Karger, SW 85:535-542.

Salama, M., W. Sandine and S. Giovannoni. 1991. Development and Application of Oligonucleotide Probes for Identification of Lactococcus-Lactis Subsp Cremoris. Appl. Environm. Microbiol. 57(5):1313-1318.

Seegers, J., S. Bron, C. Franke, G. Venema and R. Kiewiet. 1994. The majority of lactococcal plasmids carry a highly related replicon. Microbiology 140:1291-1300.

Van der Vossen, J. M., D. van der Lelie and G. Venema. 1987. Isolation and characterization of Streptococcus cremoris Wg2-specific promoters. Appl. Environm. Microbiol. 53:2452-2457.

Wells, J. M., P. W. Wilson, P. M. Norton, M. J. Gasson and R. W. F. Lepage. 1993. Lactococcus lactis: high-level expression of tetanus toxin fragment C and protection against lethal chalenge. Molecular Microbiology 8:1155-1162.

Progress in genetic research of lactic acid bacteria

Pierre Renault,
Laboratoire de Génétique Microbienne,
INRA, Biotechnologie,
78352 Jouy en Josas Cedex
France

In the last years, there has been a important increase in our knowledge of lactic acid bacteria. This progress came with the emergence of techniques allowing their manipulation although for long it has been restricted to *Lactococci*. Some key features of industrial processes have been characterized in great detail like the genes allowing the utilization of lactose and casein from milk. Knowledge on more general aspects of metabolism is now emerging and will allow the manipulation of LAB to improve important industrial traits and to use them as cell factories (Gasson, 1993).

TOOLS

In the previous course, we have described the general strategies that can be used to study LAB genomes. Techniques have been set up to detect, isolate, modify and reintroduce genes in most species of LAB. A number of vectors are available, tools like transposons based on derivatives of Tn917 or ISS1 allow genetic studies of *Lactococci*, and it is possible that these techniques could be applied to other LAB as well. Luciferase reporter genes may also allow detailed studies on gene regulation by in-line measurement during industrial processes or in the natural environment. The pGhost series of vector the replication of which is thermosensitive allow reverse genetics and stable gene integration in the chromosome in a food-compatible way.

PHYLOGENY

LAB were up to recently divided in three genera: *Streptococcus*, *Lactobacillus* and *Leuconostoc*. However, this classification was not satisfying because these genera

NATO ASI Series, Vol. H 98
Lactic Acid Bacteria:
Current Advances in Metabolism, Genetics and Applications
Edited by T. Faruk Bozoğlu and Bibek Ray

contained a number of heterogeneous species which were often transferred from a genus to another. The development of the nucleic acid sequencing methods, and more specially the catalogs of 16S RNA from LAB allowed the establishment of a new phylogeny. *Streptococcus* have been divided into three new genera, *Streptococcus*, *Lactococcus* and *Enterococcus*. *Lactobacilli* have not been separated yet in different genera, but phylogenetic trees based on 16S sequences suggest that it is composed of many very different classes which are branching deep in the tree. Lastly, new genera and species of LAB were identified.

The studies on the genetic variability among representatives of a species has also profited from the new techniques like pulse-field electrophoresis (PFGE) and random amplified polymorphic DNA (RAPD). PFGE allows to accurately differentiate strains (Le Bourgeois *et al.*, 1991), but has not been applied to large collections of bacteria. It allows also to show that major chromosomal rearrangements may occur quite frequently on repeated culture or under stress (see below). RAPD has been used as a more convenient technique to collections of more than 400 strains of one species isolated in different products at different times. Patterns of bands obtained are then digitalized, standardized and analysed by different statistical methods. This allows to show that strains of a given species in the collections can be grouped in several, but limited number of clusters of unequal size suggesting that the population is made up of different clones rather than a pool of homogeneous strains (Taillez *et al.*, 1996). The difference in the size of the clusters may reflect the preferred use of certain strains over others for starter cultures.

GENE AND GENOME ORGANIZATION

The genes characterized since the first chromosomal gene from *L. lactis* has been described in 1989 cover a few percent of the chromosome and allow to have a picture of the gene and genome organization. Genes involved in metabolic pathways are highly clustered, which contrasts with the more scattered organization observed in several other bacteria (Chopin, 1993). The presence of several genes not required for the biosynthesis within certain operons is intriguing (Renault *et al.*, 1995). Since these gene are cotranscribed with the other genes of the operon, their function might be linked to the same metabolism. Some of them have a regulatory function, and examples will be presented below.

Another interesting feature is the organization of the chromosome for which physical maps are now available (Le Bourgeois *et al.*, 1995). The size of the genomes of LAB turns around 2 to 3 megabases, a rather small size a compared to those *E. coli*, *B.*

subtilis Pseudomonas or *Myxococcus* which are ranging from 5 to 9 Mb. This reflects the limited biosynthetic and catabolic capacities of LAB as compared to these bacteria able to live in ecosystems where resources are scarce compared to milk, plant or digestive tract from where LAB are usually isolated. LAB genomes are still bigger than the chromosome of *Mycoplasma genitalium* which is only 0.58 Mb, containing only the genes corresponding to essential functions of this intracellular parasite (Fraser *et al.*, 1995). LAB isolated from nature are still able to synthetize most amino acids and vitamins, traits which tend to be lost in species growing exclusively in milk. Loss of metabolic activities have been well documented in *Lactobacilli* (Morishita *et al.*, 1981) and the events which led to gene inactivation characterized at the molecular level for histidine and branched-chained amino acids in *Lactococci* (Delorme *et al.*, 1993; Godon *et al.*, 1993).

The circular map of *L. lactis* shows that most the genes are transcribed in the same orientations than replication as in *B. subtilis* and several other bacteria. The comparison of the physical maps of several strains of lactococci revealed differences due to inversions of certain segments (Davidson*et al.*, 1995; Le Bourgeois *et al.*, 1995). These events occurred during fermentations in industrial strains, but also during cultures of laboratory strains. An instability of the genome was also described into *Streptococcus thermophilus* (Colmin *et al.*, 1991). Chromosomal genes can be amplified on plasmids, and plasmidic genes can be integrated in the chromosome. Moreover, it seems that genes might be exchanged between bacteria of different species and genera (Gasson & Fitzgerald, 1994; Guedon *et al.*, 1995). For example, the genes involved in rapid lactose utilization and casein breakdown in *Lactococci* are carried on conjugative plasmids and flanked by IS elements (van Roijen *et al.*, PhD thesis). It is likely that these genes might have been transferred on the plasmid by transposition events. More recently, it has been found that endopeptidase F gene, present as a single gene on the lactose-protease plasmid of *L. lactis*, is originated from a chromosomal cluster which contained three open reading frames (Nardi *et al.*, 1996).

In conclusion, like those of several bacterial species, the chromosome of LAB might be considered as a molecule in constant evolution, able to acquire new genes encoding for new metabolic pathways, and loosing functions which are not necessary in a rich environment.

GENE EXPRESSION

A key feature in the expression of metabolic traits is the way they are controlled. The regulation of several genes has been extensively studied and a variety of mechanisms have now been characterized. Indeed the basic mechanisms of transcriptional control found in LAB are similar to those previously characterized in other bacteria. Gene expression can also be controlled at the translational and enzymatic level.

Control of transcription

Presently, more than 50 lactococcal promoters have been characterized and their comparison shows that their -10 and -35 consensus sequences are similar to those recognized by the vegetative sigma factor in *E. coli* and *B. subtilis* (de Vos & Simons, 1994). About one half of them contain an additional conserved dinucleotide sequence one base upstream of the -10 region, a feature present rarely in *E. coli* but frequent in *Streptococci* (Sabelnikov *et al.*, 1995). The major sigma factor of *L. lactis* encoded by the *rpoD* gene has been characterized as a 38.8 kDA protein with high similarity to the vegetative sigma A factor of *B. subtilis* (Araya *et al.*, 1993). Other genes for sigma factors remain to be charaterized in *L. lactis* and LAB.

The first characterized gene encoding a regulator in LAB is *mleR*, an activator the presence of which is required to activate the transcription of the genes involved in malolactic fermentation (Renault *et al.*, 1989). MleR is homologous to the proteins of the LysR family, which bind to DNA and activate transcription in presence of a co-inducer, malate in this example. Activators belonging to the two component system family have also been found. NisK and NisR have been shown to induce the expression of the genes involved in nisin synthesis (Kuipers *et al.*, 1995).

Gene expression can be controlled negatively by repressors. The operon encoding for the lactose utilization system is transcribed divergently from *lacR* which product, represses its expression (van Roijen *et al.*, 1990). LacR belongs to the DeoR family of repressors. In absence of lactose, LacR was shown to bind to an operator overlapping the *lac* promotor whereas it was inactivated by the presence of a metabolite of the tagatose pathway. This operon is also subject to catabolic repression in presence of glucose. This response could be mediated by a protein homologous to the *B. subtilis* catabolite repressor CcpA, a consensus site for its binding being present in the immediate neighbourhood of the promotor (de Vos & Simons, 1994). Another repressor, RepC controls the expression of the protein promoting the replication of a plasmid. Lastly, operons encoding pathways for saccharose utilisation, histidine and branched-chained amino acid biosynthesis (Renault *et al.*, 1995), and a gene encoding the cell wall proteinase (Marugg *et al.*, 1995), seem also to be subject to repression.

Elongation of transcripts might be arrested early by a terminator, the activity of which can be modulated by the formation of alternative mRNA secondary structures. The choice between the formation of the terminator or the antiterminator structures depends on several factors which are the translation of a short peptide rich in leucine for the *leu-ilv* operon (Godon *et al.*, 1992), the binding of a phosphorylated protein for the betaglucosidase regulator gene *bglR* (Bardowski *et al.*, 1994) and the interaction of the cognate uncharged tRNA for the *trp* and *his* operons (Renault *et al.*, 1995; Chopin, 1993).

Control of translation

Translation is initiated by the ribosomes, the 16S rRNA of which recognizes a complementary sequence present upstream the initiation codon called ribosome biding site (RBS). In most cases the complementarity of RBS and 16S rRNA measured by the free energy is higher in *L. lactis* than in *E. coli*, explaining the poor translation of most gram negative bacteria genes in *Lactococci* and more generally in LAB. In LAB operons, many genes are overlapping, suggesting that translational coupling may occur and contribute to the control of gene expression. Such coupling has been recently evidenced by the polarity of a small deletion in a non essential gene present in the *his* operon of *L. lactis* (C. Delorme, personal communication) and by *lacZ* gene fusions downstream of the translated ORF (van de Guchte *et al.*, 1991). Secondary structures can also reduce the access to the RBS and lower the translation rate of some genes (Renault et al. 1995).

Low translation rate might be a barrier to the production of high amounts of heterologous proteins. This is often due to the presence of rare codons in the coding sequences. Some other factors may also affect translation as the bases flanking the codons which can interact with the tRNA anticodon loop. Modification of the coding strand might thus be required in order to optimize expression of heterologous genes. To realize the suitable changes, it is necessary to know the tRNAs present in the cell, their sequence and relative abundance.

46 tRNAs, with 26 different anticodons have been characterized by our group in *Lactococci*. Most of them are grouped in four clusters, three of them being linked to ribosomal operons (Figure 1). The promoters of the tRNA genes have some common features with those of two ribosomal operons, suggesting a coordinate regulation for tRNA genes. Interestingly, the last tRNA gene (Leu-UAA) of *trnE* might be controlled by attenuation of transcription in *L. lactis* subsp. *lactis* but not in *cremoris*.

Figure 1: Organisation of the tRNA genes in *Lactococci*.

* *rrnB-trnB* ... - 23S - 5S - **Val**$_{UAC}$ - **Tyr**$_{GUA}$ - **Trp**$_{CCA}$ - **His**$_{GUG}$ - **Gln**$_{UUG}$ - **Leu**$_{CAA}$ - Ter - Ter

* *rrnE-trnE* ... - 23S - 5S - **Val**$_{UAC}$ - **Asp**$_{GUC}$ - **Lys**$_{UUU}$ - **Leu**$_{UAG}$ - **Thr**$_{UGU}$ - **Gly**$_{GCC}$ - ter - **Leu**$_{UAA}$ - Ter - P32 -

* *rrnF-trnF* ... - 23S - 5S - **Val**$_{UAC}$ - **Asp**$_{GUC}$ - **Lys**$_{UUU}$ - **Leu**$_{UAG}$ - **Thr**$_{UGU}$ - **Gly**$_{GCC}$ -
-**Arg**$_{ACG}$ - **Pro**$_{UGG}$ - **Met**$_{AUG}$ - **Ile**$_{CAU}$ - **Phe**$_{GAA}$ - **Gly**$_{UCC}$ - **Ile**$_{GAU}$ - **Ser**$_{GCU}$ - Ter - P - *dnaE* - *rpoD*

* *trnA* --- P - **Glu**$_{UUC}$ - **Ser**$_{UGA}$ - (f)**Met**$_{AUG}$ - **Phe**$_{GAA}$ - **Gly**$_{UCC}$ - **Ile**$_{GAU}$ - 5S - **Asn**$_{GUU}$ - Ter---

* *rrnA-asn* --- P - 16S - **Ala**$_{UGC}$ - 23S - 5S - **Asn**$_{GUU}$ - Ter --

* *rrnB, C, D, E, F* --- P - 16S - **Ala**$_{UGC}$ - 23S - 5S ...

* --- P - **Ser**$_{CGA}$ - Ter -- * --- P - **Leu**$_{AAG}$ - Ter -- * --- P - **Arg**$_{ACG}$ - Ter?

* --- P - **Gln**$_{UUG}$ - Ter -- * --- P - **Leu**$_{AAG}$ ...?

* additionnal tRNAs detected by hybridisation: * ? - **Glu**$_{UUC}$ - ? * ? - **Ser**$_{UGA}$ - ?

trna (Nilson & Johansen, 1994), Arg$_{ACG}$ and Gln$_{UUG}$ (Dickely *et al.*, 1995) *rrnA* (Chiarutini *et al.*, 1993), Ala$_{UGC}$ (Davidson *et al.*, 1995)

Although some tRNAs remain to be identified, the absence of certain tRNA species might explain some features of the codon bias in *L. lactis*. For example, *L. lactis* may not have tRNA alaGGC, leuCAG, valCAC, serCGA, proGGG, thrGGU like *Mycoplasma* (Muto *et al.*, 1990) and *B. subtilis* for some of them (Green & Vold, 1993). This would imply that the UNN anticodons of the unique isoacceptor for these species translate synonymous codons through four-way wobbling. tRNAleuAAG has not been found yet in other bacteria. Further studies are required to learn whether the first anticodon base A is modified to Inosine and recognizes three synonymous codons (as tRNAargACG) or is not modified (as tRNAthrAGU from *Mycoplasma*) and allows translation of the CUU codon only, a codon very frequently used by *L. lactis*.

Change of the rare CTG codon to the frequent TTG codon at position 10 and 46 of the CAT gene improves the translation of this gene 24 and 1.4 fold respectively (Bojovic et al., 1994). This result suggests that the translational efficiency of the CTG codons varies according to secondary interactions of the tRNA anticodon loop with bases surrounding the codon as presented in figure 2.

Figure 2: Codon-anticodon interaction between the two CTG codons of the *cat-86* gene and the *L. lactis* cognate tRNA.

```
                             (      )                 (      )
                              \    /                   \    /
tRNA                          GGAUU                    GGAUU
                               .||.                    |||.|
codons           pos. 10     -UCUGC->       pos. 46   -CCUGA->

                          Poor interaction          Good interaction
```

METABOLISM

Sugar and pyruvate catabolism

A key metabolic point of LAB is their glycolytic pathway leading to pyruvate which is then catabolized to lactate, ethanol, acetate, formate, acetoin and several other products involved in the development of the flavor in fermented products. The genes involved in lactose utilization by the PTS pathway were identified and characterized at the molecular level in *Lactococci*. These genes clustered in one operon allow the uptake of lactose with its concomitant phosphorylation to lactose 6-phosphate, the breakdown of this molecule to glucose and galactose 6-phosphate and finally, the transformation of galactose 6-phosphate to the intermediate of glycolysis glyceraldehyde 3-phosphate. Three genes encoding key enzymes of the glycolytic pathways are known. Two of them are transcribed with the gene encoding lactate dehydrogenase (Llanos *et al.*, 1993). The genes involved in pyruvate transformation to acetolactate (AL) and acetoin are also known. Interestingly, in *L. lactis* the gene encoding the acetolactate decarboxylase (*aldB*) and catabolic acetolactate synthase (*alsS*) are not clustered like in *B. subtilis* and *Klebsiella terrigena* (Marugg *et al.*, 1994; Renault *et al.*, 1995). *aldB* is the last but one gene of the anabolic operon for branch-chained amino acids and may play a dual role in the regulation of the AL pool generated during pyruvate catabolism or leucine and valine biosynthesis (Goupil *et al.*, 1996b; Renault *et al.*, 1995).

The Acetoin pathway was extensively studied in LAB, because one of its by-product, diacetyl is a flavour compound essential in many dairy products such as butter, cream and certain cheeses. Its production was shown to depend on the strains used as starters and the conditions of fermentation like pH, oxygen or temperature (Hugenholz, 1993). Although changing the conditions of fermentation can lead to a

Figure 3

Sugar and pyruvate metabolism in *L. lactis*

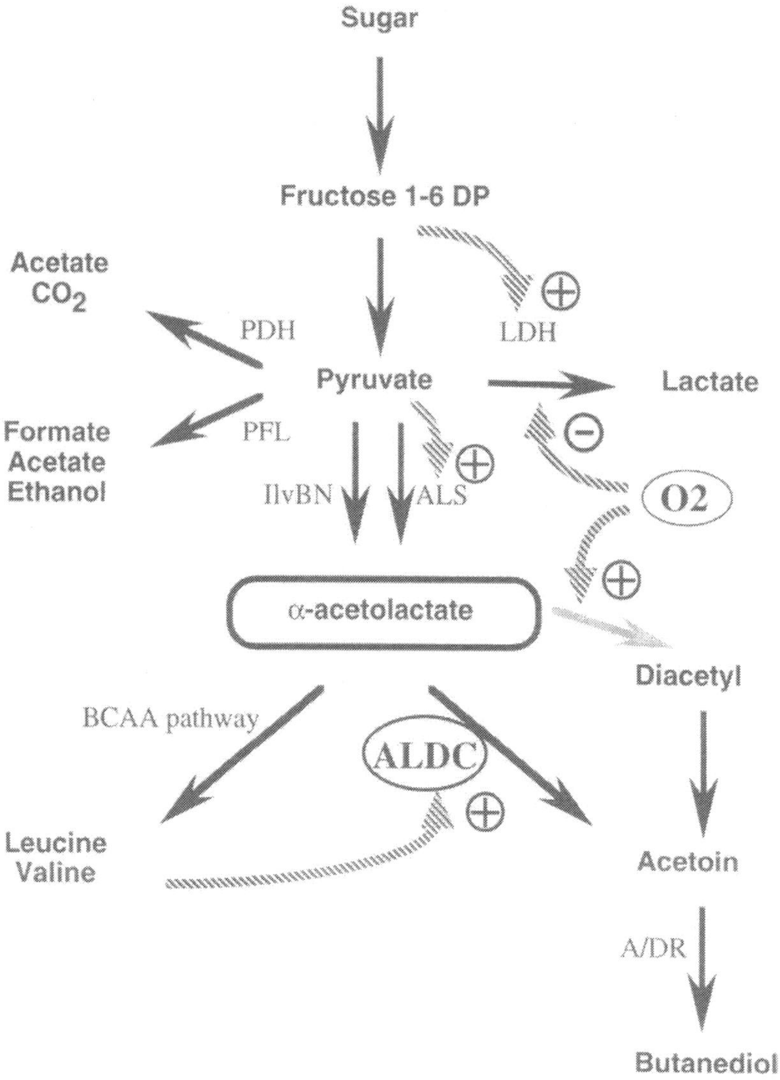

very significant increase in diacetyl concentration, it is sometimes impossible to apply the optimal conditions in real factory processes. For example, high oxygenation which might lead to a considerable increase in diacetyl production might also totally inhibit the growth of oxygen sensitive strains present in starters. Shaking of the milk during the process which allows high oxygenation also affects the texture of the product. Obtaining diacetyl overproducing strains is thus desirable, to alleviate some of the constraints inherent in factory process.

Diacetyl is formed from the chemical oxidative decarboxylation of AL. During citrate utilization, AL is the result of the activity of AL synthase on pyruvate. The key enzyme, AL synthase, which condenses two molecules of pyruvate to give AL and CO_2, has been purified (Snoep *et al.*, 1992) and its gene characterized (Marugg *et al.*, 1994). The allosteric properties of this enzyme suggest that when the intracellular concentration of pyruvate is high, part of it is converted to AL instead of L-lactate (Snoep *et al.*, 1992). AL is then decarboxylated to acetoin by the AL decarboxylase (ALDC) so that only a small amount of diacetyl is produced. The low amount of AL in the medium seems to be the bottle-neck for high diacetyl production. To increase this amount, AL production should be enhanced, or alternatively, its degradation rate to acetoin reduced (Hugenholz, 1993).

Several pairs of strains consisting of a wild type *L. lactis* subsp. *lactis* biovar diacetylactis and its isogenic constructed mutant for ALDC activity have been studied (Goupil *et al.*, 1996a). The wild type strains accumulate acetoin in the medium whereas the mutants produce AL instead of acetoin during exponential growth. Consequently, the ALDC defect allows a 10 fold increase in the production of diacetyl which confirms that ALDC activity has a negative effect on diacetyl synthesis. This approach based on the elimination of an enzymatic step appears to be more efficient than stimulating AL production. Indeed, overproduction of AL synthase, encoded by *alsS*, increases only 7.3% the amount of pyruvate diverted into the AL pathway, probably because this enzyme requires a high concentration of pyruvate to be active (Snoep *et al.*, 1992; Platteeuw *et al.*, 1995). Moderate expression of the *ilvBN* genes encoding the anabolic AL synthase increases the carbon flux to this pathway by 50% (Benson *et al.*, 1995; Swindell *et al.*, 1996). This enzyme has indeed an higher affinity for pyruvate than the catabolic one. Mutation of the *ldh* gene leads to a drastic change in the fermentation products, and to the transformation of a significant amount of pyruvate in AL derivatives. However, in these cases, AL is rapidly converted to acetoin and butanediol and the yield of diacetyl production is very low (Platteeuw *et al.*, 1995). Moreover, *ldh* defective mutants produce a significant amount of formate, acetate and ethanol instead of lactate, which leads to a complete change in the growth and acidification curves.

Proteolysis

Studies on proteolysis have received particular attention in LAB because it limits LAB growth in milk. Caseins should be cleaved in oligopeptides and especially in di and tri peptides for an optimal growth (Juillard *et al.*, 1995a). Furthermore, peptides and amino acid derivatives may play a important role in the development of cheese flavour during ripening.

Early studies showed that the cell-wall proteinase was responsible for the first step of casein hydrolysis (Kok & de Vos, 1994). The peptide are then internalized by several permeation systems, two of which have been characterized as a di and tri peptide permease and an oligopeptides permease (Juillard *et al.*, 1995b). Peptides are then cleaved to amino acids in the cells by a series of peptidases exhibiting different specificities listed in Table 1. The role of these peptidases in nitrogen supply is now studied of mutants expressing different level of each peptidase and/or combinations of these mutations (Mierau *et al.*, 1995). A specific course by J. Kok is covering this topic in this issue.

Table 1: Characterized peptidases in *L. lactis*.

Peptidases	**Specificity**	**Sequenced**	**Peptidases**	**Specificity**	**Sequenced**
Aminopeptidases			*Proline specific peptidases*		
Aminopeptidase N	X I Y-Z....	yes	X-PDP	X-Pro I Z	yes
Aminopeptidase C	X I Y-Z....	yes	Prolidase	X I Pro	
Aminopeptidase A	asp(Glu) I Y-Z....		Iminipeptidase	Pro I Y-Z	
			Aminopeptidase P	X I Pro-Z....	
Di-peptidases					
			Endopeptidases		
Dipeptidase 1	X I Y				
Dipeptidase 2			LEPI (PepF)	....W-X I Y-Z....	yes
Dipeptidase 3			LEPII (PepO)	....W-X I Y-Z....	yes
Tripeptidase 1	X I Y-Z	yes			

Stress response

In the environment, cells are subjected to a number of stresses. During these conditions which do not allow an optimal growth, the bacteria exhibit a response that manifests itself by the production of new proteins. The best studied response is the one induced by an increase of temperature above the "normal". Indeed, a wide variety of stresses have been evidenced to induce a response like oxidation, nutrient starvation, low temperature, alcohol, radiations.... The set of proteins induced is adapted to the change of the environmental conditions the bacteria have to face. However, several proteins are present in most cases, like GroEL which appears to serve as a chaperone.

The major heat shock genes, *groEL* and *dnaK* have been cloned using a PCR-based approach (Eaton *et al.*, 1993). The *groEL* is part of an operon which also contains a gene homologous to *groES* (Batt, 1995). By a similar approach, another operon containing four open reading frames *orf1-grpE-dnaK-orf4* have been characterized. GrpE and DnaK exhibit homology with the same proteins of other bacteria, the *orf1* product is a protein of 39 kDa found in gram positive bacteria, and the *orf4* product does not share strong homology with any known protein. Finally, *dnaJ* which is usually present in the *dnaK* operon is on a separate unit of transcription elsewhere in the chromosome (van Asseldonk *et al.*, 1993).

Interestingly, a DNA conserved motif in many gram positive bacteria is found between the -10 and the start of the first open reading frame. This motif, termed CIRCE for controlling inverted repeats of chaperone expression, has been shown to have a regulatory role for heat shock operons. Recently, evidence have been presented that Orf39 of *B. subtilis* (homologous to Orf1 of *L. lactis*) encodes a protein which is a negative regulatory factor for the expression of the *groE* and *dnaK* operons (Yuan & Wong, 1995). It is believed that it might function as a CIRCE-specific repressor. In *L. lactis*, GroEL seems to be induced 100-1000 fold 15 minutes after a thermal shift from 30 to 42°C, whereas DnaK and DnaJ are induced only 3 fold.

Recently, a surprising effect of *recA* null mutations was described (Duwat *et al.*, 1995). In addition to its role in DNA metabolism, *recA* was shown to be also involved in response to oxygen and heat stresses. The role of RecA in oxidative stress would be to repair DNA damage resulting from the inefficient elimination of oxygen radicals in *Lactococci*. Moreover, a *recA* strain does not grow at high temperature as a result of a deficiency in the expression ofheat-shock proteins (DnaK, GroEL and GrpE). In contrast, the novel heat shock protein HflB which down-regulates the heat shock response in *E. coli* has an elevated level of expression. This suggests that in *L. lactis*, these two major pathways of stress response are coregulated by the means of RecA. Other examples of cross-protection have also been shown to occur under carbon-starvation which enhances resistance to heat, ethanol, acid, osmotic, and oxidative stresses (Hartke *et al.*, 1994).

Several other genes involved in stress response have been characterized like the superoxide dismutase gene (Sanders *et al.*, 1995) and a two-component regulatory system responding to SOS signals (Huang *et al.*, 1995).

INTERACTIONS LAB-ENVIRONMENT

Phages

Infection by bacteriophages of starter cultures is one of the major problems that might occur during lactic fermentation. It has a considerable economic impact since it may lead to complete loss of vat and the arrest of a factory for days. Numerous research projects have been carried out to find strategies minimizing the effect of bacteriophages on fermentation, processes like culture rotations with strains having different sensitivities to phages and selection of phage resistant strains.

Genomic organization of LAB phages

In the future, rational approaches to devise a phage resistance system based on naturally existing or novel mechanisms could be considered, provided that we have a good insight in phage physiology and genetics. This information may be used for other purposes, like the use of the lysin and holin genes which liberate the cell content and accelerate cheese ripening. Tightly regulated promotors can provide new expression systems (Nauta *et al.*, 1995). Integrative vectors have been obtained with the integration system of lysogenic phages (Dupont *et al.*, 1995). Suicide vectors can also be conceived.

Several phage genome sequencing projects have been undertaken and the complete nucleotide sequence of certain phages is now available (Schouler *et al.*, 1994; Schouler *et al.*, 1996). The function of some open reading frames has been established. The genes encoding integration-excision systems, DNA metabolism and replication, cell death and lysis, phage structural proteins and regulators have been identified. Even if many open reading frame have still unknown functions, studies on the temporal expression of phage genes provide first hand information on their potential role. Phages containing non-sense mutations corresponding to known suppressor tRNA in unassigned open reading frames can be considered. Moreover, comprehensive studies on the interaction of phage defence systems interfering with different steps of the phage life cycle seem to be a promising approach to understand phage genetics.

LAB phage genomes have a modular organization as is the case for phages from many bacteria (Fig. 4). Each domain is involved in a step of the life cycle, and it is thought that phages may exchange these domains generating by this mean new virulent phages. Certain studies provided information on how the phages could evolve after acquisition of new functions by recombination. Phage phi50 acquired a methylase gene from a plasmid restriction-modification system to overcome the host defence mechanism (Hill *et al.*, 1991b). Another phage, ul36, acquired a chromosomal fragment of its host allowing to overcome an abortive infection mechanism (Moineau *et al.*, 1994). Sequence data from virulent phages and prophages will allow to understand the modular organization and plasticity of the phage genomes.

phage resistance mechanisms

Molecular and genetic studies have given insight into phage resistance mechanisms and opened the way to construct strains with improved defence capacities against phages. It is worthy to note that less than one percent of the isolated strains have both phage resistance and proper industrial properties. Spontaneous phage resistant mutants that might be obtained easily by selection often loose some of their pertinent industrial traits, and do not resist to other phages that appear during their subsequent industrial use.

Some lactococcal strains are naturally able to resist at various degrees to phages, and they have been shown to possess phage resistance mechanisms which may act on a broad range of phages and at different steps of their cycle. These mechanisms might also be combined together on plasmids to confer a very high degree of resistance (Hill *et al.*, 1993; Klaenhammer & Fitzgezrald, 1994).

Up to date, mechanisms which interfere with phage adsoption, attack DNA upon injection (restriction-modification systems), and do not allow phage development have been selected (Klaenhammer & Fitzgerald, 1994). The first two mechanisms either are not very efficient by themselves or may be rapidly overcome by phage methylation and cannot probably constitute the backbone for an efficient constructed phage defence system. The third, also called phage abortive infection mechanism (Abi) might be more promising and considerable research has been undertaken to

Figure 4
MODULAR GENOMIC ORGANIZATION OF PHAGES

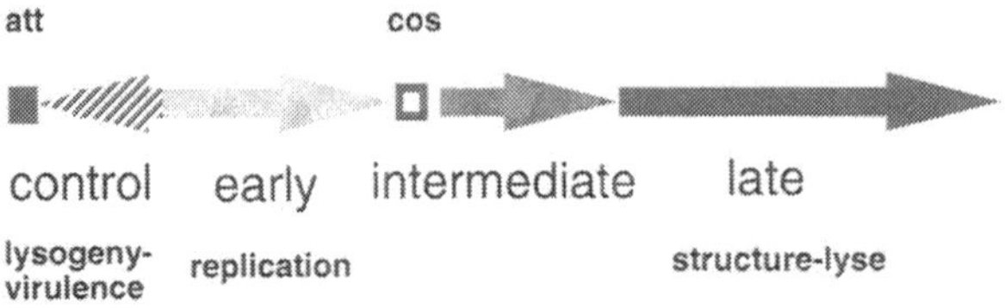

Figure 5
EXAMPLE OF PHAGE RESISTANCE MECHANISMS

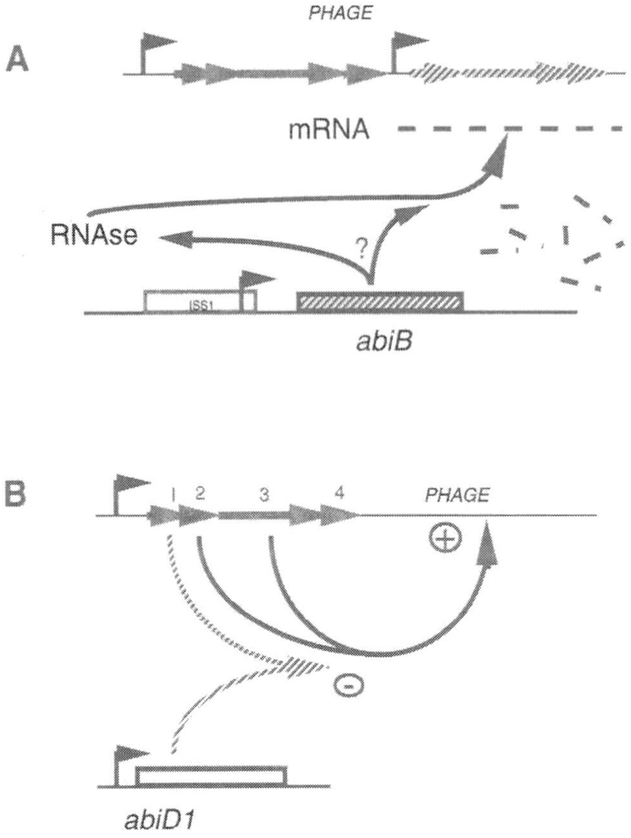

characterize it. In the presence of an Abi mechanism, the efficiency of phage adsoption remains the same but number of phage particles released by the cell has considerably decreased which allows to minimize their number in a culture.

Among the characterized genes encoding an Abi mechanism, at least four different types exist based on the absence of similarity among them. For simplicity, these mechanism were named *abiX*, X being a letter attributed in function of the date of characterization. The effect of several Abi mechanisms was investigated. In presence of *abiA*, no phage DNA replication was observed although an intact copy of the injected phage could be detected in the cell (Hill *et al.*, 1991a). *abiC* seems to interfere with the expression of phage proteins but it is not known if this occurs at the level of transcription, translation or the packaging process (Durmaz *et al.*, 1992; Geis *et al.*, . *abiC* , 1992). The *abiB* determinant reduces the yield of phage particles 100 to 1000 fold by preventing phage growth. It was shown to prevent phage development by promoting a dramatic decay of phage transcripts which appear normally 10-15 minutes after infection. It was proposed that in AbiB+ cells, the sensitive phages induce the synthesis or stimulate an RNase activity (Fig. 5; Pareira *et al.*, 1996). AbiD1 is a mechanism active on both prolate and small isometric headed phages (Anba *et al.*, 1995). The AbiD1 phage target has been characterized in an operon called M-operon composed of four open reading frames (Bidnenko *et al.*, 1996). Mutation and deletion analysis indicate that the first and the third open reading frames are involved in the AbiD1 mechanism. Analysis of this operon in *E. coli* suggested it encodes an endonucleolytic activity required for a normal phage life cycle. AbiD1 would decrease the amount of endonuclease below the level required for phage development (Fig. 5). These results confirm that Abi systems acting on different targets are available and the combination of them in a single strain might provide higher levels of phage resistance.

Indeed some naturally occurring plasmids are already carrying several mechanisms for phage resistance. The introduction of such plasmid allows to render industrial strains resistant to phages during long periods of time, but phages that can overcome these mechanisms appeared spontaneously in industrial conditions (Hill *et al.*, 1991b; Klaenhammer & Fitzgerald, 1994). The construction of a set of "super resistant" strains will require a further comprehensive study of phage resistance mechanisms in order to combine them in a rational way and optimize their use. Lastly, improved knowledge of phage genetics could provide new methods to prevent phage infection, for example by the expression of phage products able to interefere with the phage life cycle, or the production of antisense mRNA inhibiting the expression of essential genes (Chung *et al.*, 1992).

Bacteriocins

LAB produce many different molecules with antimicrobial properties, and some of them might have a potential for food preservation. In recent years, a number of these molecule have been purified and many of them were shown to be proteins of small size (Dodd & Gasson, 1994). Cloning the genes encoding these antimicrobial proteins was often succesfully done by the design of probes deduced from the N-terminal amino acid sequence. The genes for biosynthesis of bacteriocins are often organized in operons, the first gene of which is usually encoding the structural protein (Klaenhammer, 1993). The structural protein may be processed. Lantibiotic nisin is one of the best studied examples of such a system (de Vos *et al.*, 1995).

The spectrum of action of bacteriocins is also a key point for their use as food preservative. Their resistance to acid pH, proteolitic clivage and several other enviromnemtal conditions will determine the persistance of their action upon production.

The mode of action of antimicrobial molecules has also been investigated and its comprehensive study will allow the design of new molecule with improved industrial properties. A special course on this topic is presented by W. Konings in this issue.

Human health

A number of studies addressing the possible health benefits associated with the consumption of LAB have been carried out. LAB are believed to stimulate the immune system, lower the risk for cancer, inhibit potential pathogens and limit the blood cholesterol level. However no convincing prove of such effects has so far been provided yet (Marteau & Rambaud, 1993). Most LAB are not able to colonize the human digestive tract. The fact is that several enzymatic activities of interest and expressed by LAB, including lactose assimilation and conjugated bile salt hydrolases.

The most promising aspect in the health field might come from the use of LAB as cell factories for usefull molecules. They might be used as delivery vectors for new antimicrobials, antigenic proteins (Wells *et al.*, 1993), molecules stimulating the immune system (Steidler *et al.*, 1995), and enzymes.

CONCLUSION

Understanding LAB genetics and optimizing the techniques to modify LAB may allow, in the near future to improve the properties of these bacteria. However, it is hard to imagine that a single strain will exhibit all together optimum technological traits like their phage resistance, the texture, the aroma, the flavor, the nutritional quality and conservation of food and their "probiotics" properties. Several lactococcal strains modified for better phage resistance, or producing increased flavor, modify texture, or express molecules potentially benefical for health are already been tested for applications but more studies are still required for a broader application, to LAB other than *Lactococci* (Mercenier *et al.*, 1994).

AKNOWLEDGEMENTS

I am indebted to Christine Delorme, Nathalie Goupil, Marie-Christine Chopin, Catherine Schouler, Jamila Anba, Ricardo Pareira, Elena Bidnenko, Emanuelle Maguin, Philippe Langella, Yves Leloir, Patrick Taillez, Patrick Duwat and Michael Gasson for communicating results prior to publication. I thank Costa Anagnostopoulos for critical reading of the manuscript.

REFERENCES

Anba, J., E. Bidnenko, A. Hillier, D. Ehrlich and M. C. Chopin. 1995. Characterization of the lactococcal *abiD1* gene coding for phage abortive infection. J. Bacteriol. 177:3818-3823.

Araya, T., N. Ishinashi, S. Shimamura, K. Tnaka and H. Takahashi. 1993. Genetic and molecular analysis of the *rpoD* gene from *Lactococcus lactis*. Biosc. Biotech. Biochem. 57:88-92.

Bardowski, J., S. D. Ehrlich and A. Chopin. September 1994. BglR protein, which belongs to the BglG family of transcriptional antiterminators, is involved in beta-glucoside utilization in *Lactococcus lactis*. Journal of Bacteriology 176:5681-5685.

Batt, C. A. 1995. Stress response in *Lactococcus lactis*. *in* Genetics of *Streptococci, Enterococci* and *Lactococci*. Ferretti JJ., Gilmore M.S. Klaenhammer T.R. and F. Brown (eds) Dev. Biol. Stand. Basel, Karger, SW 85:449-454.

Benson, K. H., J.-J. Godon, P. Renault, H. G. Griffin and M. J. Gasson. 1995. Effect of *ilvBN*-encoded alpha-acetolactate synthase expression on diacetyl production in Lactococcus lactis. Appl. Microbiol. Biotechnol. *in Press.*

Bidnenko, E., D. Ehrlich and M. C. Chopin. 1995. Phage operon involved in sensitivity to the *Lactococcus lactis* abortive infection mechanism AbiD1. J. Bacteriol. 177:3824-3829.

Bojovic, B., G. Djordjevic, A. Banina and L. Topisirovic. 1994. Mutational analysis of *cat-86* gene expression controlled by lactococcal promoters in *Lactococcus lactis* subsp. *lactis* and *Escherichia coli*. J. Bacteriol. 176:6754-6758.

Chiaruttini, C. and M. Milet. 1993. Gene Organization, Primary Structure and RNA Processing Analysis of a Ribosomal RNA Operon in *Lactococcus lactis*. J. Mol. Biol.230:57-76.

Chopin, A. September 1993. Organization and Regulation of Genes for Amino Acid Biosynthesis in Lactic Acid Bacteria. FEMS Microbiology Reviews 12:21-38.

Chung, D. K., S. K. Chung and C. A. Batt. 1992. Antisense RNA directed against the major capsid protein of Lactococcus-lactis Ssp cremoris bacteriophage-F4-1 confers partial resistance to the host. Appl. Environm. Microbiol. 37:79-83.

Colmin, C., M. Pebay, J. M. Simonet and B. Decaris. 1991. A Species-Specific DNA Probe Obtained from *Streptococcus salivarius* Subsp *thermophilus* Detects Strain Restriction Polymorphism. FEMS Microbiology Letters 81:123-128.

Davidson, B. E., N. Kordias, N. Baseggio, A. Lim, M. Dobos and A. J. Hillier. 1995. Genomic organization of *Lactococci.* in Genetics of *Streptococci, Enterococci* and *Lactococci.* Ferretti JJ., Gilmore M.S. Klaenhammer T.R. and F. Brown (eds) Dev. Biol. Stand. Basel, Karger, SW 85:411-422.

de Vos, W. M., O. P. Kuipers, J. R. van der Meer and R. J. Siezen. 1995. Maturation of nisin and other lantibiotics: post-translationally modified antimicrobial peptides exported by Gram-positive bacteria. Molecular Microbiology 17:427-437.

de Vos, W. M. and S. F. M. Simons. 1994. Gene cloning and expression systems in *Lactococci.* in Genetic and biotechnology of lactic acid bacteria, (Ed.) Gasson M.J. and de Vos W.M., Blackie Academic & Professional, London. :pp. 52-105.

Delorme, C., J.-J. Ehrlich, S.D. Godon and P. Renault. 1993. Gene inactivation in *Lactococcus lactis*: Histidine biosynthesis. J. Bacteriol. 175(14):4391-4399.

Dickely, F., D. Nilsson, E. B. Hansen and E. Johansen. 1995. Isolation of *Lactococcus lactis* nonsense suppressors and construction of a food-grade cloning vector. Molecular Microbiology 15:839-847.

Dodd, H. M. and M. J. Gasson. 1994. Bacteriocins of lactic acid bacteria. in Genetic and biotechnology of lactic acid bacteria, (Ed.) Gasson M.J. and de Vos W.M., Blackie Academic & Professional, London. :pp. 211-251.

Dupont, L., B. Boizetbonhoure, M. Coddeville, F. Auvray and P. Ritzenthaler. 1995. Characterization of genetic elements required for site-specific integration of *Lactobacillus delbrueckii* subsp *bulgaricus* bacteriophage *mv4* and construction of an integration-proficient vector for *Lactobacillus plantarum*. J. Bacteriol. 177:586-595.

Durmaz, E., D. L. Higgins and T. R. Klaenhammer. 1992. Molecular Characterization of a 2nd Abortive Phage Resistance Gene Present in *Lactococcus-lactis* subsp *lactis* ME2. J. Bacteriol. 174:7463-7469.

Duwat, P., S. D. Ehrlich and A. Gruss. 1995. The *recA* gene of *Lactococcus lactis*: characterization and involvment in oxidative and thermal stress. Molecular Microbiology 17:1121-1131.

Eaton, T., C. Shearman and M. Gasson. 1993. Cloning and sequence analysis of the *dnaK* gene region of *Lactococcus lactis*. J. Gen. Microbiol. 139:3253-3264.

Fraser et al. 1995. The minimal gene complement of Mycoplasma genitalium. Science 270:397-446.

Gasson, M. J. 1993. Progress and Potential in the Biotechnology of Lactic Acid Bacteria. FEMS Microbiology Reviews 12:3-20.

Gasson, M. J. and G. F. Fitzgerald. 1994. Gene transfer systems and transposition. in Genetic and biotechnology of lactic acid bacteria, (Ed.) Gasson M.J. and de Vos W.M., Blackie Academic & Professional, London. :pp. 1-51.

Geis, A., T. Janzen, M. Teuber and F. Wirsching. 1992. Mechanism of Plasmid-Mediated Bacteriophage Resistance in *Lactococci*. FEMS Microbiology Letters 94:7-14.

Godon, J. J., C. Delorme, J. Bardowski, M.-C. Chopin, S. D. Ehrlich and P. Renault. 1993. Gene inactivation in *Lactococcus lactis*: Branched-chain amino acid biosynthesis. J. Bacteriol 175:4383-4390.

Goupil, N., G. Corthier, S. D. Ehrlich and P. Renault. 1996a. Unbalance of leucine flux in *Lactococcus lactis* and its use for the isolation of diacetyl overproducing strains. Submitted

Goupil, N., J.-J. Godon, M. Cocaign-Bousquet, S. D. Ehrlich and P. Renault. 1996b. Dual role of alpha-acetolactate decarboxylase in *Lactococcus lactis* subsp. *lactis*. Submitted

Green, C. and B. S. Vold. 1993. Transfer RNA, tRNA processing, and tRNA synthetases. *in Bacillus subtilis* and other gram-positive bacteria (Ed. Sonenshein A.L., Hoch J.A. and Losick R.) ASM, Washington, DC :p.683-698.

Guedon, G., F. Bourgoin, M. Pebay, Y. Roussel, C. Colmin, J. M. Simonet and B. Decaris. 1995. Characterization and distribution of two insertion sequences, *IS1191* and *iso-IS981*, in *Streptococcus thermophilus*: Does intergeneric transfer of insertion sequences occur in lactic acid bacteria co-cultures? Molecular Microbiology 16:69-78.

Hartke, A., S. Bouche, X. Gansel, P. Boutibonnes and Y. Auffray. 1994. Starvation-induced stress resistance in *Lactococcus lactis* subsp *lactis* IL1403. Appl. Environm. Microbiol. 60:3474-3478.

Hill, C. 1993. Bacteriophage and Bacteriophage Resistance in Lactic Acid Bacteria. FEMS Microbiology Reviews 12:87-108.

Hill, C., I. J. Massey and T. R. Klaenhammer. 1991a. Rapid Method to Characterize Lactococcal Bacteriophage Genomes. Appl. Environm. Microbiol. 57:283-288.

Hill, C., L. A. Miller and T. R. Klaenhammer. 1991b. In vivo genetic exchange of a functional domain from type II methylase between lactococcal plasmid pTR2030 and a virulent bacteriophage. J. Bacteriol. 173:4363-4370.

Huang, X. F., D. C. Huang, G. Novel and M. Novel. 1995. Two *Lactococcus lactis* genes, including *lacX*, cooperate to trigger an SOS response in a *recA*-negative background. J. Bacteriol.177:283-289.

Hugenholz J. 1993. Citrate metabolism in lactic acid bacteria. FEMS Microbiol. Reviews 12:165-178.

Juillard, V., H. Laan, E. R. S. Kunji, C. M. Jeronimusstratingh, A. P. Bruins and W. N. Konings. 1995a. The extracellular P-I-type proteinase of *Lactococcus lactis* hydrolyzes beta-casein into more than one hundred different oligopeptides. J. Bacteriol.177:3472-3478.

Juillard, V., D. Le Bars, E. R. S. Kunji, W. N. Konings, J. C. Gripon and J. Richard. 1995b. Oligopeptides are the main source of nitrogen for *Lactococcus lactis* during growth in milk. Appl. Environm. Microbiol. 61:3024-3030.

Klaenhammer, T. R. 1993. Genetics of Bacteriocins Produced by Lactic Acid Bacteria. FEMS Microbiology Reviews 12:39-86.

Klaenhammer, T. R. and G. F. Fitzgerald. 1994. Bacteriophages and bacteriophage resistance. in Genetic and biotechnology of lactic acid bacteria, (Ed.) Gasson M.J. and de Vos W.M., Blackie Academic & Professional, London. :pp. 106-168.

Kok, J. and W. M. de Vos. 1994. The proteolitic system of lactic acid bacteria. in Genetic and biotechnology of lactic acid bacteria, (Ed.) Gasson M.J. and de Vos W.M., Blackie Academic & Professional, London. :pp. 169-210.

Kuipers, O. P., M. M. Beerthuyzen, P. Ruyter, E. J. Luesink and W. M. de Vos. 1995. Autoregulation of nisin biosynthesis in *Lactococcus lactis* by signal transduction. J. Biol. Chem. 270:1-6.

Le Bourgeois, P., M. Lautier, L. Vandenberghe, M. J. Gasson and P. Ritzenthaler. 1995. Physical and genetic map of the *Lactococcus lactis* subsp *cremoris* MG1363 chromosome: Comparison with that of *Lactococcus lactis* subsp *lactis* IL 1403 reveals a large genome inversion. J. Bacteriol. 177:2840-2850.

Le Bourgeois, P., Mata, M. and Ritzenthaler, P. 1991. Pulsed-fied gel electrophoresis as a tool for studying the phylogeny and genetic history of lactococcal strains. 140-145: Dunny, G.M.//Cleary, P.P.//McKay, L.L.

Llanos, R. M., C. J. Harris, A. J. Hillier and B. E. Davidson. 1993. Identification of a Novel Operon in *Lactococcus lactis* Encoding 3 Enzymes for Lactic Acid Synthesis - Phosphofructokinase, Pyruvate Kinase, and Lactate Dehydrogenase. J. Bacteriol.175:2541-2551.

Marteau, P. and J. C. Rambaud. 1993. Potential of Using Lactic Acid Bacteria for Therapy and Immunomodulation in Man. FEMS Microbiology Reviews 12:207-220.

Marugg, J. D., D. Goelling, U. Stahl, A. M. Ledeboer, M. Y. Toonen, W. M. Verhue and C. T. Verrips. 1994. Identification and Characterization of the alpha-Acetolactate Synthase Gene from *Lactococcus lactis* subsp. *lactis* biovar Diacetylactis. Appl. Environm. Microbiol. 60:1390-1394.

Marugg, J. D., W. Meijer, R. Vankranenburg, P. Laverman, P. G. Bruinenberg and W. M. Devos. 1995. Medium-dependent regulation of proteinase gene expression in *Lactococcus lactis*: Control of transcription initiation by specific dipeptides. J. Bacteriol. 177:2982-2989.

Mercenier, A., P. H. Pouwels and B. M. Chassy. 1994. Genetic engineering of *Lactobacilli, Leuconostocs* and *Streptococcus thermophilus*. in Genetic and biotechnology of lactic acid bacteria, (Ed.) Gasson M.J. and de Vos W.M., Blackie Academic & Professional, London. :pp. 252-294.

Mierau, I., A. J. Haandrikman, K. Leenhouts, P. S. T. Tan, W. N. Konings, G. Venema and J. Kok. 1995. Genetics and peptides degradation in *Lactococcus*: gene cloning, construction and analysis of peptidase negative mutants. in Genetics of *Streptococci, Enterococci* and *Lactococci*. Ferretti JJ., Gilmore M.S. Klaenhammer T.R. and F. Brown (eds) Dev. Biol. Stand. Basel, Karger, SW 85:503-508.

Moineau, S., S. Pandian and T. R. Klaenhammer. 1994. Evolution of a lytic bacteriophage via DNA acquisition from the *Lactococcus lactis* chromosome. Appl. Environm. Microbiol. 60:1832-1841.

Morishita, T., Y. Degushi, M. Yajima, T. Sakurai and T. Yura. 1981. Multiple nutritional requirement of *Lactobacilli*: genetic lesions affecting amino acid biosynthetic pathways. J. Bacteriol. 148:64-74.

Muto, A., Y. Andachi, H. Yuzawa, F. Yamao and S. Osawa. September 11, 1990. The Organization and Evolution of Transfer RNA Genes in *Mycoplasma capricolum*. Nuc. Acids Res. 18:5037-5043.

Nardi, M., P. Renault, J. C. Gripon and V. Monnet. 1996. Duplication of *pepF* gene encoding an oligopeptidase in *Lactococcus lactis*. Submitted

Nauta, A., D. van Sinderen, H. Karsen, E. Smit, G. Venema and J. Kok. 1995. Inducible gene expression mediated by a repressor-operator system isolated from *Lactococcus lactis* bacteriophage r1t. In press

Nilsson, D. and E. Johansen. 1994. A conserved sequence in tRNA and rRNA promoters of *Lactococcus lactis*. Biochim. Biophys. Acta 1219:141-144.

Parreira, R., S. D. Ehrlich and M.-C. Chopin. 1996. Dramatic decay of phage transcripts in lactococcal cells carrying the abortive infection determinant AbiB. Molecular Microbiology :in press.

Platteeuw, C., J. Hugenholz, M. Starrenburg, I. Van Alen-Boerrigter and W. M. de Vos. 1995. Metabolic engineering of *Lactococcus lactis*: Influence of the overproduction of a-acetolactate synthase in strains deficient in lactate dehydrogenase as a function of culture conditions. Appl. Environm. Microbiol. 61:3967-3971.

Renault, P., C. Gaillardin and H. Heslot. 1989. Product of the *Lactococcus lactis* gene required for malolactic fermentation is homologous to a family of positive regulators. J. Bacteriol. 171:3108-3114

Renault, P., J.-J. Godon, N. Goupil, C. Delorme, G. Corthier and S. D. Ehrlich. 1995. Metabolic operons in *Lactococci*. in Genetics of *Streptococci, Enterococci* and *Lactococci*. Ferretti JJ., Gilmore M.S. Klaenhammer T.R. and F. Brown (eds) Dev. Biol. Stand. Basel, Karger, SW 85:431-441.

Sabelnikov, A. G., B. Greenberg and S. A. Lacks. 1995. An extended -10 promoter alone directs transcription of the DpnII operon of *Streptococcus pneumoniae*. J. Mol. Biol. 250:144-155.

Sanders, J. W., K. J. Leenhouts, A. J. Haandrikman, G. Venema and J. Kok. 1995. Stress response in *Lactococcus lactis*: Cloning, expression analysis, and mutation of the lactococcal superoxide dismutase gene. J. Bacteriol. 177:5254-5260.

Schouler, C. 1996. Genomic organization of lactic acid bacteria phages. Le Lait :in press.

Schouler, C., S. D. Ehrlich and M.-C. Chopin. 1994. Sequence and organization of the lactococcal prolate-headed bIL67 phage genome. Microbiology 140:3061-3069.

Snoep, J. L., M. J. T. Demattos, M. J. C. Starrenburg and J. Hugenholtz. 1992. Isolation, Characterization, and Physiological Role of the Pyruvate Dehydrogenase Complex and alpha-Acetolactate Synthase of *Lactococcus lactis* subsp *lactis* BV diacetylactis. J. Bacteriol. 174:4838-4841.

Steidler, L., J. M. Wells, A. Raeymaekers, J. Vandekerckhove, W. Fiers and E. Remaut. 1995. Secretion of biologically active murine interleukin-2 by *Lactococcus lactis* subsp *lactis*. Appl. Environm. Microbiol. 61:1627-1629.

Swindell, S. R., K. H. Benson, H. G. Griffin, P. Renault, Ehrlich S.D. and M. J. Gasson. 1995. Genetic manipulation of the pathway for diacetyl metabolism in *Lactococcus lactis*. Submitted

Taillez, P., Ehrlich, S.D. and A. Chopin. 1996. Manuscript in preparation.

van Asseldonk, M., A. Simons, H. Visser, W. M. de Vos and G. Simons. 1993. Cloning, nucleotide sequence, and regulatory analysis of the *Lactococcus lactis dnaJ* gene. J. Bacteriol. 175:1637-1644.

van de Guchte, M., J. Kok and G. Venema. 1991. Distance-Dependent Translational Coupling and Interference in *Lactococcus lactis*. Mol. Gen. Genet. 227:65-71.

van Rooijen, R. J. and W. M. de Vos. 1990. Molecular Cloning, Transcriptional Analysis, and Nucleotide Sequence of *lacR*, a Gene Encoding the Repressor of the Lactose Phosphotransferase System of *Lactococcus lactis*. J.Biol. Chem. 265:18499-18503.

Wells, J. M., P. W. Wilson, P. M. Norton, M. J. Gasson and R. W. F. Lepage. 1993. *Lactococcus lactis*: high-level expression of tetanus toxin fragment C and protection against lethal chalenge. Molecular Microbiology 8:1155-1162.

Yuan, G. and S.-L. Wong. 1995. Isolation and characterization of Bacillus subtilis *groE* regulatory mutants: evidence for *orf39* in the *dnaK* operon is a repressor gene on regulating the expression of both *groE* and *dnaK*. J. Bacteriol. 177:6462-6468.

Cloning and Expression Vectors for Lactococci

Jeremy M. Wells and Karin M. Schofield
Department of Pathology
University of Cambridge
Tennis Court Rd
Cambridge CB2 1QP
U.K.

Introduction

In last decade there has been considerable interest in developing recombinant lactic acid bacteria with novel properties for applications in the health care industry and for their more traditional use in food fermentation and preservation. It is envisaged that genetically modified lactic acid bacteria will eventually be used to increase the production of certain flavour compounds or enzymes involved in various food processes or as cell factories for the production of antimicrobials, high value proteins and primary or secondary metabolites. A more recent development has been the growing interest in the use of non-pathogenic lactic acid bacteria as vaccine delivery vehicles for mucosal immunisation. A key component in the development of many of these applications is the requirement for efficient expression systems for the production of heterologous proteins in different species of lactic acid bacteria. This chapter summarises the progress of work on the development of general cloning vectors and inducible and constitutive expression systems for the lactococci. Emphasis has been given to our own work on gene expression with

NATO ASI Series, Vol. H 98
Lactic Acid Bacteria:
Current Advances in Metabolism, Genetics and Applications
Edited by T. Faruk Bozoğlu and Bibek Ray

reference to the many important contributions made by other research groups in this field.

General Cloning Vectors for the Lactococci

Plasmid vectors currently developed for use in the lactococci have been based on both homologous and heterologous replicons. The majority of vectors based on homologous replicons are derived from the small cryptic plasmids pWVO1 and pSH71 and have been genetically marked with antibiotic resistance genes which are known to be expressed in several Gram-positive and Gram-negative bacteria. Plasmids pWVO1 and pSH71 are members of a large family of Gram-positive plasmids which have a rolling circle replication (RCR) system. It is not unexpected therefore that plasmids based on these replicons have a broad Gram-positive host range.

One of the first lactococcal cloning vectors to be reported (pGK12) was generated by incorporating the chloramphenicol acetyl transferase gene from pC194 (*cat-194*) and erythromysin gene from pE194 (*ery-194*) into the 2.1 kb cryptic lactococcal plasmid pWVO1. Plasmid pGK12 and the derivatives which have been subsequently generated from it are able to replicate in *E. coli,* a wide range of Gram-positive bacteria, and exist at low copy number in *L. lactis* and *B. subtilis* (Table 1; Kok et al., 1984).

Genetically marked derivatives of pSH71 were first constructed by ligating *Taq*I restriction endonuclease fragments of pSH71 to a DNA fragment from plasmid pBD64 containing the *S. aureus cat-194* and *knt-110* genes (Anderson and Gasson 1985). One of the resulting derivatives designated pCK1 (5.7 kb) contained a 1.7 kb fragment of pSH71 and was shown to replicate in *Bacillus, E. coli* and *L. lactis* at higher copy number than pGK12 and its derivatives. Similar studies carried out in parallel with pSH71 resulted in the construction

Table 1. Vectors for cloning and expression in *Lactococcus*.

	Replicon	Size (kb)	Selection	Copy number	Features	Reference
General cloning vectors						
pGK12	pWVO1	4.4	Erm^R Cat^R	Low	*E. coli* shuttle vector	Kok et al., 1984
pCK1	pSH71	5.5	Cat^R Kan^R	High	*E. coli* shuttle vector	Anderson & Gasson, 1985
pNZ12	pSH71	4.3	Cat^R Kan^R	High	*E. coli* shuttle vector	de Vos et al., 1986a-c
pNZ121	pSH71	4.3	Cat^R Kan^R	Low	*E. coli* shuttle vector	de Vos et al., 1986a-c
pNZ123	pSH71	2.8	Cat^R	High	Derivative of pNZ12; MCS	de Vos et al., 1992
pMIG1	pSH71	4.8	Cat^R Kan^R	High	*E. coli* shuttle vector; MCS	Wells et al., 1993
pMIG2	pSH71	4.0	Cat^R	Low	Smaller derivative of pMIG1	Wells et al., 1993
pMIG2H	pSH71	4.2	Cat^R	High	High-copy-number version of pMIG2	Wells et al., 1993
pMIG4	pSH71	4.5	Erm^R	High	Derivative of pMIG1; MCS	Wells, unpublished
pFX3	pDI25	4.5	Cat^R	High	Derivative of pFX1; MCS	Xu et al., 1991
pIL252	pAMβ1	4.7	Erm^R	Low	MCS	Simon & Chopin, 1988
pIL253	pAMβ1	5.0	Erm^R	High	High copy derivative of pIL252; MCS	Simon & Chopin, 1988
pTRKL1	pAMβ1/ p15A	8.3	Erm^R Cat^R	Low	*E. coli* shuttle vector; insertional inactivation of Cat^R in *E. coli*	O'Sullivan and Klaenhammer, 1993
pTRKH1	pAMβ1/ p15A	11.0	Erm^R Cat^R Tet^R	High	*E. coli* shuttle vector; insertional inactivation of Cat^R in *E. coli*	O'Sullivan and Klaenhammer, 1993
pTRKH2	pAMβ1/ p15A	6.9	Erm^R	High	Blue/white screening of clones in *E. coli*	O'Sullivan and Klaenhammer, 1993
pTRKH3	pAMβ1/ p15A	7.8	Erm^R Tet^R	High	Insertional inactivation of Tet^R in *E. coli*	O'Sullivan and Klaenhammer, 1993

Table 1 continued.

pSA3	pIP501	10.2	Cat^R	Low	*E. coli* shuttle vector	Dao & Ferreti, 1986
	pACYC184		Erm^R			
pCI374	pCI305	4.9	Cat^R	Low	Narrow host range	Hayes et al., 1990
pWVO2E	pWVO2	4.1	Erm^R		MCS	Waterfield, 1996
Expression vectors						
pGKV2	pWVO1	4.7	Erm^R	Low	Contains Bacillus SPO2 promoter	van der Vossen et al., 1985
pMG36e	pWVO1	3.6	Erm^R	Low	Contains lactococcal chromosomal promoter P32	van de Guchte et al., 1989
pNZ2014	pSH71	3.6	Cat^R	High	Contains *L.lactis lac* promoter, inducible[1]	de Vos and Simons, 1994
pTREX1	pAMβ1	5.2	Erm^R	High	Contains *L.lactis* chr. promoter P1	Schofield et al., 1996
pTREX7	pAMβ1	5.6	Erm^R	High	Contains *L.lactis* chr. promoter P7	Schofield et al., 1996
pTREX11	pAMβ1	5.2	Erm^R	High	Contains *L.lactis* chr. promoter P11	Schofield et al., 1996
pTREX14	pAMβ1	5.1	Erm^R	High	Contains *L.lactis* chr. promoter P14	Schofield et al., 1996
pTRCX1	pAMβ1	5.0	Cat^R	High	*cat-194* containing derivative of pTREX1	Schofield et al., 1996
pILPol	pAMβ1	8.3	Erm^R	Low	Contains *L.lactis lac* promoter & T7 RNA polymerase, inducible[1]	Wells et al., 1993
pLET1	pSH71	4.9	Cat^R	High	Intracellular expression; gene10 fusion	Wells et al., 1993
pLET2	pSH71	5.0	Cat^R	High	Secretion vector; *usp45* SL	Wells et al., 1993
pLET3	pSH71	4.9	Cat^R	High	Secretion vector; *prtP* SL and TIR	Wells et al., 1993
pLET4	pSH71	5.4	Cat^R	High	Cell-wall anchoring vector	Wells et al., unpublished
pLET5	pSH71	4.9	Cat^R	High	Intracellular expression vector	Wells et al., unpublished

[1] Expression from these vectors is inducible in hosts containing the lactose operon.
Erm^R, Cat^R, Kan^R and Tet^R indicate antibiotic resistance to erythromycin, chloramphenicol, kanamycin and tetracycline respectively.
MCS: multiple cloning site; SL: signal leader.

of high and low copy number plasmids pNZ12 and pNZ121 respectively (de Vos et al., 1986a-c). A smaller high copy number derivative of pNZ12, designated pNZ123 (Table 1), containing chloramphenicol resistance gene from *S. aureus* (*cat-194*) and a multiple cloning site has been used for cloning a variety of genes in *Lactococcus* (de Vos et al., 1992). Similarly, we have generated improved derivatives of pCK1 which have been instrumental in the development of the lactococcal T7 RNA polymerase based expression system. A smaller derivative of pCK1 was constructed by incorporating a multiple cloning site between two unique restriction sites in pCK1 to generate pMIG1 (Fig 1; Wells et al., 1993a). A smaller replicating derivative of pMIG1 was generated by the deletion of an internal fragment of the *knt-110* gene and insertion of a second cloning site linker. The resulting plasmid designated pMIG2 replicates at a lower the copy-number than pMIG1 in *L. lactis*. The reasons for this are not clear but it may involve uncontrolled transcription from the *knt-110* promoter through neighbouring plasmid sequences as the insertion of a transcription terminator downstream of the *knt-110* promoter resulted in an increase in the copy number of the plasmid in *L. lactis*. The resulting plasmid is designated pMIG2H to indicate that it is a higher copy-number derivative of pMIG2 (Fig 1; Wells et al., 1993a). In order to facilitate the use of the pMIG series of vectors with other compatible cloning vectors we have recently constructed pMIG4 in which the *cat-194* gene present in pMIG1 has been replaced with the *S. aureus ery-194* gene from pE194 (Fig 1). The segregational stability of vectors based on the pWVO1/pSH71 replicons appears to be high in *Lactococcus* although derivatives containing large inserts in the order of 8kb show decreased stability (Kok et al., 1984; de Vos and Simons, 1987; Kiewiet et al., 1993b). Additional derivatives of the pMIG, pGK and pNZ vectors have also been developed for other specialised applications some of which are mentioned in more detail in the following sections. (Wells et al., 1993b, Wells et al., 1993c, van der Vossen et al., 1985, de Vos, 1987, van Asseldonk et al., 1993, Platteeuw et al., 1994).

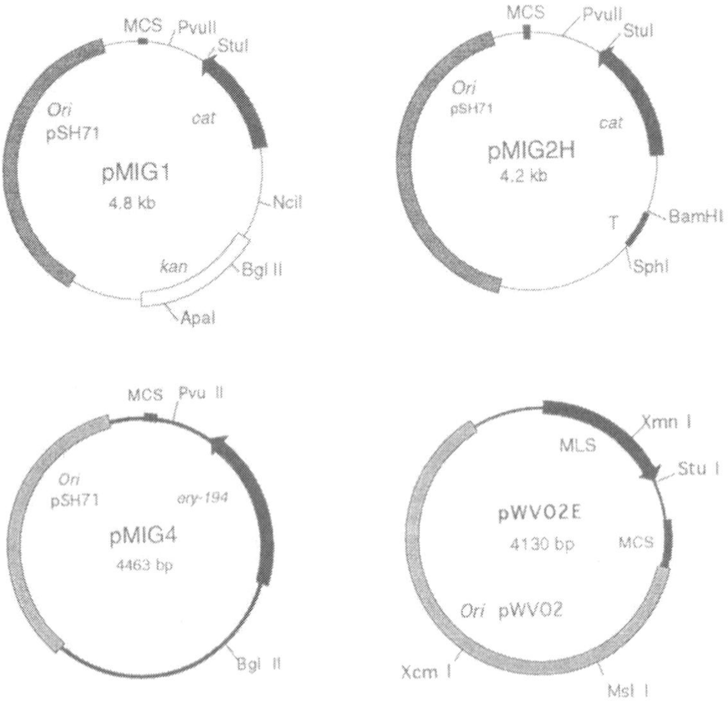

Fig 1. Physical and genetic maps of cloning vectors pMIG1, pMIG2, pMIG4 and pWVO2E. The coding sequences of the determinants encoding resistance to chloramphenicol (*cat*), erythromysin (*ery-194*) or macrolides, lincosamides and streptogramin B (MLS) and the replicons (*ori* pSH71 & *ori* pWVO2) are depicted as thick lines or arrows on which the the direction of transcription is indicated. T: terminator; MCS: multiple cloning site.

Another cryptic lactococcal plasmid pDI25 which is highly related to pWVO1 and pSH71 has also been used to generate the cloning vectors pFX1 and pFX3, the latter being a polylinker containing derivative of the former (Table 1; Xu et al., 1991).

Most of the plasmids for the lactococci which are based on heterologous replicons have been derived from the conjugative enterococcal plasmid pAMβ1 (Clewell et al., 1974) or the streptococcal conjugative plasmid pIP501 (Horodniceanu et al., 1976). Both of these plasmids are relatively large (24 - 34 kb), self-transmissible, replicate via a theta-type mechanism and encode resistance to macrolides, lincosamides and streptogramin B (MLS) (Brantl et al., 1990). Together these two theta-type plasmids form another replication family separate to the pWVO1/pSH71 group.

Two small cloning vectors designated pIL252 and pIL253 which are based on pAMβ1 and which are compatible with the pWVO1/pSH71 replication group of plasmids have proved to be highly useful for gene cloning in the lactic acid bacteria (Table 1; Simon and Chopin, 1988). Plasmids pIL252, pIL253 and their derivatives are not self transferable but can be mobilised from *L. lactis* to other Gram-positive bacteria at low frequency by *tra*$^+$ derivatives of pIP501 (Gruzza et al., 1993) Unlike the pWVO1/pSH71 derived vectors, both pIL252 and pIL253 are likely to support the cloning of large inserts up to 22kb in length (Simon and Chopin, 1988). Although these vectors will not replicate in *E. coli*, the pAMβ1 replicon on which they are based shows a very broad host range within the Gram-positive bacteria (Brantl and Behnke, 1992). Recently, improved derivatives of pIL252 and pIL253 were constructed which contain the *E. coli* p15A plasmid origin of replication allowing them to also be used in *E. coli* (Table 1; O'Sullivan and Klaenhammer, 1993). Fewer vectors have been described based on pIP501 from *Streptococcus agalactiae* and those which have been developed are all large and exist in low copy number in *Lactococcus*.

More recently the theta mode of replication has been identified in a number of lactococcal plasmids including pCI374, pWVO2, pJW563 (encoding a type II restriction/modification system) and pUCL22 (a lactose-proteinase plasmid) (Kiewiet et al., 1993a; Gravesen et al., 1995; Frere et al., 1993). Few cloning vectors have been reported which are based on homologous theta-type replicons one exception being pCI374 which is derived from the cryptic plasmid pCI305 (Hayes et al., 1990). pCI374 replicates at low-copy-number in *L. lactis*, is highly stable both structurally and segregationally and has a narrow host range which appears to be restricted to certain species of *L. lactis*. A genetically marked derivative of the narrow host range plasmid pWVO2, designated pWVO2E, has been constructed in our laboratory by joining a DNA fragment containing the pWVO2 origin of replication to a DNA fragment from pMIG4 containing a multiple cloning site and erythromycin resistance gene *ery-194* (Fig 1; Waterfield, 1996). The general ability of plasmids which replicate by a theta-mode to stably maintain large inserts is likely to encourage the future development of homologous theta-type vectors for the genetic modification of lactic acid bacteria, especially with respect to applications in food grade systems.

Constitutive expression vectors

All of the expression vectors which have been published for use in lactococci are based on the pWVO1/pSH71 replication group of plasmids. One of the first lactococcal expression vectors to be described (pGKV2) utilises the SPO2 promoter from *Bacillus subtilis* (van der Vossen et al., 1985). Subsequently, there have been a number of studies reporting the isolation and characterisation of lactococcal chromosomal promoters and their use in constitutive expression vectors (van der Vossen et al., 1987; Koivula et al., 1991; Waterfield et al., 1995). One of the lactococcal promoters identified by van der Vossen et al., was used to construct pMG36e (van de Guchte et al., 1989). This expression plasmid incorporates the P32 promoter, the first 12 amino acids of the open reading frame naturally present downstream of promoter P32 (ORF32) into which a polylinker has been inserted, and the transcription terminator of the lactococcal proteinase gene *prtP*. Genes to be expressed can be inserted in frame with the first 12 amino acids of the P32 open reading frame in pMG36e. The expression of the hen egg white lysozyme and bacteriophage T4 lysozyme genes has been studied in *L. lactis* by cloning these genes in frame with ORF32 in the expression vector pMG36e. Immunoblotting experiments indicated that the amounts of lysozyme expressed in *L. lactis* were low and only in the case of T4 bacteriophage lysozyme was the active enzyme detected in cell extracts of expressing cells.

Using pMG36e the *Bacillus subtilis* neutral protease (*nprE*) gene has been expressed in *L. lactis* under control of its own translation initiation signals and secretion signal. The mature form of the protease was found in the culture supernatant of the expressor cells and its activity detected on indicator plates and *in vitro* enzyme assays indicating that the protein had been properly secreted and processed in *L. lactis* (van de Guchte et al., 1990). A derivative of pMG36e has been used to study distance-dependent translational coupling of genes to ORF32 in *L. lactis* (van de Guchte et al., 1991). The lactococcal *prtP* gene promoter has been used to construct the expression vector pNZ337 a derivative of which was used to study the expression and processing of bovine prochymosin in *L. lactis* (Simons et al., 1992).

Although expression plasmids based on the pWVO1/pSH71 replicons should have a broad Gram-positive host range they are considered to be inherently less stable structurally than those which replicate via a theta-type mechanism due to the formation of single-stranded DNA intermediates and high molecular weight plasmid multimers (Gruss and Ehrlich 1989; Kiewiet et al., 1993b). In addition plasmids which replicate via this mechanism do not support large plasmid inserts as compared to plasmids which replicate via a theta type mechanism. A further disadvantage with the lactococcal expression vectors described by other workers is that the different expression signals influencing gene expression such as the 5' untranslated sequences (including RNA stabilising sequences), translation initiation region, promoter, and terminator cannot be readily manipulated to study their effects on gene expression. In order to overcome some of the limitations of the existing expression plasmids we have constructed a new series of constitutive expression vectors based on pTREX (for theta replicating expression vectors) which incorporate the replicon of pAMβ1 (Schofield et al., 1996). This theta type replicon has been transferred to various *Streptococcus, Lactobacillus* and *Bacillus* species as well as *Clostridium acetobutylicum*, (Oultram and Young, 1985; Gibson et al., 1979; LeBlanc et al., 1978) indicating that the pTREX vectors are also likely to be useful for studies of gene expression in several other organisms.

The expression cassette in pTREX (Fig. 2) incorporates the following *E. coli* bacteriophage T7 expression signals: (1) a putative RNA stabilising sequence, (2) the translation initiation region (TIR) of *gene10* which has been modified at one nucleotide position to increase complementarity of the Shine Dalgarno (SD) sequence to the ribosomal 16sRNA of *L. lactis* and (3) the T7 RNA polymerase terminator. Unique cloning sites are present upstream of the putative RNA stabilising sequence and downstream of the translation initiation codon to facilitate insertion of promoters and target genes respectively. To increase the versatility of these vectors pTRCX was constructed in which the MLS determinant from pAMβ1 has been replaced with the chloramphenicol resistance gene (*cat-194*) from *S. aureus* (Fig 2). The pTREX and pTRCX vectors are structurally stable in *L. lactis*, *Lactobacillus johnsonii* VPI11088, *Lactobacillus gasseri* VPI6033, *Lb. gasseri* ATCC33323 and *Bacillus thuringiensis*. The plasmid copy number of pTREX vectors in *L. lactis* is estimated to be similar to the pIL253 vector from which they are derived i.e. 45 - 85 copies per cell. The frequency of loss of the resistance marker in the absence of antibiotic selection (pTREX1) was 2.5 x 10^{-4} per generation (tested

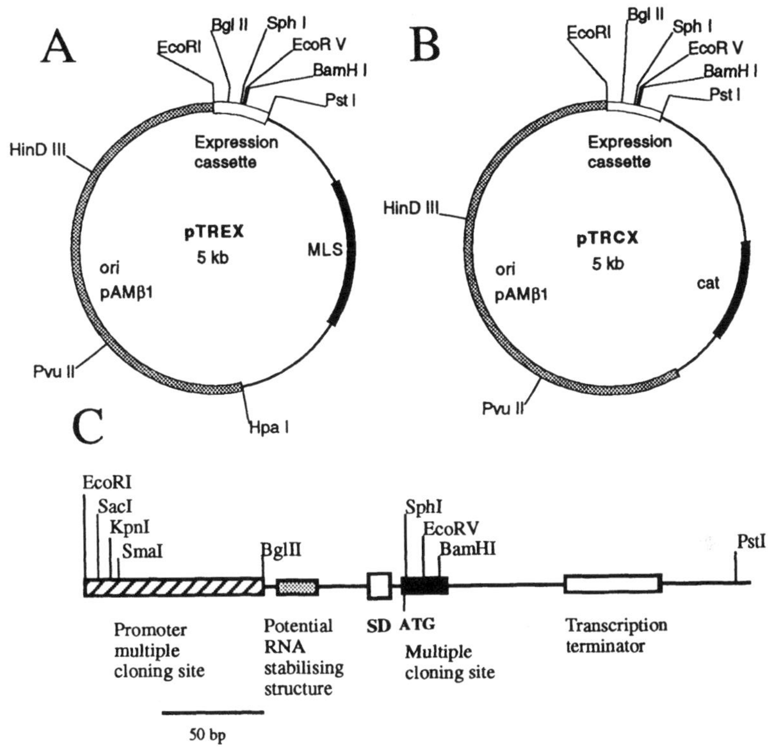

Fig. 2. (A & B) Physical and genetic maps of the broad Gram-positive host range vectors pTREX and pTRCX. The expression cassette, chloramphenicol acetyl transferase gene (*cat-194*), macrolides, lincosamides and streptogramin B (MLS) resistance determinant and the replicon (*Ori*-pAMβ1) are depicted as thick lines. (C) Schematic representation of the expression cassette showing the various sequence elements involved in gene expression, the location of unique restriction endonuclease sites, Shine Dalgarno (SD) motif and translation initiation start codon (ATG).

over 40 generations) in *L. lactis* which is slightly lower than the value previously reported for pIL253 (Chopin et al., 1988). The segregational stability of pTRCX in *L. lactis* was slightly lower. In the *L. gasseri* strains pTREX1 was structurally stable but segregationally unstable with complete loss of the resistance marker occurring within 40 generations in the absence of antibiotic selection. Segregational instability of plasmids in *Lactobacillus* has been noted previously by other workers (Claasen et al., 1994) and in this case may be due to the absence of a resolvase (*res* gene) present in pAMβ1 which reduces plasmid multimerisation and increases segregational stability of the plasmid. Incorporation of the pAMβ1 *res* gene into pIL252 significantly increased its segregational stability in *L. lactis*; a similar effect might be expected with *res* containing derivatives of pTREX (Kiewiet et al., 1993b).

A series of expression vectors were constructed using pTREX and four of the seventeen lactococcal promoters recently isolated in *L. lactis* using the *Vibrio fischeri* luciferase gene (*luxAB*) as a reporter (Waterfield et al., 1995). Four *L. lactis* chromosomal DNA promoter fragments of varying activity, designated P1, P7, P11 and P14, were amplified by PCR and cloned between the EcoRI and BglII sites in pTREX to generate pTREX1, pTREX7, etc. (Schofield et al., 1996). The SD motif present downstream of the transcription start sites of these promoters was deliberately excluded from the PCR amplified promoter fragment to prevent translation initiation at sites other than the start codon indicated in the expression cassette (Fig 2)

The relative activity of the pTREX1, 7, 11 and 14 expression vectors in *L. lactis* was assessed by cloning the *V. fischeri* luxAB reporter genes in the multiple cloning site to generate pTREX1-*luxAB*, pTREX7-*luxAB* etc. The relative promoter strengths were in the order P1> P11> P14> P7. A variety of other genes have also been expressed in *L. lactis* using pTREX1, including other reporter enzymes such as β-galactosidase, β-glucuronidase, antigens such as tetanus toxin fragment C (TTFC), glutathione transferase of *S. mansoni* and also secreted proteins such as neutral protease, α-amylase and xylanase from *Bacillus* sp. (Table 2).

The *Bacillus* neutral protease and α-amylase have been included as model proteins for expression in *L. lactis* and *Lactobacillus* using the pTREX series

Table 2. Examples of heterologous genes expressed in lactic acid bacteria using pTREX1.

Protein (Gene)	Source	Secreted	Expression host	Comments	Reference
β-galactosidase (*lacZ*)	*E. coli*		*L. lactis*	Visual plate assay	Schofield et al., unpublished
β-glucuronidase (*gusA*)	*E. coli*		*L. lactis*	Visual plate assay	Schofield et al., unpublished
Luciferase (*luxAB*)	*Vibrio fischeri*		*L. lactis*	Bioluminesence *in vivo*.	Schofield et al., 1996
α-amylase (*amyL*)	*Bacillus licheniformis*	Yes[a]	*L. lactis*	Activity on starch plates; secreted	Wells et al., unpublished
		Yes[a]	*Lb. johnsonii*	Activity on starch plates; secreted	Wells et al., unpublished
Neutral protease (*nprE*)	*Bacillus subtilis*	Yes[a]	*L. lactis*	Actvity on milk plates; secreted	Wells et al., unpublished
Streptavadin	*Streptomyces avidinii*	Yes	*L. lactis*		Wells et al., unpublished
Xylanase	*Bacillus* sp.		*L. lactis*	Active enzyme formed	Bailey et al., unpublished
Tetanus toxin fragment C	*Clostridium tetani*		*L. lactis*	1-3% of total cell protein, intracellular accumulation	Schofield et al., 1996
			Lb. gasseri	Intracellular accumulation	Schofield et al., 1996
			Lb. johnsonii	Intracellular accumulation	Schofield et al., 1996
			Lb. paracasei	Intracellular accumulation	Mercenier & Kleinpeter pers. comm.
Glutathione S-transferase	*Schistosoma mansonii*		*L. lactis*	Intracellular accumulation	Schofield et al., unpublished

[a] Evidence for protein secretion based on the detection of the active protein in culture supernatants and on indicator plates.

of vectors as the activity of the secreted protein can be easily visually detected on indicator plates and because both proteins have previously been expressed and secreted by *L. lactis* using other expression systems (van de Guchte et al., 1990; van Asseldonk et al., 1993). The full length *Bacillus* neutral protease gene (*nprE*) was cloned into pTREX1 so that the native TIR and signal leader would be utilised for expression. The neutral protease (Npr) expressor strains of *L. lactis* were evident from the halo of digested milk protein surrounding colonies which were grown on glycerophosphate-milk agar containing 0.005% bromocresol purple. Npr activity in the culture supernatant of overnight cultures of expressor strains of *L. lactis* was quantitatively determined using the azocoll assay and shown to be similar to that previously reported using the expression vector pMG36e to produce Npr in *L. lactis* (van de Guchte et al., 1990). These findings are in keeping with the results of other workers who provided further evidence that the *Bacillus* neutral protease is properly processed and secreted by *L. lactis*.

The α-amylase of *Bacillus lichenformis* has also been expressed in *L. lactis* using a derivative of pTRCX (Table 1) containing the signal leader of the lactococcal *usp45* gene. The mature form of the amylase was cloned in frame with the signal leader by cloning into a unique restriction site (NaeI) which overlaps the last codon of the signal peptide. Bacteria expressing α-amylase could be visualised by the production of halos around colonies growing on starch plates after staining with iodine. Similarly, strains of *Lactobacillus* containing pTREX1-*amyL* but not the control strains transformed with the vector alone produced halos around colonies grown on starch plates. These results indicated that the lactococcal P1 promoter is active in *Lb. gasseri* and *Lb. johnsonii* and that the heterologous translation initiation signals also function in these hosts. Although it seems likely that α-amylase is properly processed and secreted in *Lactobacillus* and *Lactococcus* when expressed in pTREX or pTRCX these results need to be confirmed by further experiments involving the detection and measurement of amylase levels in sub-cellular fractions of expressing bacteria.

The gene encoding tetanus toxin fragment C (TTFC) has been cloned into pTREX1, pTREX7, pTREX11 and pTREX14 to generate pTREX1-TTFC etc. A crude estimate of the quantity of TTFC expressed in *L. lactis* MG1363 was obtained by a comparison of the intensity of bands detected on Western blots of cell extracts of the TTFC expressor strains and known amounts of purified recombinant TTFC. The amount of TTFC produced by a particular pTREX-TTFC construct correlated with the activity (as predicted by luciferase assay) of the promoter used (Schofield et al., 1996). The quantity of TTFC produced in *L. lactis* MG1363 harbouring pTREX1-TTFC was estimated to be in the range of 1-3% of total cell protein. Expression of TTFC has also been detected in *Lb. gasseri*, *Lb. johnsonii* (Schofield, unpublished data) and *Lb. paracasei* (Mercenier & Kleinpeter, personal communication) which were transformed with pTREX1-TTFC. Similar amounts of TTFC were expressed in the *Lactobacillus* and *L. lactis* as deduced from the intensity of bands detected on Western blots of cell extracts from the different expressor strains.

The use of constitutive promoters which have a relatively low basal level of activity usually avoids the problems that can occur due to the high level constitutive expression of foreign proteins, such as slow growth of the cells, the accumulation of plasmid mutations and deletions, and the accumulation of cells which have lost the plasmid. Such problems are exasperated by the expression of proteins which have a harmful effect on the cell. There have been a number of anecdotal reports of the failure to stably clone strong lactococcal promoters into plasmids in *L. lactis* (van der Vossen et al., 1987 Eaton et al., 1993 Wells unpublished) which may have been related to the physiological stresses which are imposed on cells constitutively expressing certain proteins in high amounts. These problems can be overcome by using promoters which are regulated.

Inducible expression systems

The lactococcal *lac* promoter which controls the expression of the lactose-specific enzymes of the lactose phosphotransferase system (PTS), and the enzymes of the tagatose-6-phosphate pathway has been used for lactose-inducible expression in strains of *L. lactis* which carry the *lac* genes either on a plasmid or integrated into the chromosome (van Rooijen and de Vos 1990; van Rooijen et al., 1992). The *lac* promoter can also be used as a strong constitutive promoter in strains lacking the genes required for growth on lactose. However, in *L. lactis* the instability of plasmids containing only the *lac* promoter region has been reported by some workers (Eaton et al., 1993, Schofield, unpublished results). In a series of studies on the regulation of the *lac* promoter it has been shown that the LacR repressor which is divergently transcribed from the *lac* promoter represses operon expression by binding to the operator sites lacO1 and lacO2 located in the *lac* promoter and between the lac promoter and the start of the LacR gene respectively. The model proposed for the regulation of *lac* gene expression is as follows. Binding of LacR repressor to lacO1 during growth on glucose activates transcription of *lacR*. As the concentration of LacR increases within the cell the second lower affinity operator site (lacO2) becomes occupied by the repressor. When both operator sites are occupied transcription from the *lac* promoter is repressed. Expression is induced during growth on lactose by the binding of tagatose-6-phosphate to the *lac* repressor. None of the other phosphorylated intermediates formed during the catabolism of lactose or galactose have been shown to bind LacR (van Rooijen and de Vos, 1993).

The *lac* promoter has been used to construct lactose inducible expression vectors in plasmids with both RCR and theta-modes of replication. The expression vector pNZ2014 is based on pNZ123 and contains the *lac* promoter and a downstream terminator from the *L. lactis pepN* gene in pNZ123 (Table

1). This expression vector has been used for the inducible over-expression of various genes in *L. lactis* (Simons and de Vos 1994). Similarly, an inducible expression vector was constructed by cloning the *lac* promoter into pTREX to generate pTREX*plac* (Table 1).

The *lac* promoter has also been used to develop a regulated expression system which exploits the properties of the *E. coli* T7 bacteriophage RNA polymerase (Wells et al., 1993b). In this expression system the T7 RNA polymerase has been placed under control of the *lac* promoter in low-copy-number plasmid pIL227 to generate pILPol. An expression host strain which would permit lactose induction of T7 RNA polymerase expression was generated by transforming *L. lactis* strain MG1820 with pILPol. Strain MG1820 carries a large 23.7 kb plasmid containing the *lac* operon genes. A series of expression vectors which are compatible with pILPol have been constructed based on pMIG1 (Fig 1). These vectors designated the pLET (for lactococcal expression by T7 RNA polymerase) series of vectors each contain a different version of a T7 expression cassette cloned into the multiple cloning site (MCS) of pMIG1 (Fig 3). The most recently constructed derivative pLET4 has been designed to allow proteins of interest to be expressed as a fusion to the C-terminal cell-wall spanning and membrane anchoring domain (nt 6518 to 6913) of PrtP (Fig 3). Target genes are cloned into the unique BamHI site which is in frame with usp45 signal leader and C-terminal anchoring domains of PrtP.

The pLET series of vectors have been used for the expression of a variety of genes in *L. lactis* (Table 3). Most of the genes expressed in the lactococcal T7 system have encoded antigens (e.g. TTFC) which are of interest to us in studies relating to the use of *L. lactis* as an oral vaccine delivery vehicle (Wells et al., 1995). *L. lactis* has also been evaluated as a potential production micro-organism for high value proteins such as the interleukins. Murine interleukin 2 (*mIL-2*) has been expressed and secreted in *L. lactis* using the lactococcal T7

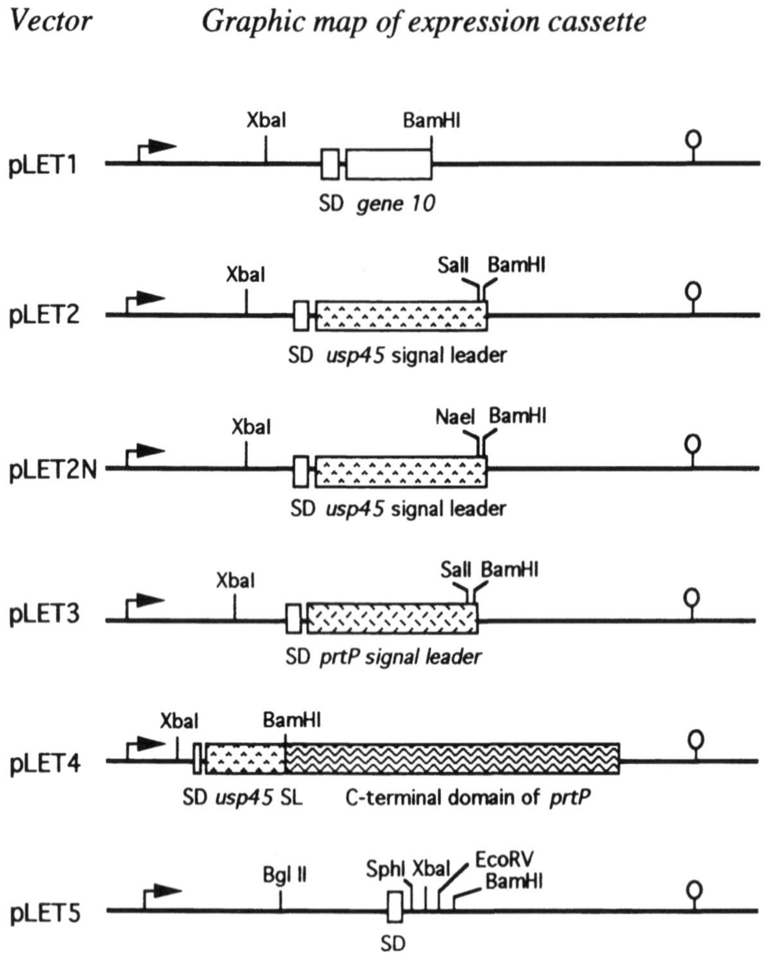

Fig.3. Schematic representation of the expression cassettes present in the pLET series of vectors. The coding sequences are boxed and labelled accordingly. The position of the T7 promoter and T7 terminator is shown. The endonuclease restriction sites indicated are unique in the vector. SL: signal leader; SD: Shine Dalgarno motif. Further information regarding the different sequences and features of the pLET cassettes can be found in the text and references therein.

Table 3. Examples of heterologous genes expressed in *L. lactis* using the pLET series of vectors.

Protein	Source	Vector	Comments	Reference
Tetanus toxin fragment C	*Clostridium tetani*	pLET1	Approx. 20 % of soluble cell protein	Wells et al., 1993
		pLET3	Secreted - 3 mg/L	Wells et al., 1993
		pLET4	membrane anchored protein	Wells et al., unpublished
Glutathione S-transferase	*Schistosoma mansoni*	pLET1	5% of total cell protein	Chamberlain et al., 1995
Diphtheria toxin frag. B	*Corynebacterium diphtheria*	pLET1	1-2% of total cell protein	Wells et al., unpublished
		pLET2	Secreted	Wells et al., unpublished
		pLET3	Secreted	Wells et al., unpublished
Cholera toxin subunit B	*Vibrio cholerae*	pLET1	No detectable expression	Hildebrand & Wells, unpublished
		pLET2	No detectable expression or secretion	Hildebrand & Wells, unpublished
		pLET3	No detectable expression or secretion	Hildebrand & Wells, unpublished
P69	*Bordetella pertussis*	pLET1	Trace amounts of expression	Bailey et al., unpublished
V3 loop of gp120	HIV-1 type MN	pLET1	Intracellular accumulation of protein	Litt et al., unpublished
Interleukin 2 (mIL2)	Murine	pLET2N	Secreted 3 mg/L	Steidler et al., 1995
Xylanase	*Bacillus* sp	pLET1	Active protein formed intracellularly	Bailey et al., unpublished
Streptavidin	*Streptomyces avidinii*	pLET1	Secretion via streptavidin signal leader	Steidler et al., unpublished
Insect toxin (Cry1A(a))	*Bacillus thuringiensis*	pLET1	Insoluble; 20-25% of total cell protein, toxin active	Wells et al., unpublished

system in yields of up to 3 mg/L (Steidler et al., 1995). The IL-2 secreted by *L. lactis* was 100% biologically active and correctly processed to generate a recombinant protein which is identical to the mature form of the natural protein. The insect toxin Cry1A(a) of *Bacillus thuringiensis* HD-1 is another example of a protein which has been efficiently expressed in *L. lactis* (in amounts up to 20-25% of total cell protein; Wells et al., unpublished). The Cry1A(a) protein expressed in *L. lactis* was insoluble as it is in *B. thuringiensis* and toxic when given in the form of a cell extract or as live expressor cells to the larvae of tobacco hornworm (*Manduca sexta*)

Although some proteins such as TTFC, a glutathione S-transferase and the Cry1A(a) insect toxin have been expressed at high levels, only low or trace amounts of certain other proteins are formed. For example, no expression of *Vibrio cholera* toxin subunit B (CTB) was detected in *L. lactis* when the gene lacking its own signal sequence was cloned into pLET1 (for intracellular expression) or in the expression secretion vectors pLET2 and pLET3 despite the fact that high amounts of CTB were formed in *E. coli* using the same CTB expression cassettes present in *L. lactis* (Hildebrand and Wells unpublished). This apparent failure of the expression system cannot be explained by plasmid instability, lack of transcription or mRNA stability indicating that the problem lies with the translation of CTB in the different bacterial hosts or that CTB is rapidly degraded by proteases in *L. lactis*. The reasons for the failure to express efficiently express certain proteins in *L. lactis* using the T7 RNA polymerase system is currently the subject of investigation.

Concluding Remarks

The extensive work carried out over the past decade by a several laboratories on the genetics of lactococci and their plasmid biology has enabled a variety of cloning plasmids and specialised vectors to be constructed for use in LAB. Several constitutive expression vectors are now available for expression in lactococci which are either based on the pWVO1/pSH71 replication group of cloning vectors or on derivatives of pAMβ1. Both types of plasmid replicate in a wide range Gram-positive bacteria and but those which contain the pAMβ1 replicon (i.e. pTREX and pTRCX) can be used for the cloning of large DNA fragments and provide more structural stability to recombinant plasmids. The pTREX series of expression vectors have been designed to the study of the effects of different expression signals on gene expression in LAB and in these vectors the various sequence elements influencing gene expression can be easily manipulated. The suitability of pTREX for studies of gene expression in LAB has been demonstrated by constructing a series of vectors which express genes encoding various reporter enzymes and antigens at varying levels in *L. lactis*. The pTREX vectors were also shown replicate and to be structurally stable three species of *Lactobacillus* and also *B. thuringiensis* indicating that these vectors will have a potentially broad application for studies of gene expression in the LAB. Studies on the lactose operon of *L. lactis* led to the identification of the regulated lac promoter which has been instrumental in the development of the inducible expression systems described for the lactococci. Expression vectors containing the lac promoter such as pNZ2014 and pTREX-*plac* can used for lactose-inducible expression in strains of *L. lactis* which carry the *lac* genes either on a plasmid or integrated into the chromosome. The *lac* promoter has also been used as the basis for the development of a high level inducible expression system which utilises the *E. coli* T7 bacteriophage RNA polymerase to drive expression of antigen genes cloned in the pLET series of vectors in *L. lactis*. Using the lactococcal T7 system a number of heterologous proteins have been expressed at high levels (2-20% total soluble cell protein) in *L. lactis*. We have prepared pLET vectors which can be used (a) to produce the protein intracellularly in amounts up to 25% of soluble protein (b) to secrete the protein into the growth medium or (c) to anchor the protein in the wall and membrane. The expression systems currently under development have already enabled us to explore the feasibility of using recombinant *L. lactis* as a vaccine delivery vehicle to deliver antigens via the digestive tract. It is also envisaged that the continued development of optimised expression systems for *L. lactis* could enable these organisms to be used for the production of a number of enzymes, compounds, antimicrobials, and heterologous proteins of value to the food, pharmaceutical and agricultural industry.

Acknowledgements

The authors gratefully acknowledge the financial support of the Biotechnology and Biological Sciences Research Council, UK. We wish to thank Nina Hey (University of Cambridge) for her invaluable help in constructing pTRCX. We thank Todd Klaenhammer and Gwen Allison (University of North Carolina, USA) for their expert help and advice with the manipulation of *Lactobacillus*

References

Anderson, P.H. & Gasson, M.J. (1985) High copy number plasmid vectors for use in lactic streptococci. *FEMS Microbiology Letters* 30: 193-196.

Brantl S., and Behnke D. (1992). Characterisation of the minimal origin required for replication of the streptococcal plasmid pIP501 in *Bacillus subtilis*. Mol. Microbiol. 6: 3501-3510.

Brantl S., Behnke D., and Alonso J.C. (1990). Molecular analysis of the replication region of the conjugative *Streptococcus agalactiae* plasmid pIP501 in *Bacillus subtilis*. Comparison with plasmids pAMβ1 and pSM19035. Nucl. Acids Res. 18: 4783-4790.

Chamberlain, L.M., Robinson, K., Wells, J.M., and Le Page, R.W.F. (1996) Immune response to whole glutathione S-transferase (P28) and its epitopes from *Schistosoma mansoni* expressed in *Lactococcus lactis*. Immunology 86, Supplement 1, 27.

Claassen E., Kottenhagen M.J., Pouwels P.H., Posno M., Boersma W.J.A., and Lucas C.J. (1994). Use of Lactobacillus, a GRAS (Generally Recognised As Safe) organism; as a base for a new generation of "oral" live vaccines. (G.P. Talwar et al., eds.). pp. 407-412.

Clewell D.B., Yagi Y., Dunny G.M., and Schultz S.K. (1974). Characterisation of three deoxyribonucleic acid molecules in a strain of *Streptococcus faecalis*: identification of a plasmid determining erythromycin resistance. J. Bacteriol. 117: 283-289.

de Vos W.M. (1986a). Genetic improvement of starter streptococci by the cloning and expression of the gene coding for a non-bitter proteinase. In "Biomolecular Engineering in the European Community: Achievements of the Research Programme (1982-1986): Final Report." (E. Magnien, ed.). pp. 465-472. Martinus Nijhoff Publishers, Dordrecht.

de Vos W.M. (1986b). Gene cloning in lactic streptococci. Neth. Milk Dairy J. 40: 141-154.

de Vos W.M. (1986c). Sequence organisation and use in molecular cloning of the cryptic lactococcal plasmid pSH71. Second Streptococcal Genetics Conference, Miami Beach. American Society for Microbiology. P223.

de Vos W.M. (1987). Gene cloning and expression in lactic streptococci. FEMS Microbiol. Rev. 46: 281-295.

de Vos W.M., Siezen R.J., and Kuipers O.P. (1992). Lantibiotics similar to nisin A. PCT Patent Application WO 92/18633

de Vos W.M., and Simons A.F.M. (1987). Method for preparing proteins using transformed lactic acid bacteria. European Patent 0 228 726

Eaton T.J., Shearman C.A., and Gasson M.J. (1993). The use of bacterial luciferase gene as reporter genes in *Lactococcus*: regulation of the *Lactococcus lactis* subsp. *lactis* lactose genes. J. Gen. Microbiol. 139: 1495-1501.

Frere J., Novel M., and Novel G. (1993). Identification of the theta-type minimal replicon of the *Lactococcus lactis* CNRZ270 lactose-protease plasmid pUCL22. Current Microbiology. 27: 97-102.

Gibson E.M., Chace N.M., London S.B., and London J. (1979). Transfer of plasmid-mediated antibiotic resistance from streptococci to lactobacilli.

Gravesen A., Josephsen J., von Wright A., and Vogensen F.K. (1995). Characterisation of the replicon from the lactococcal theta-replicating plasmid pJW563. Plasmid. 34: 105-118.

Gruss A., and Ehrlich S.D. (1988). Insertion of foreign DNA into plasmids from gram-positive bacteria induces formation of high molecular weight plasmid multimers. J. Bacteriol. 170: 1183-1190.

Gruss A., and Ehrlich S.D. (1989). The family of highly interrelated single-stranded deoxyribonucleic acid plasmids. Microbiol. Rev. 53: 231-241.

Gruzza M., Langella, P., Duval-Iflah, Y., Ducluzeau, R. (1993). Gene transfer from engineered *Lactococcus lactis* strains to *Enterococcus faecalis* in the digestive tract of gnotobiotic mice. Microb. Releases 2: 121-125.

Hayes F., Daly C., and Fitzgerald G.F. (1990). Identification of the minimal replicon of *Lactococcus lactis* subsp. *lactis* UC317 plasmid pCI305. Appl. Environ. Mirobiol. 56: 202-209.

Horodniceanu T., Bouanchaud D., Biet G., and Chabbert Y. (1976). R plasmids in *Streptococcus agalactiae* (group B). Antimicrob. Agents Chemother. 10: 795-801.

Kiewiet R., Bron S., de Jonge K., Venema G., and Seegers J.F.M.L. (1993a). Theta replication of the lactococcal plasmid pWVO2. Molec. Microbiol. 10: 319-327.

Kiewiet R., Kok J., Seegers J.F.M.L., Venluna G., and Bron S. (1993b). The mode of replication is a major factor in segregational plasmid instability in *Lactococcus lactis*. Appl. Environ. Microbiol. 59: 358-364.

Koivula T., Sibakov M., and Palva I. (1991). Isolation and characterisation of *Lactococcus lactis* subsp. *lactis* promoters. Appl. Environ. Microbiol. 57: 333-340.

Kok J., van der Vossen J.M.B.M., and Venema G. (1984). Construction of plasmid cloning vectors for lactic streptococci which also replicate in *Bacillus subtilis* and *Escherichia coli*. Appl. Environ. Microbiol. 48: 726-731.

Le Blanc D.J., Hawley R.J., Lee L.N., and St. Martin E.J. (1978). "Conjugal" transfer of plasmid DNA among oral Streptococci. Proc. Natl. Acad. Sci. USA. 75: 3484-3487.

O'Sullivan D.J., and Klaenhammer T.R. (1993). High- and low-copy-number *Lactococcus* shuttle cloning vectors with features for clone screening. Gene. 137: 227-231.

Oultram J.D., and Young M. (1985). Conjugal transfer of plasmid pAMβ1 from *Streptococcus lactis* and *Bacillus subtilis* to *Clostridium acetobutylicum*. FEMS Micro. Lett. 27: 129-134.

Platteeuw C., Simons G., and de Vos W.M. (1994). Use of the *Escherichia coli* β-glucuronidase (*gusA*) gene as a reporter for analyzing promoters in lactic acid bacteria. Appl. Environ. Microbiol. 60: 587-593.

Schofield K.M., Wilson P.W., Waterfield N.R., Le Page R.W.F., and Wells J.M. (1996). Novel and versatile vectors for studies of gene expression in lactic acid bacteria. Submitted.

Simon D., and Chopin A. (1988). Construction of a vector plasmid family and its use for molecular cloning in *Streptococcus lactis*. Biochimie. 70: 559-567.

Simons G., Buys H., Hogers R., Koehnen E., and de Vos W.M. (1990). Construction of a promoter-probe vector for lactic acid bacteria using the *lacG* gene of *Lactococcus lactis*. Dev. Ind. Microbiol. 31: 31-39.

Simons G., and de Vos W.M. (1994). Gene cloning and expression systems in *Lactococci* In: Genetics and Biotechnology of Lactic Acid Bacteria (eds. Gasson M.J. and de Vos W.M.) Blackie, Glasgow UK.

Steidler L., Wells J.M., Raeymaekers A., Vandekerckhove J., Fiers W., and Remaut E. (1995). Secretion of biologically active murine interleukin-2 by *Lactococcus lactis* subsp. *lactis*. Appl. Environ. Microbiol. 61: 1627-1629.

Swinfield T.J., Oultram J.D., Thompson D.E., Brehm K.J., and Minton N.P. (1990). Physical characterisation of the replication region of the *Streptococcus faecalis* plasmid pAMβ1. Gene. 87: 79-90.

van Asseldonk M., de Vos W.M., and Simons G. (1993). Functional analysis of the *Lactococcus lactis usp45* secretion signal in the secretion of a homologous proteinase and a heterologous α-amylase. Mol. Gen. Genet. 240: 428-434.

van de Guchte M., Kok J., and Venema G. (1989). Construction of a lactococcal expression vector: expression of hen egg white lysozyme in *Lactococcus lactis* subsp. *lactis*. Appl. Environ. Microbiol. 55: 224-228.

van de Guchte M., Kodde, J., van der Vossen, J.M.B.M., Kok, J. and Venema, G. (1990). Heterologous gene expression in *Lactococcus lactis* subsp. *lactis*: synthesis, secretion, and processing of the *Bacillus subtilis* neutral protease. Appl. Environ. Microbiol. 56: 2606-2611.

van de Guchte M., Kok J., and Venema G. (1991). Distance-dependent translational coupling and interference in *Lactococcus lactis*. Mol. Gen. Genet. 227: 65-71.

van der Vossen J.M.B.M., Kok J., and Venema G. (1985). Construction of cloning, promoter screening and terminator screening shuttle vectors in *Bacillus subtilis* and *Streptococcus lactis*. Appl. Environ. Microbiol. 50: 540-542.

van der Vossen J.M.B.M., van der Lelie D., and Venema G. (1987). Isolation and characterisation of *Streptococcus cremoris* Wg2-specific promoters. Appl. Environ. Microbiol. 53; 2452-2457.

van Rooijen R.J., and de Vos W.M. (1990). Molecular cloning, characterisation and nucleotide sequence of *lacR*, a gene encoding the repressor of the lactose phosphotransferase system of *Lactococcus lactis*. J. Biol. Chem. 265: 18499-18503.

van Rooijen R.J., and de Vos W.M. (1993). Purification of the *Lactococcus lactis LacR* repressor and characterisation of its DNA binding sites. R.J. van Rooijen, Ph.D. thesis pp 101-119.

van Rooijen R.J., Gasson M.J., and de Vos W.M. (1992). Characterisation of the promoter of the *Lactococcus lactis* lactose operon: contribution of flanking sequences and LacR repressor to its activity. J. Bacteriol. 174: 2273-2280.

Waterfield N.R., Le Page R.W.F., and Wells J.M. (1995). The isolation of lactococcal promoters and their use in investigating bacterial luciferase synthesis in *Lactococcus lactis*. Gene. 165: 9-15.

Waterfield N.R., (1996) Ph.D. thesis, University of Cambridge, UK.

Wells J.M., Norton P.M., and Le Page R.W.F. (1995). Progress in the development of mucosal vaccines based on *Lactococcus lactis*. Int. Dairy Journal 5: 1071-1079.

Wells J.M., Wilson P.W., and Le Page R.W.F. (1993a). Improved cloning vectors and transformation procedure for *Lactococcus lactis*. J. Appl. Bacteriol. 74: 629-636.

Wells J.M., Wilson P.W., Norton P.M., Gasson M.J., and Le Page R.W.F. (1993b). *Lactococcus lactis*:high level expression of tetanus toxin fragment C and protection against lethal challenge. Molec. Microbiol. 8: 1155-1162.

Wells J.M., Wilson P.W., Norton P.M., and Le Page R.W.F. (1993c). A model system for the investigation of heterologous protein secretion pathways in *Lactococcus lactis*. Appl. Environ. Microbiol. 59: 3954-3959.

Xu F., Pearce L.E., and Yu P.-L. (1991). Construction of a family of lactococcal vectors for gene cloning and translational fusions. FEMS Microbiol. Lett. 77: 55-60.

Expression of Human and Murine Interleukins in *Lactococcus lactis*

Lothar Steidler, Walter Fiers and Erik Remaut
Laboratory for Molecular Biology
Universiteit Gent
KL. Ledeganckstraat 35
B-9000 Gent
Belgium

The need for recombinant cytokines

In recent years there has been an increasing interest in the development of systems for the high yield production of biologically active proteins. One of the most tempting domains, because of the requirement for high analogy to the natural products, economic and scientific value of these end products and fundamental difficulties encountered along the way, is the synthesis of recombinant cytokines.

Cytokines, or lymphokines when strictly related to lymphoid cells, are a series of freely diffusible secretory proteins which modulate different aspects of the immune system, such as growth stimulation of lymphoid cells, gene expression, cell mobility, immunosuppression etc. For practical and economical reasons it is almost impossible to perform preparative purification of these proteins from their natural source, as they are very often only present in trace amounts. Furthermore, because of the requirement for uncompromised, highly purified lymphokine preparations for medical and research purposes, a lot of attention has been put on the use of recombinant expression systems for the production of these scientifically and commercially important proteins.

Although recombinant gene technology has enabled eukaryotic proteins to be efficiently produced in micro-organisms, in only a few cases has this led to the development of elegant production procedures for lymphokines. A general feature of the high level intracellular production of foreign proteins in *Escherichia coli*, the most established and well studied of the microbial production organisms, is the formation of insoluble inclusion bodies (Schein 1989). This often results in the formation of an

NATO ASI Series, Vol. H 98
Lactic Acid Bacteria:
Current Advances in Metabolism, Genetics and Applications
Edited by T. Faruk Bozoğlu and Bibek Ray

inactive protein which is sometimes impossible to refold in vitro. Unfortunately this phenomenon is not restricted to the cytoplasm and may still occur when proteins are secreted into the periplasm of *E. coli* (Cleary 1989). In the most favourable cases soluble protein can be recovered from the periplasm by selective disruption of the outer membrane (Steidler et al., 1994 ; Robbens et al. 1995).

Another recognised problem with the use of *E. coli* as a production organism is the presence in its cell wall of lipopolysaccharides (LPS). These compounds are known to be extremely pyrogenic and prove to be difficult to remove during purification of the desired protein. When present as contaminants, they can strongly mask any lymphokine effect or lead to the death of lab animals by the induction of systemic inflammatory responsive syndrome. The use of *Lactococcus lactis* as a production host is an elegant by-pass of this problem since the start material for purification can be produced LPS free.

A *lac* controlled T7 RNA polymerase driven expression system in *L. lactis*.

There is now a growing interest in the development of recombinant expression systems for the food grade Gram-positive lactic acid bacteria as these organisms are already widely used in a variety of industrial fermentation processes. Most progress has been made with the lactococci which have a Generally Regarded As Safe (GRAS) status and the capacity to secrete proteins into the growth medium, making them potentially useful organisms for the safe production of commercially important proteins. The chromosomal derived promoter P32, and the expression signals of the regulated *lac* operon, nisin operon and cell wall associated proteinase *prtP* gene have all been used to express heterologous proteins in *L. lactis*. The secretion of heterologous proteins into the growth medium has also been achieved by making in frame fusions to the signal secretion leaders of the lactococcal *usp*45 and *prt*P genes (reviewed by de Vos and Simons, 1994).

Two examples of the expression of eukaryotic proteins in. *L. lactis* are hen egg white lysozyme and prochymosin. The hen egg white lysozyme gene was fused to the open reading frame downstream of the P32 promoter (van de Guchte et al, 1989). Lysozyme could be detected by western blotting but no lysozyme activity was observed in cell extracts of *L. lactis*. In the second study a series of in-frame fusions were made between the coding sequence of bovine prochymosin and the lactococcal

prtP gene. The fusion to the first 62 amino acids of the *prtP* gene resulted in the secretion of a processed PrtP-prochymosin fusion into the growth medium which was detected by western blotting (Simons et al., l992).
The lactococcal *lac* promoter region (Van Rooijen et al., 1992) has also been used to develop a regulated expression system for *L. lactis* (figure 1). This system makes

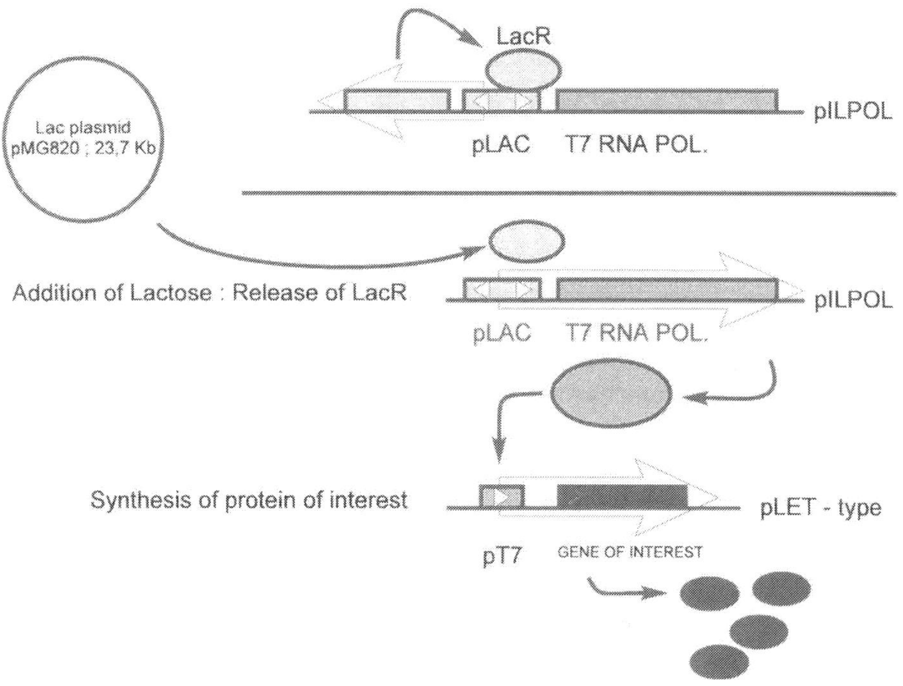

Figure 1. : *lac* controlled T7 RNA polymerase driven expression system (Wells et al., 1993b). Following switching from glucose to lactose containing growth medium, the *lacABCDFEGX* promotor is derepressed and drives the synthesis of the T7 RNA polymerase. This polymerase is active upon the T7 promotor and leads to the expression of the gene of interest.

use of the *E. coli* T7 bacteriophage RNA polymerase to drive expression of genes cloned in the pLET series of vectors (Wells et al., 1993b). The *lacR* and *lacABCDFEGX* promotors are arranged divergently. The LacR protein acts as a repressor upon the *lacABCDFEGX* promotor and autoregulates itself (Eaton et al., 1993). A vector pILPOL has been constructed on which the T7 RNA polymerase gene is cloned downstream of inducible *lacABCDFEGX* promoter. The same plasmid holds the LacR expression cassette, so in this way, in the absence of the inducer the expression of the T7 RNA polymerase is repressed. When the carbon source is switched from glucose to lactose, tagatose 6-phosphate, a lactose metabolite which prevents the repressor from binding the *lac* promotor and thus derepresses transcription of the T7 RNA polymerase gene, is formed in the cell. Once formed in the cell, the T7 RNA polymerase initiates transcription from its cognate promoter present on the pLET series vectors and is able to drive the expression of a gene of interest, suitably cloned on this plasmid. Vectors have been prepared which can be used to produce recombinant protein intracellular in amounts up to 22% of soluble protein (Wells et al, 1993c) or alternatively to secrete the protein into the growth medium (Wells et al l993b).

Construction of a Murine Interleukin 2 expression unit

We made use of the lactococcal T7 RNA polymerase based system to express murine interleukin 2 (mIL2 ; Steidler et al. 1995). IL2 or T-cell growth factor is a 149 aa. lymphokine which allows the long-term proliferation of activated T-cells (Morgan et al., 1976; Rees, 1990). Furthermore IL2 enhances the generation of cytotoxic T lymphocytes and is involved in B cell growth (Wagner et al. 1980; Howard et al., 1983; Rees 1990).

A part of the *mIL2* gene, which codes for the mature protein, has been fused to the lactococcal *usp*45 secretion signal leader (van Asseldonck et al., 1992). The construction of the mIL2 expression plasmid is depicted in figure 2. From the plasmid pLET2 (Wells et al.,1993a) the 0.3 Kb. *Eco*RI - *Hind*III fragment was isolated and cloned in the plasmid pMa58 (Stanssens et al.1989). This plasmid was subjected to site specific mutagenesis according to Stanssens et al. (1989) by the use of the synthetic oligonucleotide GGATCCGTCGCCGGCGTAAAC. The mutant *Eco*RI - *Hind*III fragment was cloned into the *Eco*RI - *Hind*III opened plasmid pMIG1 (Wells et al.,1993a) and the resulting plasmid was called pLET2N. This plasmid contains a *Nae*I restriction site at the processing site of the *usp*45 leader peptide, while retaining the amino acid sequence of this leader. A 551 bp. *Fsp*I - *Spe*I fragment encompassing the sequence for mature mIL2 was isolated from plasmid pSP64mIL2m (De Molder et al. 1992). This fragment was ligated to the larger fragment of *Nae*I and *Bgl*II digested pLET2N and the smaller fragment of pLET2N, digested with *BglII* and *XbaI*. The resulting plasmid contained the fusion between the *usp45* leader and the part of the *mIL2* gene coding for the mature protein under the control of the T7 promotor and was called pL2MIL2.

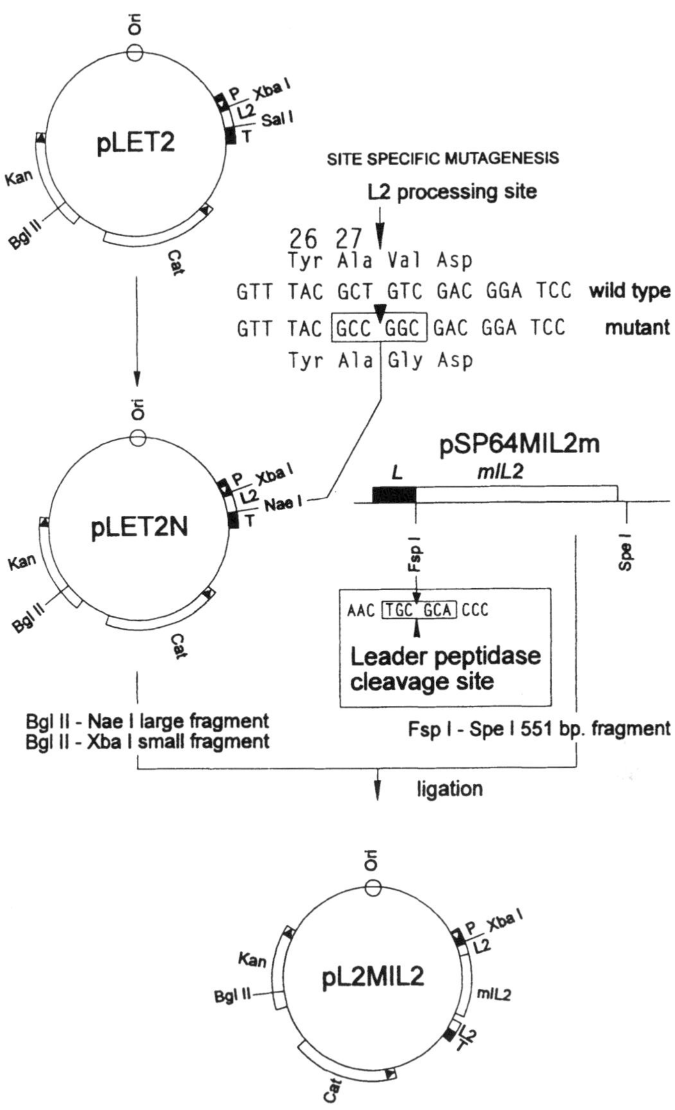

Figure 2. : Flow scheme for the construction of the mIL2 expression plasmid pL2MIL2. L represents the *mIL2* leader and L2 represents the *usp45* leader. P stands for the T7 promotor and T stands for terminator.

Synthesis of Murine Interleukin 2

The plasmid pL2MIL2 was transferred to MG1820 (pILPOL). This *L. lactis* strain holds the *lac* plasmid pMG820 and the inducible T7 expression unit. A preculture of the obtained strain MG1820 (pILPOL, pL2MIL2) was grown in growth medium supplemented with glucose. Switching from glucose to lactose containing growth medium induces the T7 RNA polymerase. This leads (figure 1) to the synthesis of the USP45-mIL2 fusion protein. We observed the accumulation of mature mIL2 into the growth medium. Prior to the preparation of total protein extracts from the growth medium, all insoluble material had been removed from the culture supernatant by centrifugation at 100.000 g for 1 hour. This step assures that all protein material used in further analysis is completely soluble.

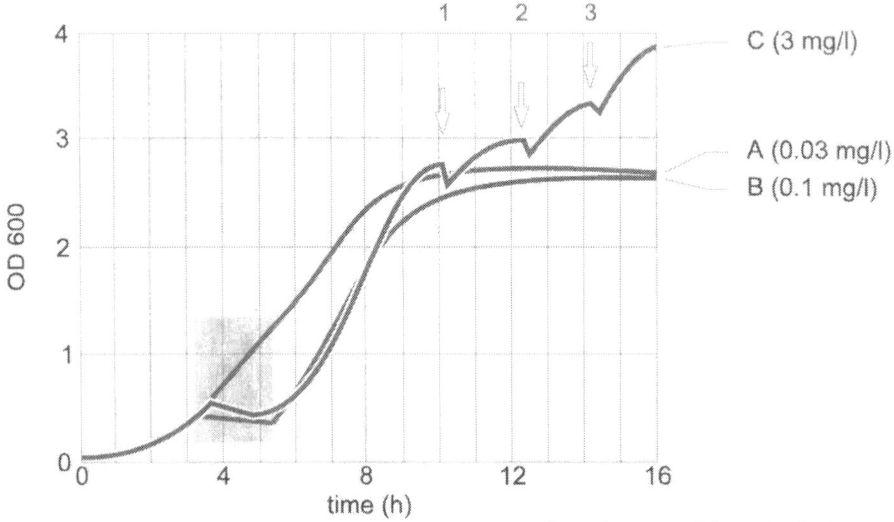

Figure 3. : OD profiles of mIL2 producing batch cultures. The shaded area represent the switching of growth medium from GM17 to LM9. mIL2 yields are indicated.

Initial studies on the production of mIL2 were carried out using M17 (Difco), a growth medium for the cultivation of Streptococci. However, when the culture supernatant was analysed by SDS-PAGE and coomassie staining, a dense smear was observed on the gels. This smear was likely to mask any protein band produced by the recombinant organism. To address this problem we changed the strategy for growth of production batches (standard 10L standing cultures in a New Brunswick Scientific Bio-Flo IV fermentor at 30°C). Initial growth was still performed in M17 supplemented with 0.5% glucose and selective antibiotics but at an OD_{600} of 0.5 the cells were harvested in a Heraeus contifuge, the cell pellet was washed once with 0.5 l of LM9 and resuspended in the same volume of LM9. LM9 is M9 (Miller, 1972) containing 0.5% lactose. This growth medium is virtually a phosphate buffer. To this we added casitone, an oligopeptide mixture and the only proteinaceous compounds present, as a source of amino acids. The use of this growth medium allowed the detection of the recombinant protein in the cell supernatant by coomassie staining. Figure 3 shows OD_{600} profiles of batch cultures of MG1820 (pILPOL,pL2MIL2). From graph A it is apparent that cell growth does not stop when cells were collected, as the resuspended culture had gained one OD_{600} unit. When only a part of the collected cells was resuspended (figure 3, graph B) so that the LM9 culture started at an OD_{600} of approximately 0.5, we observed a final increase in the production of mIL2 from 0.03 mg/l to 0.1 mg/l. This indicates that production of mIL2 occurs in the active growth phase. Hence we concluded that a further extension of this growth phase could possibly augment the final yield.

This finding redirects the question for an improved yield to that of which parameters are restricting for the further growth of the culture to a higher density. In all cultures, represented by graphs A and B, figure 3, we observed a plateau value in the OD_{600} profile of 2.5. At this plateau the culture had reached pH 4.5, whereas pH was around neutral at the beginning of the LM9 culture. Also at this point starvation of the culture was to be expected. The crucial role of these two factors, acidity and nutrient concentration, was made clear in a third type of batch culture (figure 3, graph C). In this, we adjusted the pH of the culture by the addition of 150 ml of 2 M NaOH to 6.0 when reaching the first stationary phase (1). This resulted in an immediate increase in OD. The slight initial drop of the density is caused by the dilution of the culture. At the second plateau, pH had reached 5.5. At this point (2) addition of nutrients (lactose and casitone were added up to 0.5%.), rather than adjusting pH gave rise to a second growth stimulation. The inverse was observed at point 3, where the addition of extra nutrients did not have any effect but the adjustment of pH, which had again reached 4.5 gave rise to an immediate increase in the OD., up to 4. This leads us to the conclusion that both pH and the availability nutrients are key features in this type of production batch. Crude growth medium, collected at the end of the experiment was analysed and showed the production of 3mg/l mIL2 (figure 4). From this we conclude that the mIL2 is synthesised and secreted during active growth.

Evidently, the next step in the optimalization of this production process will comprise the automation of alkaline and nutrient addition, in which we would aim for a constant pH of 5.5 and lactose and casitone concentration of 0.5%. Since the culture immediately responds to the added compounds, we think that the capacity of this type of production-culture is far from exhausted.

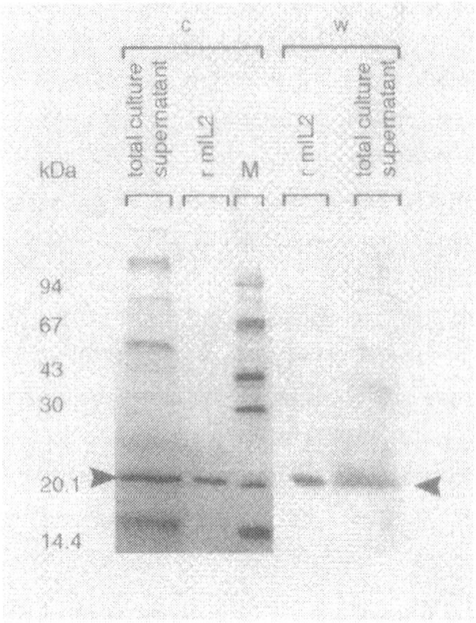

Figure 4. : Total protein fractions from culture supernatant, analysed by SDS 15% PAGE, revealed with coomassie brilliant blue staining (c) or western blot immunodetection using rabbit anti mIL2 (w) and compared to purified recombinant mIL2 (r mIL2). The equivalent of 1 ml of growth medium was loaded in each lane. Reference r mIL2 represents 1 µg.

Qualitative and quantitative analysis of the obtained product

In order to eliminate all insoluble material, part of the culture supernatant was centrifuged at 100,000 x g for 1 hour and the supernatant was used in further studies.

The full protein profile of the growth medium was compared to purified mIL2 and was visualised by SDS 15% PAGE and coomassie brilliant blue staining or alternatively by western blot using rabbit anti mIL2 (figure 4). The equivalent of 1 ml of culture medium was analysed and compared with 1 µg of purified mIL2. This shows that mIL2 is one of the major bands in the crude cell supernatant.

To estimate the mIL2 concentration in the crude supernatant, serial two-fold dilutions of this fraction and of a mIL2 standard of known concentration were separated on a SDS 15% PAGE gel and stained with coomassie brilliant blue. The image obtained was scanned with a Hewlett-Packard Scanjet IIcx. The bands corresponding to mIL2 were densitometrically quantified using Image-Quant™ software (Molecular Dynamics, Sunyvale, CA, USA). The range of results where consecutive steps in the dilution series gave rise to density values which related 1:2 was used to calculate the mIL2 concentration. This was done by comparison with the values obtained from the reference mIL2 standard. This experiment showed the presence of 3mg of mIL2 per litre of culture supernatant. These results were confirmed by direct ELISA, using the same recombinant mIL2 reference as a standard.

Biological activity was assayed as described (Gillis et al., 1978). Essentially the proliferation of T-cells by a serial dilution of crude mIl2 producing cell supernatant was compared with and expressed relative to an international hIL2 standard (1.13 x 10^7 u/mg). The biological activities of hIL2 and mIL2 are the same in this type of assay (Mosmann et al., 1987). The assay showed the presence of 3.5×10^4 u per ml of the crude cell supernatant from *L. lactis* (pILPol ; pL2MIL2), obtained after lactose induction as described. From these results we conclude the full biological activity of the recombinant mIL2, secreted by *L. lactis*.

To investigate whether the USP45-mIL2 precursor had been correctly cleaved, a total protein fraction of the growth medium was run on a SDS 15% PAGE gel and electroblotted onto a Pro Blott membrane (Bauw et al., 1988). The band corresponding to the recombinant mIL2 was cut out and the N-terminal amino acid sequence was determined by automated Edman degradation on a 470A gas-phase sequenator coupled to a 120 A on-line phenylthiohydantoin amino acid analyzer (Applied Biosystems, Foster City, CA, USA). This revealed the N-terminal sequence : Ala Pro Thr Ser Ser, which is exactly the same as in natural, mature mIL2 (Kashima et al., 1985).

From these data we conclude that the recombinant mIL2, produced from *L. lactis* is completely soluble. Amino terminal amino acid sequencing revealed that the Usp45-mIL2 had been correctly processed. Assessment of the biological activity in a T-cell proliferation assay showed no significant difference in biological activity, when compared to a native standard. This is the first description of the use of *L. lactis* to produce a fully active and correctly processed recombinant protein of eukaryotic origin in high yield.

The expression of various lymphokines from human and murine origin

The encouraging results obtained with the mIL2 brought us to investigate the possibilities for expression of other lymphokine genes in *L. lactis*. In recent work we have subcloned several of these genes in the pLET type expression vector.

	product	yield
A	murine interleukin 2	100 ng/ml
	murine interleukin 4	3 ng/ml
	murine interleukin 6	100 ng/ml
	humane interleukin 2	50 ng/ml
	murine interferon-β	0 ng/ml
	murine 55 K soluble TNF receptor	50 ng/ml
B	*S. aureus* protein A	1500 ng/ml
	streptavidin	3000 ng/ml

Table 1. : Synthesis levels of lymphokines (panel A) and proteins from prokaryotic origin (panel B).

Constructs were designed in such way that insertions into the expression plasmid were made in the same way as had been done for the *mIL2* gene : all were designed as fusion genes, in which the *usp*45 leader sequence was fused in frame to the mature part of the lymphokine genes. We have made such expression plasmids for the murine interleukins 4 and 6 (mIL4, mIL6), murine interferon-β (mIFN-β), murine 55K soluble TNF receptor (mSR55) and human interleukin 2 (hIL2). All of these constructs were verified by sequence analysis and showed exact fusions between the *usp*45 leader and the sequences encoding the mature proteins. The expression plasmids were transferred to *L. lactis* MG1820 (pILPOL) and the various expression strains were evaluated for the production of the respective lymphokines. For this, 10 ml tube cultures were incubated at 30°C without shaking. Cells were pregrown in glucose-M17, collected at OD_{600} of 0.5 and washed and resuspended in lactose-M9. Soluble fractions were assayed for the concentration and biological activities of the respective lymphokines. The bioassays performed were proliferation of myeloid cells (mIL4, Vandenabeele et al., 1990), T-cells (hIL2, Gillis et al., 1978) and a B-cell hybridoma (mIL6, Van Snick et al., 1986), antiviral activity (mIFNβ, as in Derynck et al., 1980, but performed on L929R2, (Vanhaesebroeck et al, 1991) cells) and inhibition of TNF activity (mSR55, Espevik and Nissen-Meyer, 1986)). The results obtained show a wide variety of synthesis levels (table 1, note that no growth optimisation of the cultures has been applied, as can be seen from the mIL2 expression level). Comparison of concentrations of the various lymphokines with the values obtained from assays on the biological activities show that all expressed lymphokines displayed 100 % identical activities as compared to the natural proteins.

Approaches for the understanding of highly variable expression levels

The very broad range of expression levels observed (0 to 100 ng ml^{-1} for the cytokines, 0 to 3000 ng ml^{-1} when other secreted proteins were used in the comparison) was somewhat surprising to us. Indeed, divers factors which others have reported to influence the level of expression to a considerable extent, seem to be identical to all underlying constructs. It is mainly for this reason that we believe that a profound study for the underlying causal events of synthesis differences in this series of proteins could help to gain insight in the critical aspects which have to be taken into account when considering *L. lactis* as a host for heterologous gene expression. Furthermore, cytokines prove to be very attractive as tools for this type of studies since they can be detected at fairly low concentrations and the eventual optimisation of any of the expression constructs would result in an interesting end product.

Firstly it is important to point to the very analogous way in which the different interleukin expression constructs were made. Essentially, it can be stated that, apart from the actual sequence encoding the mature proteins and some 3' untranslated regions, all of these plasmids are identical. In particular they contain the same promotor region and an identical ribosome binding site. This renders unlikely a number of possible reasons to which differences in expression levels often can be attributed. Numerous reports mention that, apart from the actual strength of the promotor as such, the region around the promotor can contain various stimulatory and silencing motifs, e.g. depending on methylation or on the binding of activating complexes. Sequence variations in the ribosome binding site can strongly affect expression levels. Of particular interest in this is the distance between a certain Shine Dalgarno sequence and any downstream open reading frame. Furthermore, initiation of translation is often influenced by the nature of sequences around the ATG, such as high or low GC content.

As mentioned, for some of the expression modules investigated the 3' untranslated regions are not identical. To verify in an experimental way whether this could influence expression levels, we exchanged these regions between mIL2 and mIL4, produced at 100 and 3 ng ml^{-1} respectively. For this we introduced *Spe*I restriciton sites immediately adjacent to the stopcodons of both genes, as they were present in the pLET type plasmids. Assessment of interleukin concentrations showed that the introduction of the mutations did not alter the levels of expression. When 3' untranslated regions were interchanged and concentrations were determined again, we noticed no difference in production of mIL2 or mIL4, so it could be ruled out that these 3' regions have either a stimulatory or a restraining activity.

To address the possible role of codon usage in the difference in expression, we compared the lymphokine genes for the degree to which frequently used codons are present. For this, we made use of the table on preferential codon usage, by Van Der Guchte and Venema (1992). Based on the data given in this table, codons were given quotations ranging from 0 to 1. The values are, for a particular codon, encoding amino acid A, ((percentage of particular codon for A) x (percentage of most frequently used codon for $A)^{-1}$). These numbers enable us to compare the degree of "optimality" for different codons. If one would e.g. consider the hypothetical amino acid A, encoded by four codons, all used in 25% of the cases, all codons would have the value 1, since no one of them is preferred over any one other. Similarly considering the hypothetical amino acid B, encoded by two codons, the first one of which occurs at a frequency of 75%, the second one at a frequency of 25%, one obtains the values 1 and 0.33 respectively. This example shows that all four codons for A are as optimal for their amino acid as is the first codon for B, although a percentage plot would not show this. A second consideration about the ratios used is that one could always obtain a value of 1 for all codons of an open reading frame. We compared all lymphokines in our study by making plots per codon position of the described values. This gives a "hills and valleys" profile, showing possible particular blocks of unfavourable codon usage. By means of this type of plot we observed a 12 consecutive CAG (rated 0.31) Gln codon stretch in mIL2. This stretch is absent from hIL2, which is for the rest highly homologous to mIL2. However hIL2 is expressed only at half the level of mIL2. If we sort the values for all codons of a certain gene according to decreasing value, giving a "slope" plot, we can compare different genes for their overall codon usage. All mouse genes gave nearly identical plots, possibly showing their genetic relatedness. When the bacterial genes involved in the comparison were taken into account, we could not even conclude the direction of a possible relation between codon usage and expression level : protein A (1500 ng ml^{-1}) had an overall much higher value, streptavidin (3000 ng ml^{-1}) scored strikingly lower than the mouse genes. From this we conclude that it is not possible to make a priori predictions for the improvement in expression of these genes based on optimisation of the codon usage.

A comparable type of analysis was performed regarding the GC content of the genes under study. For this, for any position the sum of values (A or T = 0, G or C = 1) for nucleotides ranging from -5 to +5 were plotted. This gave rise to values between 0 and 11, expressing the GC content of the surroundings of a certain position. The comparison of the obtained "hills and valleys" and "slope" graphs however, did not point to any obvious correlation between GC content and expression levels.

From these findings we do not think it is possible to design an optimisation strategy to acquire enhanced expression, merely based on the location of disadvantageous structures or assessment of the overall composition of either the codon usage or GC content for any of the lymphokine genes.

Discussion

We presented a high yield, semi large scale production method for a fully active eukaryotic protein, based on its synthesis and secretion by the GRAS organism *L. lactis*.

The fragment of the *mIL2* gene, coding for the mature part of the protein was fused to the sequence coding for the secretion signal leader of *usp*45. This fusion gene is under the control of the T7 promotor. This expression plasmid forms a part of a lactose inducible, T7 RNA polymerase driven expression system (Wells et al.1993)

We describe culture conditions by which a standing culture of L. lactis MG1820 (pILPol ; pL2MIL2) is able to grow in a defined growth medium and secrete mIL2. Upon addition of lactose to an exponentially growing culture, we could observe the appearance of a new protein band in the culture medium, which had the same mobility as mIL2, purified from E. coli. This protein band showed reaction with anti mIL2 antiserum in a western blot immunoassay. Amino terminal sequence analysis showed that the recombinant protein had the same N-terminus as had been reported for mIL2. In a T-cell proliferation assay we could observe the same specific biological activity as for natural mIL2. From these results we conclude that the strain *L. lactis* MG1820 (pILPol ; pL2MIL2) is capable of synthesising a USP45-mIL2 fusion protein. Furthermore this strain correctly cleaves of the USP45 signal sequence and releases soluble, fully active mIL2 into the growth medium. The recombinant lymphokine accumulates in the culture supernatant to one of the major protein bands. The synthesis of the recombinant mIL2 occurs predominantly during exponential growth phase. Parallel experiments in which saturated cultures were left for prolonged time (in one case up to 72 hours) showed no increase in the mIL2 concentration, which further substantiates the above statement

In an experimental fermentation we could show that it is possible to increase the bacterial density of a stationary culture by addition of nutrients and by stabilisation of the pH. Both of these appeared to be essential. Indeed, an initial addition of lactose and casitone could only give full effect after the addition of NaOH, which brought the pH from 4.5 to 5.5, but the addition of NaOH to a culture in late log phase could only give rise to an other increase in OD_{600} when nutrients were added.

This type of production offers several advantages in the downstream processing of the desired end product. The synthesis occurs in a cheap, clear growth medium, which is virtually devoid of any protein. The producing organism is food grade and LPS free, which are features that meet requirements on the quality of therapeutic proteins. The crude cell supernatant, in which the recombinant product is present prominently, contains few other proteins. The initial mIL2 concentration is 10 times higher than from an E. coli expression system and comparable to the level of expression from *Pichia pastoris.* The product is made fast and is present in a phosphate buffer, which can be applied directly on colums for hydrophobic interaction.

We subcloned a series of other lymphokine genes in the same way as had been done for *mIL2.* Concentration assessment of the synthesised products showed that a wide range of expression levels was obtained. We made use of this series of constructs to compare some possible effects on the synthesis of heterologous proteins in *L. lactis.* This comparison made clear that no a priori and general statements could be made on the effects of 3' untranslated regions, optimisation of codon usage or GC content. Obviously one could think of particular cases in which the various parameters considered here play an important role. These should however be approached in an experimental way. Our future research will consist of detailed analysis of topics regarding gene structure, which might also include adaptation of codon usage and GC content.

References

Bauw, G., Van den Bulcke, M., Van Damme, J., Puype, M., Van Montagu, M. and Vandekerckhove, J. 1988. Protein-electroblotting on polybase-coated glass-fiber and polyvinylidene difluoride membranes : an evaluation. J. Protein Chem. **7**:194-196

Cleary, S., Mulkerrin, M.G. and Kelley, R.F. (1989) Purificatin and Characterisation of Tissue Plasminogen Activator Kringle-2 Domain Expressed in Escherichia coli. Biochemistry **28** : 1884-1891

De Molder, J., Fiers, W. and Contreras, R. 1992. Efficient synthesis of secreted murine interleukin-2 by *Saccharomyces cerevisiae* : influence of 3'-untranslated regions and codon usage. Gene **111**:207-213

de Vos, W.M. and Simons, G.F.M. 1994. Gene cloning and expression systems in lactococci. in : Genetics and biotechnology of lactic acid bacteria. 52-105 de Vos, W.M. and Gasson, M. (eds.) Blackie Acad. Prof. London

Derynck, R., Remaut, E., Saman, E., Stanssens, P., Declerck, E., Content, J. and Fiers, W. (1980) Expression of human fibroblast interferon gene in *Escherichia coli* Nature, **287**, 193-197

Eaton, T. J., Shearman, C.A. and Gasson, M.J. (1993) The use of bacterial luciferase genes as reporter genes in Lactococcus: Regulation of the Lactococcus lactis ssp. lactis lactose genes. J. Gen. Microbiol. 139(7): 1495-1501

Espevik, T. and Nissen-Meyer, J.(1986) A highly sensitive cell line, WEHI 164 clone 13, for measuring cytotoxic factor/tumor necrosis factor from human monocytes. J. Immunol. Methods **95**:99-105

Gillis, S., Fern, M.M., Ou, W. and Smith, K.A. 1978. T cell growth factor : parameters of production and quantitative microassay for activity. J. Immunol. **120**:2027-2032

Howard, M., Matis, L., Malek, T.R., Shevach, E., Kell, W., Cohen, D., Nakanishi, K. and Paul, W. 1983. Interleukin 2 induces antigen-reactive Tcell lines to secrete BCGF-1. J. Exp. Med. **158**:2024-2039

Kashima, N., Nishi-Takaoka, C., Fujita, T., Taki, S., Yamada, G., Hamuro, J. and Taniguchi, T. 1985. Unique structure of murine interleukin-2 as deduced from cloned cDNAs. Nature **313**:402-404

Miller, J. H. 1972. Experiments in Molecular Genetics. Cold Spring Harbor Laboratory Press, Cold Spring Harbor NY.

Morgan, D.A., Ruscetti, F.W. and Gallo, R.C. 1976. Selective in vitro growth of T lymphocytes from normal human bone marrows. Science **193**:1007-1008

Mosmann, T.R., Yokota, T., Kastelein, R., Zurawski, S.M., Arai, N and Takebe, Y. 1987. Species-specificity of T cell stimulating activities of IL 2 and BSF-1 (IL 4) :

comparison of normal and recombinant, mouse and human IL 2 and BSF-1 (IL4). J. Immunol **138**:1813-1816

Rees, R.C. (Ed.) 1990. The biology and clinical applications of interleukin-2. IRL Press, Oxford

Robbens, J., Raeymaeckers, A., Steidler, L., Fiers, W. and Remaut, E. (1995), Production of Soluble and Active Recombinant Murine Interleukin-2 in Escherichia coli : High Level Expression, Kil-Induced Release, and purification. Prot. Expr. Purif. **6** : 481-486

Schein, C.H. 1989. Production of soluble recombinant proteins in bacteria. Bio/Technology **7**(11):1141-1149

Simons, G., Rutten, G., Hornes, M., Nijhuis, M. and van Asseldonk, M. 1992. Production of prochymosin in lactococci. Adv. Exp. Med. Biol. **306**:115-120

Stanssens, P., Opsomer, C., McKeown, Y. M., Kramer, W., Zabeau, M. and Fritz, H.-J. 1989. Efficient oligonucleotide-directed construction of mutations in expression vectors by the gapped duplex DNA method using alternating selectable markers. Nucleic Acids Res. **17**:4441-4453.

Steidler, L., Fiers, W., and Remaut, E. 1994 Efficient specific release of periplasmic proteins from *E. coli* using temperature induction of the cloned *kil* gene of pMB9. Biotech. Bioeng. **44**(9): 1074-1082

Steidler, L., Wells, J.M., Raeymaeckers, A., Vandekerckhove, J., Fiers, W. and Remaut, E. (1995) Secretion of Biologically Active Murine Interleukin-2 by *Lactococcus lactis* subsp. *lactis*. Appl. Env. Microbiol. 61(4) : 1627-1629

Struhl, K. 1985. A rapid method for creating recombinant DNA molecules. Biotechniques **3**:452-453.

van Asseldonk, M. Rutten, G., Oteman, M., Siezen, R.J., de Vos, W.M. and Simons, G. 1990. Cloning of *usp45*, a gene encoding a secreted protein from *Lactococcus lactis* subsp. *lactis* MG1363. Gene **95**:155-160

van de Guchte, M. Kok, J. and Venema, G. (1992) Gene expression in Lactococcus lactis. FEMS REV. 88(2): 73-92

van de Guchte, M., van der Vossen, J.M.B.M., Kok, J. and Vennema, G. 1989. Construction of a lactococcal expression vector. Expression of hen egg white lysozyme in *Lactococcus lactis* ssp. *lactis*. Appl. Env. Microbiol. **55**(1):224-228

Vanhaesebroeck, B., Van Bladel, S., Lenaerts, A., Suffys, P., Beyaert, R., Lucas, R., Van Roy, F. and Fiers, W. (1991) Two discrete types of tumor necrosis factor-resistant cells derived from the same cell line. Cancer Res. **51**(9): 2469-77

van Rooijen, R.J., Gasson, M.J. and de Vos, W.M. (1992) Characterization of the Lactococcus lactis lactose operon promoter: Contribution of flanking sequences and LacR repressor to promoter activity. J. Bacteriol. 174(7): 2273-2280

Van Snick, J., Cayphas, S., Vink, A., Coulie, P.G. and Simpson, R.G. (1986) Purification and NH_2-terminal amino acid sequence of a T-cell-derived lymphokine with growth factor activity for B-cell hybridomas. Procl. Natl. Acad. Sci. USA **83** : 9679

Vandenabeele, P., Guisez, Y., Declercq, W., Bauw, G., Vandekerckhove, J., Fiers, W. (1990) Response of murine cell lines to an IL-1/IL-2-induced factor in a rat/mouse T hybridoma (PC60): differential induction of cytokines by human IL-1 alpha and IL-1 beta and partial amino acid sequence of rat GM-CSF. Lymphokine-Res. **9**(3): 381-9

Wagner, H., Hardt, C., Heeg, K., Pfizenmaier, K., Solbach, W., Barlett, R., Stockinger, H. and Röllinghoff, M. 1980. T cell interactions during cytotoxic T lymphocyte (CTL) responses : T cell derived helper factor (interleukin 2) as a probe to analyze CTL responsiveness and thymic maturation of CTL progenitors. Immunol. Rev. 215-255

Wells, J.M., Wilson, P. W. and le Page, R.W.F. (1993a) Improved cloning vectors and transformation procedure for Lactococcus lactis. J. Appl. Bacteriol. **74**(6): 629-636

Wells, J.M., Wilson, P.W., Norton, P.M. and Le Page, R.W.F. (1993b). A model system for the investigation of heterologous protein secretion pathways in *Lactococcus lactis*. App. Env. Microbiol. **59**(11)3954-3959

Wells, J.M., Wilson, P.W., Norton, P.M., Gasson, M.J. and Le Page, R.W.F. (1993c). *Lactococcus lactis* : high-level expression of tetanus toxin fragment C and protection against lethal challenge. Mol. Microbiol **8**(6):1155-1162.

Lactic acid bacteria involved in food fermentations and their present and future uses in food industry

Friedrich-Karl Lücke
Mikrobiol. Lab., FB Haushalt & Ernährung,
Fachhochschule Fulda,
P.O. Box 1269
D-36012 Fulda
Germany

(1) Introduction

Fermented foods may be defined as foods that have been modified, in a desired way, by the activity of micro-organisms and/or enzymes. Yeasts, moulds and lactic acid bacteria, alone or in combination, are involved in food fermentations; other bacteria may also play a role in certain products. In most cases, enzymes present in the raw material or added to it contribute to the desired changes.

The main purpose of food fermentations dominated by yeasts (manufacture of baked goods and beverages) and moulds (Asian-type fermentation of rice and soybeans) is to improve digestibity, nutritional value, texture and aroma. In contrast, lactic fermentation of foods probably evolved as a preservation method, taking (empirically) advantage of the bacteria to form large amounts of lactic acid in short time. However, it also permitted the production of a large variety of foods with different aromas, flavours and consistencies, and there is evidence that lactic fermentation of certain foods may also improve their wholesomeness (see Hammes and Tichaczek, 1994, for a recent review). Table 1 lists the major food commodities fermented by lactic acid bacteria.

This paper deals with the role of lactic acid bacteria in various food fermentations and discusses the composition of the fermentation flora, their important metabolic activities and the use of starters. It is not the purpose of this paper to provide a comprehensive review of all indige-

NATO ASI Series, Vol. H 98
Lactic Acid Bacteria:
Current Advances in Metabolism, Genetics and Applications
Edited by T. Faruk Bozoğlu and Bibek Ray

nous foods prepared with the aid of lactic acid bacteria; readers with more detailed interest in such foods are referred to Wood (1985) and the monograph by Campbell-Platt (1987).

Table 1: Food commodities fermented by lactic acid bacteria (LAB)

Raw material	Product	Main fermentation agents	Additional fermentation agents in some products
Milk	sour milks	LAB	yeasts
Milk	cheeses	LAB	*Propionibacterium; Brevibacterium; Penicillium*
Meat	fermented sausages	LAB + *Staphylococcus* or *Micrococcus*	yeasts; *Penicillium*
Doughs, grain mashes	sourdough, sour mash	LAB + yeasts	
Vegetables	sauerkraut, various pickles	LAB	
Olives	table olives	LAB	
Soybeans, peanuts	tempeh, ontjom	*Rhizopus*, LAB	
Soy mash	soy sauce	yeasts, LAB	
(Grape) musts	wine	yeasts	LAB
Fish	lightly salted fish products; fish sauces	fish enzymes; LAB	

(2) Sources of starter bacteria

Lactic acid bacteria involved in food fermentations may enter the food by the following ways:

1. They may be present in the raw material in sufficient numbers to outgrow other microorganisms. This is the case in traditional vegetable and sausage fermentations.
2. They may be present on the equipment used to prepare and to ferment the food, for example in fine cracks and niches of wooden barrels used to „brine“ foods such as cucumbers, olives or fish.
3. Material from a previous fermentation (e.g. whey from cheesemaking) is used to inoculate a new batch.

4. „Additives" empirically known to contain the required agents are added.
5. Cultures are added that contain a complex mixture of strains that have been propagated as a mixed culture. Such „undefined mixed starters" are still common in many dairy fermentations.
6. Cultures are added that contain one or more defined strains propagated as pure cultures.

The first 5 methods were - and still are - used in farm-scale and artisanal food fermentations, particularly in areas where there is no reliable or affordable supply of high-quality starter cultures. The first defined bacterial starters were introduced about 100 years ago for the fermentation of milk but it was not until the 1950s that pure or defined mixed bacterial cultures became available for the fermentation of other foods. However, in some of today's industrial food fermentations, including the sauerkraut and sourdough production, there is still little if any use of defined starters. If the intrinsic and extrinsic factors affecting micro-organisms can be adjusted so that even small initial numbers of lactic acid bacteria outgrow their competitors, and if there is little interest to accelerate acid formation, there is little benefit from using lactic starter cultures. Furthermore, the development of the desired aroma and flavour is often due to complex interactions between strains involved in the fermentation, for example, between bacteria and yeasts in the kefir or sourdough fermentation. It is also very difficult to design *in vitro* systems to select strains producing good aroma and flavour in real food fermentations. This necessitates a lot of pilot-scale fermentation trials, which makes the selection of such starters very tedious and expensive.

(3) Dairy fermentations (reviews: Marshall, 1986; Teuber, 1986; Hunger, 1987; Varnam and Sutherland, 1994)

Of all lactic fermented foods, dairy products are the most important worldwide, by both weight and value. In Europe, more than 50 % of the total milk is processed into sourmilks, sour cream butter and cheese. All dairy fermentations include an initial lactic fermentation, only in rare cases (kefir and some indigenous sourmilks prepared with the aid of undefined inocula) accompanied by yeasts. Secondary fermentations, by certain lactic acid bacteria, propionibacteria, brevibacteria, yeasts or moulds contribute to the desired sensory properties of some products.

Fermented dairy products are broadly subdivided into sourmilk products (pourable or semi-solid; including sour cream) and cheeses, depending on whether or not the whey is separated from the curd after the initial lactic acid fermentation. Preferences differ between consumers in various countries: for example, the Finnish and Swedish populations rank first in consumption of sourmilk fermented by strains having a lower temperature optimum (*Lactococcus, Leuconostoc*) while traditional yoghurt (fermented at about 42°C) is most popular in Southeastern Europe and Turkey (review: Rašić *et al.*, 1990). In the countries of Western Europe, traditional plain yoghurt had only a small market share until yoghurt-based sweetened convenience products such as fruit yoghurt in small retail containers were developed and marketed, and starters producing milder varieties of yoghurt were introduced (see below).

Tables 2 and 3 give an overview of the fermented dairy products most common in Europe and the species of the micro-organisms that are used to produce them. For information on dairy products not listed in the Tables, see reviews by Oberman (1985), Galloway and Crawford (1985) and Rašic *et al.* (1990).

Table 2: Starter cultures for fermented milks and cultured butter

No.	Product type (example)	Starter composition[1]	Main desired metabolic activity
1	Buttermilk, sour cream and sour cream butter; Scandinavian-type sourmilks	*Lc. lactic ssp. cremoris* and *lactis* *Lc. lactis biovar diacetylactis;* *Ln. mesenteroides ssp. cremoris*	Lactic acid from lactose Diacetyl from citrate
2	Scandinavian-type ropy milk	as in (1); in addition: „ropy" strains of *Lc. lactis ssp. cremoris*	as (1) Slime formation
3	Yoghurt	*Str. salivarius ssp. thermophilus;* *Lb. delbrueckii ssp. bulgaricus*	Lactic acid from lactose; Acetaldehyde from pyruvate or threonine
4	Yoghurt „mild"	as (3), but strains of *Lb. delbrueckii* partially or completely replaced by *Lb. acidophilus*	as (3)
5	Kefir	*Lc. lactis, Lb. acidophilus, Lb. kefir, Lb. kefirgranum, Lb. parakefir* *Lb. kefiranofaciens* *Leuconostoc sp.* *Candida kefir*	Lactic acid from lactose Slime formation[2] Diacetyl from citrate Ethanol + CO_2 from lactose

[1] *Lb., Lactobacillus; Lc., Lactococcus; Ln., Leuconostoc; Str., Streptococcus*
[2] matrix of Kefir grains

Table 3: Starter cultures for cheeses

Product type (example)	Starter composition[1]	Main desired metabolic activity
1. Cheeses without eyes (Cheddar)	*Lc. lactis ssp. cremoris* and *lactis*	Lactic acid from lactose; proteolysis
2. Cheeses with few small eyes (Edam)	as (1), but with ca. 5 % *Ln. mesenteroides ssp. cremoris*	Diacetyl from citrate
3. Fresh cheese (quark, cottage cheese)	as (2), and/or "thermophilic" cultures (*Str. salivarius ssp. thermophilus; Lb. delbrueckii; Lb. helveticus*)	
4. Soft cheeses:	as (2), but with surface flora:	
Limburger-type:	*Geotrichum candidum* + *Brevibacterium linens*	Lactate oxidation proteolysis
Camembert-type:	*Penicillium camembertii*	proteolysis
5. Cheeses with small eyes (Gouda)	as (2), but with 15-20 % *Lc. lactis* biovar *diacetylactis*	Diacetyl and CO_2 from citrate
6. Cheeses with moulds in core (Roquefort)	as (5), but with *Penicillium roquefortii*	Lactate oxidation; lipolysis; proteolysis
7. Hard Italian-type cheese; Feta	*Str. salivarius ssp. thermophilus*	Lactic acid from lactose
	Lb. delbrueckii ssp. bulgaricus	Lactic acid from lactose; proteolysis
8. Hard cheeses with large eyes (Emmental)	*Str. salivarius ssp. thermophilus*	Lactic acid from lactose
	Lb. helveticus and/or *Lb. delbrueckii ssp. lactis*	Lactic acid from lactose; proteolysis
	Propionibacterium sp.	propionate and CO_2 from lactate

[1] *Lc.* = *Lactococcus*, *Lb.* = *Lactobacillus*, *Ln.* = *Leuconostoc*, *Str.* = *Streptococcus*

Lactic acid from lactose is produced by all species of lactic acid bacteria listed (genera *Lactococcus, Streptococcus, Lactobacillus, Leuconostoc*). Most species of lactic acid bacteria listed in Tables 2 and 3 are homofermentative, i.e. more than 1.7 moles of lactate are formed from one more of hexose under anaerobic conditions. The leuconostocs, ***Lactobacillus kefir*** and the recently described ***Lactobacillus parakefir*** (Takizawa et al., 1994) are heterofermentative and ferment lactose to approximately equimolar amounts of lactate and ethanol (and acetate if oxygen or some extra source of pyruvate is present). CO_2 is also formed from citrate by citrate-utilizing strains (*Lactococcus lactis* biovar *diacetylactis, Leuconostoc mesenteroides ssp. cremoris*) and forms the small holes that are typical for many cheeses. Diacetyl (butter-like) and acetaldehyde (yoghurt-like) are the main aroma compounds formed by certain lactic acid bacteria, and the desired amount of these substances may be controlled by appropriate selection of starter and fermentation conditions (review: Marshall, 1987). Proteolytic and peptolytic activities of lactic acid bacteria contribute to the characteristic consistency and flavour of cheeses, and the ability to produce slime is desirable for the manufacture of semisolid sourmilk products.

Properties that are used to distinguish the homofermentative genera *Lactococcus, Streptococcus, Enterococcus* and *Pediococcus* are given in Table 2 of the accompanying paper (Lücke, this volume). *Streptococcus salivarius ssp. thermophilus* can also be distinguished from *Lactococcus lactis* by its inability to ferment galactose. In contrast to *Lc. lactis ssp. cremoris, Lc. lactis ssp. lactis* grows at 40°C, at pH 9.2, as well as in the presence of 4 % NaCl or 0.1 % methylene blue, and produces ammonia from arginine (review: Teuber et al., 1992). *Lc. lactis* biovar *diacetylactis* only differs from *Lc. lactis ssp. lactis* by its (plasmid-encoded) ability to metabolize citrate into CO_2 and diacetyl. *Leuconostoc mesenteroides ssp. cremoris* appears to be specially adapted to milk and may be distinguished easily from other leuconostoc because it does not ferment sugars other than glucose, lactose and galactose (review: Holzapfel and Schillinger, 1992).

The optimum growth temperature of dairy lactococci and leuconostocs is below 35°C. Hence, these species are often termed „mesophilic“, and fermentation temperatures are normally adjusted to between 20°C (for sourmilks) and 30°C (for cheeses).

All lactobacilli important as starters for fermented dairy products belong to the obligately homofermentative group and do not metabolize citrate. They are designed „thermophilic“ because they grow at 45°C but not at 10°C. This is why the yoghurt fermentation takes place at about 42°C, and why „thermophilic“ strains are used in the manufacture of most hard cheeses where the curd is heated to about 52 - 55°C to allow for a more effective removal of whey and to lower the water activity. Since such cheeses are often made from raw milk, it is essential that the starter remains active during this process and controls undesired bacteria.

Ripened cheeses may contain larger numbers of facultatively or obligately heterofermentative lactobacilli (e.g. *Lactobacillus casei*) capable of growing at low water activity. However, there is no clear-cut evidence that they contribute to the desired properties of the cheese (Peterson and Marshall, 1990).

Thermophilic strains (lactobacilli, *Str. salivarius ssp. thermophilus)* are increasingly being used for fresh cheeses (e.g. quark) and soft cheeses because retailers want constant sensory properties (i.e. no further acidification or proteolysis) throughout display.

Because of their sugar fermentation spectra, *Lactobacillus delbrueckii ssp. bulgaricus* and *Lactobacillus helveticus* may be considered as typical „milk strains“: The former ferments only glucose and lactose while the latter, in addition, metabolizes galactose and is therefore used in cheeses where a low residual sugar content is desired, e.g. in Emmental-type products (review: Varnam and Sutherland, 1994).

Compared to *Lb. delbrueckii ssp. bulgaricus* and *Lb. helveticus, Lb. acidophilus* and *Lb. delbrueckii ssp. lactis* metabolize a much wider range of sugars. *Lactobacillus acidophilus* gained much interest both in gastroenterology and in the dairy industry because many strains tolerate hydrochloric acid and bile and because there is evidence that in the intestinal tract, at least some strains have probiotic effects (as outlined elsewhere in this Volume). In addition, it forms DL-lactic acid whereas *Lb. delbrueckii* only forms the D isomer. Growth of *Lb. acidophilus* in milk is rather poor unless the milk is supplemented with additional growth factors. This disadvantage, however, has been turned into an advantage in the development of milder yoghurt varieties with L-lactic acid as the major metabolic product. In traditional yoghurt, the well-known synergism of *Str. salivarius ssp. thermophilus* (providing reducing conditions and formate) and *Lb. delbrueckii ssp. bulgaricus* (providing peptides and amino acids) leads to low final pH values while in „mild“ varieties, fermentation stops at about pH 4.1, there is little if any further acidification during chill storage, and less acetaldehyde formation.

In some national markets, sourmilks containing other lactobacilli are available. The strains (e.g. *Lactobacillus* GG; *Lb. casei* Shirota) have been selected on the basis of their ability to survive the passage through the stomach and the duodenum and of some clinical evidence of beneficial actions in the gastrointestinal tract.

Traditional sources of inoculum for dairy fermentations were 'spontaneously' coagulated milk, whey, remnants from a previous batch, or undefined mixed cultures which are easily recovered and maintained (e.g. kefir grains). Preparations from the intestinal tract (e.g. from calf stomachs) apparently were used as a source of thermophilic starters, propionibacteria and rennet in traditional cheesemaking. Today, undefined mixed cultures (i.e. propagated as a mixture rather than mixed after preparation) or defined single or mixed cultures are commonly used in dairies. To overcome the problem of attack by bacteriophages, strains that differ in their phage sensitivity pattern are used as mixtures and/or successively in a 'culture rotation scheme'. While most thermophilic starters used in the dairy industry are supplied as defined cultures, the application of undefined mixed cultures of mesophilic strains (*Lactococcus, Leuconostoc*) is still common in many countries, even in highly standardized processes. This probably reflects the common ecological rule that a large variety of species adds to the stability of an ecosystem. Nevertheless, there is a trend towards the use of defined cultures, in parallel of better understanding of the physiology and genetics of the strains, the introduction of new strains and new products and the increasing pressure on the dairies towards precisely defining and documenting each step of their process. The future may see the application of defined strains

- with multiple resistance against phage attack;
- with stronger antagonism against *Listeria monocytogenes* and *Clostridium tyrobutyricum*;
- with probiotic properties;
- producing more slime in sourmilk products;
- accelerating cheese ripening;
- better adapted to special products (e.g. raw milk cheeses) and production conditions.

The advantages of defined cultures can be exploited best in highly controlled production systems with as little in-plant subcultivation and handling as possible. Hence, use of undefined mixed cultures will stay important in the production of many products, to maintain a large product diversity and to fulfil the needs of small, decentralized dairies. However, a better understanding of the microbial interactions in such starters and of the reasons for the stability of their microbial associations would be of great help in improving them.

Starters containing lactic acid bacteria are usually distributed in lyophilized or frozen forms. Their further propagation in the dairy is still common but improvements in culture preservation methods made it possible to use culture concentrates for direct inoculation of the milk to be fermented. Such cultures are increasingly being used, especially when it is difficult to maintain the proper equilibrium between different strains during growth as mixed culture.

(3) Meat and fish fermentations (reviews: Lücke, 1991; Hammes *et al.*, 1990; Hammes and Knauf, 1994; Kröckel, 1995; Jessen, 1995).

Muscles become anaerobic very soon after exsanguination of the animals, and cannot be pasteurized without drastically changing their sensory properties. When carcasses are butchered and meat is cut, some lactic acid bacteria adapted to meat inevitably enter the meat. Proper formulations and adjustment of fermentation conditions are crucial to warrant the dominance of lactic acid bacteria and the inhibition of undesired and hazardous micro-organisms. Muscles also have a high buffering capacity and, after *rigor mortis,* contain only little fermentable carbohydrate (maximum 0.3 % in lean pork and beef, even lower in other tissues and in fish flesh). Therefore, fermentation of these sugars does not generally lower the pH to an extent inhibitory to pathogenic bacteria.

Raw dry sausages are the most important fermented muscle foods. Raw hams and moderately salted fish products are mainly ripened by the action of their indigenous proteolytic enzymes, with only little contribution of lactic acid bacteria. At brine concentrations above 12 % NaCl (a_w value below 0.91), activity of all lactic acid bacteria except of *Tetragenococcus halophilus* ceases; the latter organism is involved in the fermentation of some heavily salted Asian-type fish sauces.

Conditions that favour the lactic acid bacteria during sausage fermentation include a pH below 6.0 (as found in 'normal' *post-rigor* beef and pork), an initial water activity between 0.955 and 0.965 (obtained by the addition of 2.5 - 3 % salt), the addition of at least 0.3 % of a rapidly fermentable sugar and of about 100 - 150 mg sodium nitrite/kg. To exclude oxygen, the mix is then firmly stuffed into vapour-permeable casings. The lactic fermentation proceeds at about 20°C within 2 - 4 days, i.e. much more slowly than in most dairy fermentations. Sausages are subsequently dried at progressively declining relative humidities ensuring homogeneous removal of water.

The composition of the lactic flora of various types of „spontaneously" fermented sausages in various countries has been thoroughly analyzed. The studies indicate that *Lactobacillus sake* and *Lactobacillus curvatus* are most competitive at fermentation temperatures below 25°C. This also applies to sausages fermented in countries like Spain (Hugas *et al.*, 1993), Greece (Samelis *et al.*, 1994) and Turkey (Gürakan *et al.*, 1995). The main reason for this is the ability of *Lb. sake* and *Lb. curvatus* to multiply both at low temperatures (4°C) and low water activity ($a_w > 0.91$). Therefore, they can establish themselves on chilled meat and surfaces in the cutting room and thrive in sausage mixes with a_w values around 0.96. Other psychrotrophic lactic acid bacteria (such as *Leuconostoc spp., Carnobacterium spp.*) tolerate less salt, and species like *Lactobacillus plantarum, Lactobacillus pentosus, Pediococcus pentosaceus* and *Pediococcus acidilactici* grow only poorly or not at all at 7°C.

In traditional slow sausage fermentations, there is little benefit from using lactic starter cultures, especially in Italian and French-type salamis where a sour taste is to be avoided. Lactic acid bacteria were often recognized as spoilage agents interfering with the activity of nitrate-reducing bacteria which is essential for the development of curing colour and aroma when nitrate is used as curing agent. The first defined starter culture for meat fermentation were introduced in Europe in the early 1960s and consisted of „*Micrococcus* M 53" (later reclassified as *Staphylococcus carnosus*). Lactic acid bacteria were introduced as starters in parallel to the changes towards industrial processing of meat products. These changes were associated with a replacement of nitrate by nitrite, shorter ripening times and, most importantly, the need to standardize the ripening process and the safety and quality of the product. To date, almost all Central and Northern European manufacturers of fermented sausages with total ripening times of 3 weeks or below use starter cultures, and use of cultures is increasing in Mediterranean countries, too (Lücke *et al.*, 1990).

All lactic acid bacteria useful in sausage fermentation are homofermentative. By formation of lactic acid, they lower the pH to below 5.3 and thereby contribute to the inhibition of pathogenic and spoilage bacteria. At this pH, formation of curing colour is accelerated, and coagulation of soluble meat proteins facilitates drying. Lactic acid bacteria may also affect the aroma of fermented sausages by formation of acetic acid and, possibly, diacetyl or some products from amino acids. Although some strains show weak proteolysis and/or lipolysis of artificial substrates *in vitro*, there is no evidence that these activities are involved to any extent in the formation of the desired sausage aroma. In view of the large variety of factors affecting the aroma and flavour of fermented sausages, it appears unlikely that by selection of new or modification of „old" strains of lactic acid bacteria, one may obtain the aroma and flavour of a fully ripened sausage in less time.

The following species of lactic acid bacteria are available as starter cultures for sausage fermentation:

- The „mesophilic" species include *Lactobacillus plantarum* and the closely related species *Lactobacillus pentosus* as well as most strains of as well as most strains of *Pediococcus pentosaceus*. They grow sufficiently fast in sausages fermenting at 20 - 25°C to ensure reasonably short fermentation times without impairing the development of colour and aroma. They possess non-heme catalase, and some *Lb. plantarum* and *Lb. pentosus* strains reduce nitrate (review: Hammes *et al.*, 1990).
- The „psychrotrophic"species *Lactobacillus curvatus* and *Lactobacillus sake* are most effective to suppress other bacteria - including other lactic acid bacteria - at fermentation temperatures in the 20 - 22°C range but have, to date, a smaller market share than the „mesophilic" strains. This could be related to their ability to continue acid production at low temperature which may lead to oversouring unless the amount of sugar added is restricted to a minimum. In addition, they differ from *Lb. plantarum* with respect to their pyruvate and oxygen metabolism (review: Kandler, 1983; Jessen, 1995). For example, they possess either only a heme-dependent catalase or no hydrogen peroxide destroying enzyme at all (review: Hammes *et al.*, 1990).

- The „thermophilic“ species *Pediococcus acidilactici* (formerly named „*Pediococcus cerevisiae*“) was selected for the manufacture of summer sausages in the United States. These sausages are fermented at temperatures above 30°C (usually about 38°C) for a short time. Under conditions used for sausage fermentation in Europe, *P. acidilactici* grows only slowly.

In Europe, preparations containing *Lactobacillus plantarum* or *Pediococcus pentosaceus* have the greatest market share and are commonly used in conjunction with strains of *Micrococcus varians, Staphylococcus carnosus* or *Staphylococcus xylosus*. The latter three species reduce nitrate and have been found to protect the product to some degree from colour changes and rancidity, particularly during extended storage. In the United States, Sweden and The Netherlands, sausages are frequently fermented with little or no participation of catalase-positive cocci. This still results in a product of acceptable colour and sufficient shelf life for large-scale production and distribution provided that nitrite is used as the curing agent, peroxide formation is avoided and the pH is rapidly lowered to 5.0 or below.

Starter cultures for meats are grown as single strain cultures. They are distributed as pure or defined mixed cultures, most of them in lyophilized form, but several also in the frozen state. In The Netherlands, many processors use „starter sausages“ which are centrally prepared by fermentation using a strain of *Lb. curvatus* and subsequently frozen and distributed. This ensures a rapid rate of acid formation which is essential for the safe fermentation at elevated temperatures (25 - 27°C). In contrast to dairies, meat processors virtually never propagate the strains any further before use.

Lactic acid bacteria suitable for sausage fermentation (and, possibly, for biopreservation of non-fermented perishable meats) should be competitive in salted meat systems and antagonistic to pathogens. Bacteriocin-producing strains of each species used for sausage fermentation have been selected, and some of them have been shown to act against *Listeria monocytogenes*. However, bacteriocins of lactic acid bacteria have no reliable effect against Gram-negative bacteria such as salmonellae and *Escherichia coli*, and are insufficiently stable in meat systems (review: Lücke, 1992).

Starters should be of no health concern and should accumulate neither hydrogen peroxide nor biogenic amines nor compounds of objectionable aroma. As all starter cultures, they must be suited for freezing and lyophilization, i.e. they must keep their activity and become active in their substrate immediately after inoculation.

(4) Lactic fermentation of foods of plant origin

Table 4 lists the species of lactic acid bacteria involved in the fermentation of various foods of plant origin.

Table 4: Important foods of plant origin fermented by lactic acid bacteria (LAB)

Starting material	Product	Species of LAB involved	Other micro-organisms involved
shredded cabbage	sauerkraut	*Leuconostoc mesenteroides*, *Lactobacillus plantarum* and other facultatively homofer-mentative lactobacilli	
cucumbers and some other vegetables	„pickles" (brined vegetables)	*Lactobacillus plantarum*	
olives	table olives	*Lactobacillus plantarum*	
doughs (especially tho-se containing >20 % rye flour)	sourdough	*Lactobacillus sanfrancisco*, *Lb. pontis*, *Lb. brevis*	various yeasts
soybeans + wheat, pre-fermented with *Asper-gillus*	soy sauce (shoyu)	*Tetragenococcus halophilus*	osmophilic yeast
high-acid grape musts	wine	*Oenococcus oenos*	wine yeasts

4.1 Vegetable fermentations (review: Buckenhüskes *et al.*, 1990; Buckenhüskes, 1993)

To obtain good growth and competitiveness of lactic acid bacteria in vegetable material, oxygen must be removed and nutrients must be liberated from the plant cells. This is achieved by shredding and firmly packing the material (in silage and sauerkraut fermentation) or by placing it in brines containing appropriate amounts of salt (in pickle and olive fermentation). This ensures that residual oxygen is removed by tissue respiration and the access of oxygen from the air is restricted. The addition of some salt and acid also aids in the suppression of Gram-negative organisms early in fermentation although some residual activity of these bacteria is desirable to eliminate nitrate and nitrite from the material.

In the sauerkraut fermentation, *Leuconostoc mesenteroides* usually develops first. In addition to lactic acid, it forms acetic acid and CO_2 and thereby contributes to the inhibition of mould growth on the surface. *Leuconostoc* also reduces fructose to mannitol and thus forms another growth substrate for *Lactobacillus plantarum* (McFeeters and Chen, 1986) which usually takes over because of its higher tolerance to acid. It may be accompanied by other homofermentative lactic acid bacteria such as *Lactobacillus sake, Lactobacillus bavaricus, Lactobacillus curvatus* and *Pediococcus pentosaceus.* In pickle fermentation, CO_2 production is not desired, and the lower initial water activity of the brine and/or the addition of some acid tends to suppress *Leuconostoc spp.*.

Starters for vegetable fermentations are commercially available and mostly contain *Lb. plantarum* and/or related species. For sauerkraut, a strain of *Lb. bavaricus* (forming L(+)-lactic acid only) is marketed in Germany for the production of 'L(+)-sauerkraut' sold through health shops. Otherwise, there is little use of starter cultures for vegetable fermentations. Apparently, the safety and speed of commercial fermentations is sufficient, even with low numbers of cells from the raw material, the fermentation equipment or used brines. Furthermore, it is often stated that products fermented by the „natural flora“ have better aroma and flavour. This may be at least true for the sauerkraut fermentation where many different bacteria affect the sensory properties. However, there are factors which favour the use of starter cultures in the future:

- To ferment vegetables in small retail containers, the inoculum must be very homogeneously distributed which is obviously much easier with high initial numbers of lactic acid bacteria.
- Undesired metabolic activities of concomitant bacteria may be inhibited in a more reliable fashion. Such activities include formation of off-aromas, gas and biogenic amines.
- Through use of selected starters, desired metabolic activities such as removal of nitrate and nitrite and formation of compounds contributing to the wholesomeness of the product may be implemented.

- Fermentation of heat-treated vegetables and vegetable juices and development of novel products such as low-sugar vegetables for deep-frying (Slinde *et al.*, 1993) become feasible.

4.2. Sourdough fermentation (review: Spicher, 1983)

Sourdough fermentation is the most widespread method to render doughs with more than 20 % rye flour suitable for baking, i.e. crumb formation and retention of leavening gas. Sourdough breads are common in Eastern Europe, Germany and Finland. The main effect of dough acidification, in conjunction with salt, is the inhibition of amylases which would otherwise degrade the starch after its gelatinization. In addition, the micro-organisms in sourdough form free amino acids (precursors of the Maillard reaction) and other compounds that improve the aroma and flavour of the bread. There is an increasing interest to use sourdough for the manufacture of some wheat breads, too; this may positively affect the dough properties and accelerates phytate degradation in doughs prepared whith whole meal.

If suitable mixtures of flour and water are allowed to stand, spore formers and other acid-sensitive bacteria are outgrown by lactic acid bacteria and a 'spontaneous sour' is formed that can be used to inoculate the dough. Such „starters“ usually consist of *Pediococcus pentosaceus, Lactobacillus plantarum* and some related, predominantly homofermentative lactobacilli not specially adapted to doughs (Spicher *et al.*, 1987; Hochstrasser *et al.*, 1993a). Continuous „subcultivation“ by the baker (who processes the starter into a mature sourdough and saves part of this as a starter for the following batch) for longer periods under appropriate conditions, leads to a microbial association that is well adapted to doughs. Characteristic species of lactic acid bacteria are the heterofermentative species *Lactobacillus sanfrancisco* and *Lactobacillus pontis* (Böcker, 1993) which require certain growth factors present in the dough or liberated by the accompanying yeast flora, and which, apart from maltose, ferment very few sugars. Rye bread prepared from sourdough fermented by these species were ranked superior to other rye breads in sensory analyses. Unfortunately, these strains are difficult to preserve by freezing or lyophilization. Other heterofermentative species frequently found in sourdoughs include *Lactobacillus brevis, Lactobacillus fructivorans* and the „thermophilic“ species *Lactobacillus fermentum* and *Lactobacillus reuteri* (review: Spicher, 1983).

The activity of lactic acid bacteria and yeasts in sourdoughs and the properties of the dough is affected by the amount of water added to the flour (expressed as 'dough yield') and the fermentation temperature (usually in the range between 20 and 35°C). The tendency is that the molar ratio between lactic and acetic acids increases with increasing temperature and dough yield.

Most commercial starter cultures are mixtures of some of the bacterial and yeast species mentioned above and are distributed as dried doughs. The baker uses them more or less often to maintain the activity of his starter. There is considerable batch-to-batch variation of the qualitative and quantitative composition of their microflora (Spicher, 1985; Hochstrasser *et al.*, 1993b). Drying of doughs with high titrable acidity often results in low counts of starter microorganisms. Hence, large-scale cultivation of starters and, in particular, drying methods should be optimized to obtain better standardized preparations with short reactivation times. For a better theoretical base of starter improvement, the microbial interactions in sourdoughs should be studied in more detail.

4.3. Soysauce, tempeh and related foods

For the production of soy sauce, heat-treated raw materials (soybeans, wheat) are first fermented by fungi (*Aspergillus oryzae, A. sojae*) in order to hydrolyze polysaccharides and provide substrates for the subsequent fermentation carried out by lactic acid bacteria and yeasts. The latter takes place in the presence of 13 - 19 % salt. Hence, the moderately halophilic species *Tetragenococcus halophilus* and the salt-tolerant fermentative yeast, *Zygosaccharomyces baillii* are predominant in the fermentation. In tempeh fermentation, lactic acid bacteria have been shown to contribute to the safety of the product (Ashenafi and Busse, 1989; Tünçel *et al.*, 1989).

(5) Conclusions

Overall, the future for foods fermented by lactic acid bacteria looks bright because such foods fulfil two central demands of today's consumers: wholesomeness and convenience. However, researchers and product developers should keep in mind the following:

- Starter cultures are only one factor contributing to the safety of the process and the quality of the product.
- The „tools" are available to genetically modify most species of lactic acid bacteria involved in food fermentations. However, this would not help very much in cases where the biochemical basis of desired activities (e.g. aroma formation) is not known or if the desired activity is the result of complex symbiotic relationships. In addition, it will be difficult to convince regulators and consumers of the benefits of certain genetically modified starters. Hence, the impact of molecular biology on the future of lactic fermented foods should not be overestimated.
- A thorough knowledge of the microbial ecology of the food to be fermented is essential for the selection and improvement of starter cultures for food fermentations. Molecular biology can be of great help, especially as a tool to characterize technologically important properties, to investigate microbial interactions in complex starters and to assess the fate of genes and their carriers in food, in humans and the environment.
- Too much uniformity among fermented foods on the market may result in lower consumption figures. One reason for the limited acceptance of starter cultures, e.g. in some vegetable fermentations and by the Italian meat industry, appears to be that the processors fear that their product loses its 'individuality'. Therefore, a large diversity of fermented foods and of starter cultures for their production must be maintained. In particular, efforts should continue to develop starter cultures for special foods and to meet the requirements of small, decentralized food processors.

(6) References

Ashenafi M, Busse M (1989) Inhibitory effects of *Lactobacillus plantarum* on *Salmonella infantis, Enterobacter aerogenes* and *Escherichia coli* during tempeh fermentation. J Food Protect 52:169-172

Böcker G (1993) Ökologische und physiologische Charakterisierung der sauerteigtypischen Stämme *Lactobacillus sanfrancisco* und *Lactobacillus pontis* sp. nov.. Ph.D. thesis, University of Hohenheim

Buckenhüskes H (1993) Selection criteria for lactic acid bacteria to be used as starter cultures for various food commodities. FEMS Microbiol Rev 12:253-272

Buckenhüskes H, Jensen HA, Andersson R, Garrido Fernández A, Rodrigo M (1990) Fermented vegetables. In: Processing and quality of foods Vol. 2 Food biotechnology (Zeuthen P, Cheftel JC, Eriksson C, Gormley TR, Linko P, Paulus K, eds) pp. 2.162-2.187. Elsevier Applied Science, London New York

Campbell-Platt G (1987) Fermented foods of the world - a dictionary and guide. Butterworths, London

Galloway JH,, Crawford RJM (1985) Cheese fermentation. In: Microbiology of fermented foods (Wood BJB, ed.), Vol. 1, pp. 111-165. Elsevier Applied Sci. Publ., London.

Gürakan GC, Bozoglu TF, Weiss N (1995) Identification of *Lactobacillus* strains from Turkish fermented sausages. Lebensm-Wiss & Technol 28:139-144

Hammes WP, Bantleon A, Min S (1990) Lactic acid bacteria in meat fermentation. FEMS Microbiol Rev 87:165-174

Hammes WP, Knauf H (1994) Starters in the processing of meat products. Meat Sci 36:155-168

Hammes WP, Tichaczek P (1994) The potential of lactic acid bacteria for the production of safe and wholesome food. Z Lebensm Unters Forsch 198:193-201

Hammes WP, Weiss N, Holzapfel WH (1992) The genera *Lactobacillus* and *Carnobacterium*. In: The Prokaryotes, 2nd ed. (Balows A, Trüper HG, Dworkin M, Harder W, Schleifer KH, eds) pp. 1535-1594. Springer, Berlin Heidelberg New York

Hochstrasser RE, Ehret A, Geiges O, Schmidt-Lorenz W (1993a) Mikrobiologie der Brotteigherstellung II. Mikrobiologische Untersuchungen an unterschiedlich hergestellten Spontan sauerteigen aus Weizenmehl. Mitt Gebiete Lebensm-Unters & Hyg 84:356-381

Hochstrasser RE, Ehret A, Geiges O, Schmidt-Lorenz W (1993b) Mikrobiologie der Brotteigherstellung IV. Mikrobiologische Untersuchung von Backmitteln und Sauerteig-Starterkulturen. Mitt Gebiete Lebensm-Unters & Hyg 84:622-629

Holzapfel WH, Schillinger U (1992) The genus *Leuconostoc*. In: The Prokaryotes, 2nd ed. (Balows A, Trüper HG, Dworkin M, Harder W, Schleifer KH, eds) pp. 1508-1534. Springer, Berlin Heidelberg New York

Hugas M, Garriga M, Aymerich T, Montfort JM (1993) Biochemical characterization of lactobacilli from dry fermented sausages. Internat J Food Microbiol 18:107-113.

Hunger W (1987) Starterkulturen für die Lebensmittelindustrie - neuere Entwicklungen am Beispiel der milchverarbeitenden Industrie. Alimenta 26 (2): 40,43-47

Jessen B (1995) Starter cultures for meat fermentation. In: Fermented meats (Campbell-Platt G, Cook PE, eds.), pp. 130-159. Blackie Academic and Professional, Glasgow.

Kandler O (1983) Carbohydrate metabolism in lactic acid bacteria. Antonie van Leeuwenhoek 49:209-224.

Kröckel L (1995) Bacterial fermentation of meats. In: Fermented meats (Campbell-Platt, G, Cook PE, eds.), pp. 69-109. Blackie Academic and Professional, Glasgow.

Kurmann JA, Rašic JL, Kroger M (1992) Encyclopedia of fermented fresh milk products. Chapman & Hall, London

Lücke FK (1991) Fermented meats. Food Res Internat 27:299-307

Lücke FK (1992) Prospects for the use of bacteriocins against meat-borne pathogens. In: New technologie for meat and meat products (Smulders FJM, Toldrá F, Flores J, Prieto M, eds.), pp. 37-52. Audet Tijschriften Nijmegen, for ECCEAMST, Utrecht

Lücke FK, Brümmer JM, Buckenhüskes H, Garrido Fernández A, Rodrigo M, Smith JE (1990) Starter culture development. In: Processing and quality of foods Vol. 2 Food bio technology (Zeuthen P, Cheftel JC, Eriksson C, Gormley TR, Linko P, Paulus K, eds) pp. 2.11-2.36. Elsevier Applied Science, London New York

Marshall VME (1986) The microflora and production of fermented milks. In: Progress in industrial microbiology Vol. 23 (Adams MR, ed), pp 1-44. Elsevier, Amsterdam

Marshall VME (1987) Lactic acid bacteria: starters for flavour. FEMS Microbiol Rev 46: 327-336.

McFeeters RF, Chen KH (1986) Utilization of electron acceptors for anaerobic mannitol metabolism by *Lactobacillus plantarum*. I. Compounds which serve as electron acceptors. Food Microbiol 3:73-81.

Oberman H (1985) Fermented milks. In: Microbiology of fermented foods (Wood BJB, ed.), Vol. 1, pp. 167-195. Elsevier Applied Science Publ., London.

Peterson SD, Marshall RT (1990) Nonstarter lactic acid bacteria in Cheddar cheese: a review. J Dairy Sci 73:1395-1410

Samelis J, Maurogenakis F, Metaxopoulos J (1994) Characterisation of lactic acid bacteria isolated from naturally fermented Greek dry salami. Internat J Food Microbiol 23:179-196

Slinde E, Skrede G, Aukrust T, Blom H, Baardseth P (1993) Lactic acid fermentaton influence on sugar content and colour of deep-fried fermented carrot chips. Food Res Internat 26:255-260

Spicher G (1983) Baked goods. In: Biotechnology, Vol. 5 (Reed G, ed.), pp. 1-80. Verlag Chemie, Weinheim.

Spicher G (1985) Zur Frage der Bewertung von Sauerteig-Starterkulturen, Sauerteigen in Trockenform und sauerteighaltigen Fertigmehlen bzw. Fertigmehlkonzentraten anhand mikrobiologischer Kenndaten. Deutsche Lebensmittel-Rundschau 81:177-180, 205-209

Spicher G, Rabe E, Rohschenkel C (1987) Untersuchungen zur Charakterisierung und Bewertung verschiedener Verfahren zur Bereitung eines Spontansauers. 1. Mitteilung: Vergleich verschiedener Spontansauerverfahren. Getreide Mehl Brot 41:118-122

Takizawa S, Kojima S, Tamura S, Fujinaga S, Benno Y, Nakase T (1994) *Lactobacillus kefirgranum* and *Lactobacillus parakefir* sp. nov., two new species from kefir grain. Internat J System Bacteriol 44:435-439

Teuber M (1986) Mikrobiologie fermentierter Milchprodukte. Chemie Mikrobiol Technol Lebensm 9:162-172.

Teuber M, Geis A, Neve H (1992) The genus *Lactococcus*. In: The Prokaryotes, 2nd ed. (Balows A, Trüper HG, Dworkin M, Harder W, Schleifer KH, eds) pp. 1483-1501. Springer, Berlin Heidelberg New York

Tünçel G, Nout MJR, Rombouts FM (1989) Effect of acidification on the microbiological composition and performance of tempe starter. Food Microbiol 6:37-43

Varnam AH, Sutherland JP (1994) Milk and milk products - technology, chemistry and microbiology. Chapman & Hall, London

Wood BJB (ed.) (1995) Microbiology of fermented foods, Vol. 1 & 2. Elsevier Applied Sci. Publ., London

Probiotics of Lactic Acid Bacteria: Science or Myth?

Bibek Ray
Animal Science Department
University of Wyoming
P.O. Box 3684
Laramie, WY 82071, USA

SUMMARY

Many species and strains of lactic acid bacteria (LAB) have been suggested to have several beneficial effects on the health of the digestive tract of humans. These benefits are produced by the antibacterial metabolites and specific cell components of several desirable intestinal indigenous bacterial species, particularly *Lactobacillus acidophilus* (group A1) and *Lactobacillus reuteri* and several *Bifidobacterium* species, and a few species of non-intestinal LAB, such as *Lactobacillus bulgaricus, Streptococcus thermophilus,* and *Lactococcus lactis*. However, study results using fermented dairy products produced by them or consumption of live cells of these species have not been consistent. The possible reasons for such controversies are discussed. In the concluding remarks possible means to overcome the controversies are suggested.

INTRODUCTION

Consumption of several species of lactic acid bacteria (LAB), either through fermented dairy products or as live cells, has been associated with many health benefits in humans. Some of these, listed in table 1, include benefits against ailments in the gastrointestinal (GI) tracts as well as in other parts of the body (Robinson, 1991; Fuller, 1992; Wood, 1992; Marteau & Rambaud, 1993). Many species and strains of lactic acid bacteria from several genera, due to their ability to produce different types of antibacterial compounds, have been credited with these health benefits. They are consumed through many types of products available commercially (Table 2). Generally a product, depending upon the type, can contain one or more of the following species: *Streptococcus thermophilus, Lactococcus lactis, Leuconostoc*

NATO ASI Series, Vol. H 98
Lactic Acid Bacteria:
Current Advances in Metabolism, Genetics and Applications
Edited by T. Faruk Bozoğlu and Bibek Ray

Table 1. Health benefits attributed to lactic acid bacteria

A.	To combat :	growth of indigenous microflora in the intestine
	:	control infections in the intestine by enteric pathogens
	:	control infections in the urinogenital tract
	:	lactose-intolerance
B.	To reduce :	cancer/tumor in colon (and other organs)
	:	serum cholesterol and cardiac heart disease
C.	To stimulate :	immune system
	:	bowel movement

Table 2. Types of probiotics, desirable bacteria in the probiotics and antimicrobial metabolites produced by these bacteria

Probiotics (predominant)	Bacteria (most commonly used)	Metabolites (in the products)
A. Fermented dairy products (yogurt, buttermilk, acidophilus milk, etc.)	*Lab. bulgaricus* *Str. thermophilus* *Lac. lactis* *Leu. mesenteroides* *Lab. acidophilus* *Lab. casei* *Bifidobacteria* spp. *Lab. reuteri*	Lactate, acetate, diacetyl, HCO_3, H_2O_2, bacteriocins, reuterine
B. Supplemented foods (pasteurized milk, drinks)	*Lab. acidophilus* *Bifidobacterium* spp. *Lab. reuteri* *Lab. bulgaricus* *Str. thermophilus*	Can supply ß-galactosidase. The indigenous types establish and produce metabolites.
C. Pharmaceuticals (tablets, capsules, granules)	*Lab. acidophilus* *Lab. bulgaricus* *Bifidobacterium* spp.	Can supply ß-galactosidase. The indigenous types, establish and produce metabolites.
D. Health food products (liquid, capsules, powders)	*Lab. acidophilus* *Bifidobacterium* spp. (many other lactobacilli, currently not recognized as species, are also used)	None, except in liquid fermented products. Can supply ß-galactosidase.

mesenteroides, Lactobacillus bulgaricus, Lactobacillus casei, Lactobacillus acidophilus, and *Lactobacillus reuteri* and some *Bifidobacterium* species (in this article bifidobacteria are included in the general term LAB). In addition, some products can contain other *Lactobacillus* species, some are even currently not regarded as a species (such as *Lab. caucacicus* in some products sold in health food stores).

Fermented dairy products were produced accidentally and consumed by the ancient civilizations, most probably initially in the Middle East and Indian subcontinent, as far back as 6000 B.C. It followed the successful domestication of cattle, invention of potteries, recognition of milk as a human food and production of sufficient milk for storage for fermentation to occur. The transformation of milk to fermented dairy products was probably conceived by the priests as divine with magical healing power about 3000 B.C. Subsequently, most major cultures believed in the healing properties of fermented dairy products and probably other fermented foods. This notion continued until the 1860's without understanding the scientific basis of food fermentation. Then in 1970 Louis Pasteur proposed that souring of milk is associated with the growth and metabolic activities of the microscopic living entities, later grouped as lactic acid bacteria. He also was first to conceive the concept of the antagonist interactions among bacteria (Pasteur & Joubert, 1877). In 1907 Metchnikoff forwarded the hypothesis of intestinal replacement therapy by implanting lactic acid bacteria from consuming Bulgarian sour milk (which later was identified as *Lactobacillus bulgaricus*). He advocated that the putrefied products of undesirable GI-tract flora contributed to the aging process of the human body and replacement of these undesirable organisms by the microorganisms in sour milk could help in the prolongation of life (Metchnikoff, 1907).

Since the 1950's microbial types and ecology of human GI-tract have been investigated scientifically and the influence of many factors on microbial flora in this organ has been identified (Drasar & Hill, 1974). In the 1970's controlled studies were conducted on the beneficial effect of *Lab. acidophilus* and *Lab.*

casei. Later in the 1980's similar investigations were undertaken on the benefits of several *Bifidobacterium* species. In the 1990's *Lactobacillus reuteri* was included in similar studies. General interest on the health benefits from the consumption of some lactic acid bacteria has resulted in the availability of many types of commercial products in the market world wide. Currently, the health benefits of lactic acid bacteria in controlling enteric pathogens in food animals and birds are being studied.

Although many health benefits of lactic acid bacteria, as listed before, have been suggested by many investigators, the results are not always conclusive. The major objectives of this article are to determine the scientific basis of the health claims, briefly analyze some study results, determine possible reasons of controversy and finally suggest means to overcome the controversies.

MICROBIOLOGY OF HUMAN INTESTINAL TRACT

The GI-tract of humans contains an astounding number and type of microflora. Their weight collectively amounts to about 1 kg and is closely equal to the weight of some of our organs, such as the kidneys or the brain. The total numbers of intestinal microflora could be as high as 10^{16} which is higher than the average number of cells of the human body (about 10^{14}). The intestinal flora are also most metabolically active and carry on a diverse type of metabolic reactions. This in turn can profoundly influence our health and well being. Because of this, the intestinal microflora are viewed by many as a specialized organ with a wide variety of metabolic activities (Drasar & Hill, 1974; Drasar & Barrow, 1985).

It has not been possible even now to isolate and identify the different microbial types that inhabit the human GI-tract. A current estimate that a normal human GI-tract is populated by 400 to 500 bacterial species of which 95% are obligate anaerobes, especially many of those in the large intestine. However, only 30 to 40 species, among all, probably constitute 99% of the total population of the intestinal microflora. In the GI-tract of an adult the bacterial population in the small intestine, especially in the jejunum and ileum, ranges between 10^6 to 10^7 cells/g and in the large intestine between 10^{10} to 10^{11}/g of the content. Numerically, the predominant types, in the small intestine are several species of *Lactobacillus* and *Enterococcus* and in the large intestine are several genera and species of *Fusobacterium, Enterobacteriaceae, Bacteroides, Eubacterium, Enterococcus, Clostridium, Bifidobacterium* and *Lactobacillus* (Drasar & Hill, 1974; Drasar & Barrow, 1985).

The intestine of a fetus in the uterus is sterile. At birth the intestine is inoculated with vaginal and fecal flora of the mother. Subsequently, a wide variety of bacteria are introduced from the many sources in the environment. From these mixtures, the natural flora of the human digestive tract are selected. Within a day after birth, the fecal population of *Escherichia coli* and *Enterococcus* species reached to about 10^{10}/g and remained at that level for 2 to 3 days (Mitsuoka, 1989). The populations of lactobacilli and bifidobacteria then increase. In breast-fed babies the numbers reached to about 10^7 to 10^{10}/g within a week with simultaneous reduction in the population of *E. coli* and enterococci to about 10^8/g of feces. In contrast, in formula-fed babies the levels of *E. coli* and enterococci together with *Bacteroides, Clostridium* and several other species remained very high with very low population of bifidobacteria. This can result in incidence of diarrhea among formula-fed babies. During this period, *Bacteroides, Clostridium* and other species also appear in the feces of breast-fed babies but at substantially lower levels. As soon as the babies are introduced to supplementary foods, along with mothers' milk, the levels of *E. coli, Bacteroides, Clostridium, Enterococcus* and other species increase; however, the level of *Bifidobacterium* still remains high. When the breast-feeding stops and the babies are introduced completely to solid foods, the fecal microbial profile rapidly changes. *Bacteroides* and *Bifidobacterium* become predominant flora with a decline in numbers of *E. coli, Enterococcus* and *Clostridium* species. By the second year of life of a child, the different microflora have established at specific ecological niches in the intestine and the population resembles that of an adult intestine (Table 3). Although the data showed a lower population of the five main bacterial groups in the jejunum, in the ileum both lactobacilli and enterococci are high. In individuals, the level of lactobacilli in the ileum can range between 10^6 and 10^7/g. In the colon content as well as in the feces the average population of these five bacterial groups ranged between 10^6 to 10^9/g. Also, geographic locations and the nature of the diet greatly influence the relative population of these five major bacterial groups. A diet rich in foods from plant sources, generally favors higher numbers of lactobacilli and bifidobacterium (Hentges, 1983; Drasar & Barrow, 1985).

The GI-tract microflora are arbitrarily divided into indigenous (or autochthonous) and transient (or allochthonous) groups. The indigenous species are considered to be permanent inhabitants of the intestines and are able to establish, maintain in specific niches and multiply. In contrast, the transient types are either passing through following their introduction through mouth or temporarily occupying a habitat vacated by an indigenous species due to some reasons. Many strains of indigenous species are capable of adhering to the intestinal epithelia that helps them in both establishment and subsequent maintenance in specific niches. Several other inherent and environmental factors influence the establishment and maintenance of the indigenous flora in the GI-tract. Some inherent factors include: gastric acidity (pH), lysozyme and other lytic enzymes, bile acids, intestinal antibodies, mucous, oxidation-reduction (O-R) potential and intestinal motility. Several environmental factors include: the geographical locations, nature of diet, alcohol consumption, intake of antimicrobial compounds (such as antibiotics), starvation, stress and the types and numbers of microorganisms are introduced in the GI-tract from various sources. The indigenous flora established during childhood are not stable, instead they are dynamic and undergo frequent changes with the change in the inherent and environmental factors (Drasar & Hill, 1974; Hentges, 1983; Savage & Fletcher, 1985; Savage, 1987).

Table 3. Average population ($\log_{10}$) of major bacterial groups in the intestinal contents of normal people[a]

Bacterial groups	Jejunum	Ileum	Colon	Feces
Lactobacillus	3	5	6	6
Gram-positive non-sporing anaerobes[b]	2	2	5	6
Enterococcus	3	5	7	7
Bacteroides	3	3	7	9
Enterobacteriaceae	3	4	6	8

[a]Adopted from Drasar & Hill (1974)
[b]They include *Bifidobacterium* species

DESIRABLE INTESTINAL BACTERIA

Among the indigenous GI-tract microflora, several species of *Lactobacillus* and *Bifidobacterium* are considered to have beneficial effects on the health and well being of hosts. Again, there are differences in opinions as one study recently suggested that bifidobacteria may have a undesirable role in colon cancer in some populations, while some *Eubacterium* species may have a beneficial role. However, some *Lactobacillus* species have definite beneficial roles in colon cancer (Moore & Moore, 1995). The benefits of different indigenous and non-indigenous lactic acid bacteria are discussed in this section.

Benefits of lactobacilli: ***Lactobacillus acidophilus***

Several *Lactobacillus* species have regularly been isolated from the intestinal contents of humans, animals and birds. Some of these are: *Lab. acidophilus, Lab. reuteri, Lab. lactis, Lab. casei* and *Lab. fermentum*. However, it is not clearly established that all of them or only a few of them are indigenous. It has been accepted that *Lab. acidophilus* and *Lab. reuteri* are definitely indigenous to the GI-tract and inhabit predominantly in the jejunum and ileum, with higher numbers in the ileum, especially at the distal segment. Several others, among those listed above, are resistant to GI-tract environment; thus, they probably can remain in the intestine for a period of time and can be confused with those which are truly indigenous, such as *Lab. acidophilus*. In contrast, some other lactobacilli, like *Lab. delbrueckii* ssp. *bulgaricus* (referred to as *Lab. bulgaricus* in this article) used in dairy fermentation are sensitive to the GI-tract environment and can not survive long in the intestinal tract (Hallgrand, 1983; Hentges, 1983; Drasar & Barrow, 1985; Speck *et al.*, 1993).

Many bacterial strains, previously isolated from different sources and designated as *Lab. acidophilus* on the basis of biochemical reaction profile, have now been divided into five groups on the basis of DNA homology. These are: A1 (*Lab. acidophilus*), A2 (*Lab. crispatus*), A3 (*Lab. amylovorus*), A4 (*Lab. gallinarum*), B1 (*Lab. gasseri*) and B2 (*Lab. johnsonii*). They all are considered to be indigenous to the GI-tract of humans, animals and birds and some groups and strains show host specificity. Except for *Lab. acidophilus*, the roll of other groups in the intestinal health of hosts has not been well studied. In this article only the contributions of *Lab. acidophilus* of group A1 is discussed (Johnson *et al.*, 1980; Bhowmik *et al.*, 1985; Johnson *et al.*, 1987; Wood, 1992, Pot *et al.*, 1993).

Since the 1970's many studies have been focused to ascertain the benefits of *Lab. acidophilus* to the health benefits of the GI-tract in humans. Strains of this species have regularly been isolated from the contents of jejunum, ileum and colon of healthy humans of all ages. The resistance of *Lab. acidophilus* cells to low pH, bile acids and lysozyme of the digestive tract, and their facultative anaerobic nature, together with the ability of some strains to adhere to the intestinal epithelial cells enable them to survive, establish and multiply in the GI-tract. The population of *Lab. acidophilus* in the intestinal content of normal adult humans range between 10^5 to 10^7/g with the higher level in the distal ileum and colon than in the jejunum. Even with a relatively low population level they are presumed to control unrestricted growth and activities of some other undesirable bacterial species that are usually present at a level of 10^{10} to 10^{11}/g in the large intestine, such as *Bacteroides* and *Fusarium* spp. Not only *Lab. acidophilus* cells are thought to maintain a desirable microbial ecology in the intestine by controlling growth and metabolic activities of some undesirable indigenous and transient bacteria; they are also able to reduce the chances of transient enteric pathogens from establishing in the GI-tract, especially in the small intestine. This is achieved most likely through the ability of *Lab. acidophilus* to produce several antibacterial metabolites, particularly large amounts of lactic acid and also depending upon a strain, some other organic acids, hydrogen peroxide (H_2O_2), bacteriocins, and deconjugated bile acids (Table 2). In one of our studies we examined the sensitivity of four strains of enteric pathogens against *Lab. acidophilus* isolates from jejunum, ileum and colon of dairy calves (Table 4). Growth of all four strains were inhibited by several *Lab. acidophilus* isolates from all three

Table 4. In vitro inhibition of enteric pathogens by *Lactobacillus acidophilus* isolates from the intestinal contents of calves

	Number inhibited/out of number tested: isolates from[b]		
Enteric pathogens tested[a]	Jejunum	Ileum	Colon
E. coli 1	10/18	14/23	10/21
E. coli 2	12/18	18/23	13/21
Salmonella typhimurium 1	11/18	15/23	11/21
Salmonella typhimurium 2	8/18	15/23	10/21

[a] *E. coli* strains produce K 99 toxin. All four strains were obtained from the Wyoming State Veterinary Laboratory (Ray, unpublished data)

[b] Eight calves were killed at 2 to 4 weeks after birth and the contents of jejunum, ileum and colon were used to isolate and biochemically identify the *Lactobacillus acidophilus* isolates. Sensitivity is determined from the presence of ≥ 3 to 4 mm zone of growth inhibition of an enteric pathogen around the growth of *Lab. acidophilus* on agar plate assay (See text for explanations)

locations. This growth inhibition is most likely not due to low pH, as a highly buffered medium was used and some other *Lab. acidophilus* isolates failed to inhibit growth of these pathogens (Ray, unpublished data).

One of the factors important for a *Lab. acidophilus* strain to produce beneficial effects on host intestinal tract is the number of normal viable cells. Many inherent and environmental factors of the host can eliminate or reduce considerably the number of *Lab. acidophilus* cells in the intestine. Some of these, mentioned earlier, are antibiotic intake, abusive and improper food habits, diseases and surgery of the digestive tract, pelvic irradiation and stress (Drasar & Hill, 1985; Gilliland, 1990; Sandine, 1990; Hentges, 1993). These conditions not only reduce the numbers of beneficial bacteria, but also give a chance for the undesirable bacteria to flare up and enteric pathogens to invade the digestive tract. We studied the effect of stress on the populations of *Lab. acidophilus* and coliforms (*E. coli*) in the intestinal contents by weaning dairy calves 4 weeks after birth and beef calves 4 months after birth. The results of the dairy calf study are presented in Figure 1. Two dairy calves were slaughtered at 1, 2 and 4 weeks after birth but before weaning and then after 2 days, and 1, 2 and 4 weeks after weaning (total 8 week study) and the contents from the jejunum, ileum and colon were selectively enumerated for *Lab. acidophilus* and coliforms (*E. coli*). The populations of both *Lab. acidophilus* and coliforms dropped dramatically in all sites soon after weaning. Then slowly the population of both types increased in all sites, but coliforms increased at a rapid rate, especially in the small intestine. Also, 4 weeks after weaning the population of *Lab. acidophilus* (also of coliforms) remained at levels lower than at 4 weeks of age after birth and before weaning. These results indicate stress can upset the microbial ecology in the digestive tract and can adversely affect the intestinal health of the hosts.

In the event the population of *Lab. acidophilus* in the host GI-tract is reduced, daily oral consumption of a large number (about 10^9 cells/day) of healthy viable cells of *Lab. acidophilus*

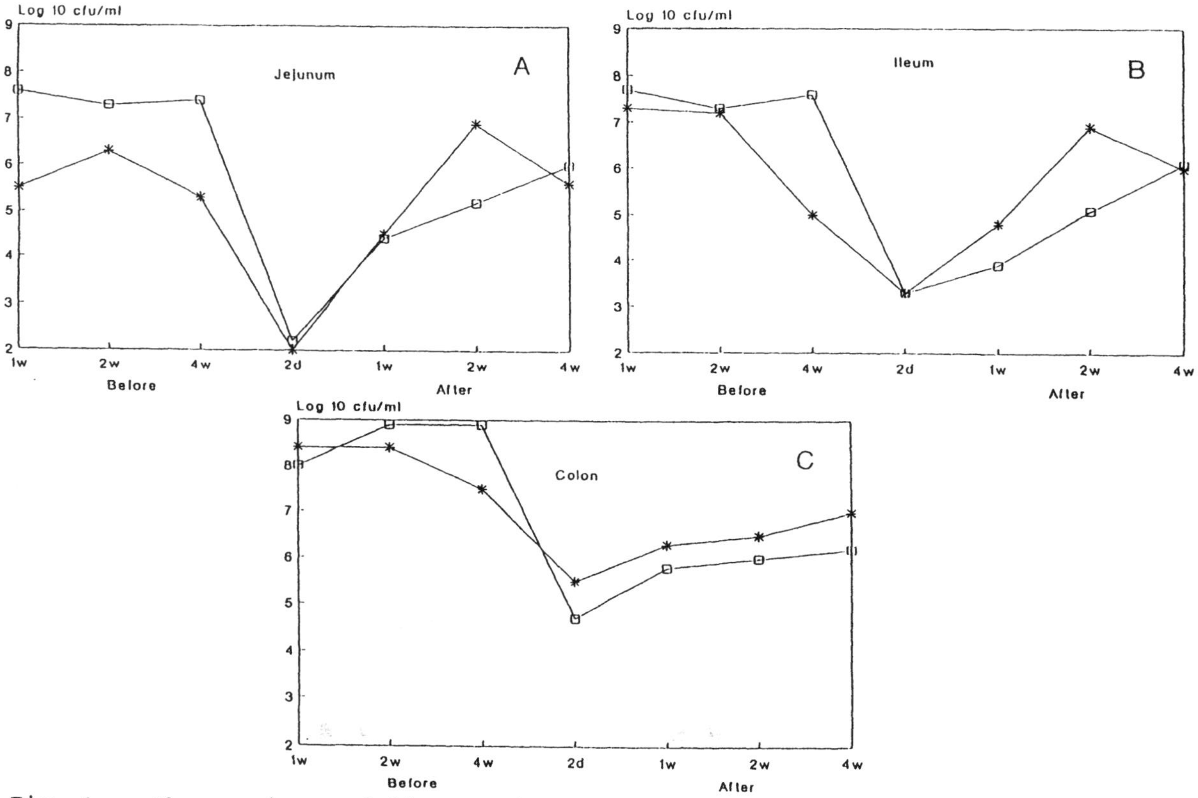

Fig. 1. Changes in population of bile resistant lactobacilli and coliforms in jejunum (A), ileum (B) and colon (C) of dairy calves before and after weaning (withdrawal from suckling) after birth (w, weeks; d, days. , lactobacilli; *, coliforms) (Ray, unpublished data).

over a period of 2 weeks or more has advocated to raise their levels (Speck, 1978). A large variety of commercial preparations containing viable *Lab. acidophilus*, many of which are dried, are now available that are intended to restore the population following oral administration. A product in which high levels of healthy viable *Lab. acidophilus* cells have been inoculated in refrigerated pasteurized milk will maintain the high population by the "sell date". However, acidic (low pH) fermented products and particularly the dried products, many of which are stored at room temperature (22 to 30°C or higher depending upon a country), most likely will not have high numbers of viable *Lab. acidophilus* cells. Also, many of the viable cells can be injured and thus will be killed by the low pH in stomach and bile salts and lysozyme in the intestine. In a study we found that the population of *Lab. acidophilus* in these products sold by the "health food store" and pharmacies have very low numbers of *Lab. acidophilus* (Table 5). Some of these products not only contained lactobacilli that are non-indigenous and sensitive to GI-environment (*Lab. bulgaricus*) and not accepted now as a species (*Lab. caucacicus*) but also had coliforms and other Gram-negative bacteria, possibly due to unsanitary practices during production and handling (Brennan *et al.*, 1983).

One reason for low viable population of *Lab. acidophilus* in the dried commercial preparations could be due to the viability loss of the cells during freezing, drying and subsequent storage, especially at higher temperatures for long times and some in the presence of air. Freezing and drying are known to cause viability loss as well as sublethal injury of *Lab. acidophilus* cells and other bacteria. Storage, even at refrigeration temperature, can reduce the viability of surviving population in the dried products rapidly. Higher storage temperatures, long storage times, and presence of oxygen (especially after opening a sealed bottle during use) can accelerate the rate of viability loss. The injured cells develop sensitivity to low pH, bile salts and lysozyme and when consumed are rapidly killed. Thus, such a product quite often is not expected to supply high numbers of healthy viable cells of *Lab.*

Table 5. Prevalence of viable *Lactobacillus acidophilus* cells in dried commercial products[a]

Product type	Claim	Viable lactobacilli enumerated	Viable *Lab. acidophilus* enumerated	Comments
<u>Health Food</u>				
A (capsules):	contains $5x10^7$ viable cells of *Lab. acidophilus* *Lab. bulgaricus* *Lab. caucasicus* per capsule	$2.1x10^3$	$3.2x10^2$	Gram-negative cells present along with coliforms. *Lab. bulgaricus* *Lab. caucacicus* are non-GI type
B (capsules):	contains $4x10^8$ viable cells of *Lab. acidophilus* per capsule	$2.0x10^3$	$<10^1$	Gram-negative cells present
<u>Pharmaceutical</u>				
A (tablets):	contains *Lab. acidophilus* *Lab. bulgaricus* (no number claimed)	$7.6x10^7$	$2x10^6$	No coliform. *Lab. bulgaricus* is not GI bacteria
B (capsules):	Contains 10^8 viable cells of *Lab. acidophilus* per capsule	$4.3x10^3$	$<10^1$	No coliform

[a]Adopted from Brennan *et al.* (1983)

acidophilus as the need dictates (Brennan *et al.*, 1986; Ray & Johnson, 1986).

Some of the *Lab. acidophilus* strains used commercially or by private laboratories may not belong to group A1. As indicated before the previously recognized *Lab. acidophilus* strains have been divided into six groups and those which are in group A1 are considered as *Lab. acidophilus*. We examined a large number of strains from all six groups and developed a procedure by which these strains can be identified by the absence and presence and molecular weight (MW) of surface layer proteins (SLP) and for the absence and presence of inducible ß-galactosidase (Table 6). Strains from groups B1 and B2 were found not to have SLP and MW of SLP of strains A1, A2, A3 and A4 were, respectively, about 46, 59, 56 and 53 KDa. Electron microscopy of these cells also revealed the presence of a SLP in A1, A2, A3, A4 but absence in B1 and B2 strains. Electron micrographs of the cell surface of a strain from group A1 and a strain from group B1 in Figure 2 show the presence and absence of SLP. Analysis of the isolates from commercial sources showed, many of those designated as *Lab. acidophilus* did not have the 46 KDa SLP (Bhowmik *et al.*, 1985; Ray & Johnson, 1986; Johnson *et al.*, 1987).

It was also found from these studies that the strains in B1 group (now designated as *Lab. gasseri*) do not have ß-galactoside trait. All strains from the other 5 groups, except for VPI 11694 in group B2, carry inducible ß-galactosidase system (Table 6; activity present only in lactose-grown strains). VPI 11694 of B2 has a constitutive ß-galactosidase system as the activity is present in both lactose and glucose grown cells. Also, the ß-galactoside activity among lactose grown *Lab. acidophilus* strains varied greatly under the same study conditions. These findings have important implications if the strains are used as probiotics for lactose intolerant individuals to overcome problems associated with consumption of milk. If the *Lab. acidophilus* cells, used in capsules, tablet, granules, etc., are grown in a broth containing glucose to obtain large cell mass, consumption of those products will not supply ß-galactosidase in the small intestine until they

Table 6. Variation in surface layer proteins and ß-galactosidase activity among strains from *Lactobacillus acidophilus* subgroups (adopted from Bhowmik *et al.*, 1985 and Johnson *et al.*, 1987)

Homology group (species designation)	VPI strains[a]	MW of surface layer protein (KDa)	ß-galactosidase activity (µm ONP/OD/min)[e] Lactose	Glucose
A1: (*Lab. acidophilus*)	VPI 6032[b]	46	1511	0
(*Lab. acidophilus*)	VPI 0330[b]	46	1545	0
(*Lab. acidophilus*)	VPI 11695	46	434	0
A2: (*Lab. crispatus*)	VPI 1799	52	1251	0
(*Lab. crispatus*)	VPI 6272	44	639	0
(*Lab. crispatus*)	VPI 1784	59	1128	0
A3: (*Lab. amylovorus*)	VPI 1754	56	1277	0
(*Lab. amylovorus*)	VPI 1756	ND[c]	1361	0
(*Lab. amylovorus*)	VPI 0824	40	1890	0
A4: (*Lab. gallinarum*)	VPI 1294	53	1042	0
(*Lab. gallinarum*)	VPI 1793	57	0	0
(*Lab. gallinarum*)	VPI 2164A	49	0	0
B1: (*Lab. gasseri*)	VPI 6033[b]	None[d]	0	0
(*Lab. gasseri*)	VPI 0333[b]	None	0	0
(*Lab. gasseri*)	VPI 6325	None	0	0
B2: (*Lab. johnsonii*)	VPI 0325	None	612	0
(*Lab. johnsonii*)	VPI 11694	None	2085	872
(*Lab. johnsonii*)	VPI 11696	None	305	0

[a]VPI strains: Virginia Polytechnic Institute Collection
[b]These strains also have ATCC numbers: VPI 6032 is ATCC 4356; VPI 0330 is ATCC 4357; VPI 6033 is ATCC 19992; VPI 0333 is ATCC 4962.
[c]Not determined.
[d]No surface layer protein was also detected on cell surface by electron microscopy
[e]ß-galactosidase activity was measured with cells grown in broth containing either lactose or glucose as carbohydrate by the method of Miller (1972) from the absorbency at 420, 550 and 600 nm.

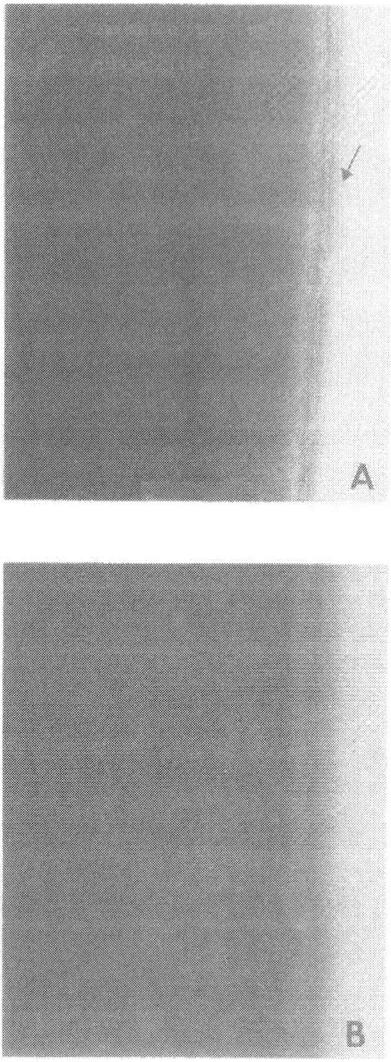

Fig. 2 Negatively stained electron micrograph of a portion of cell surface of two strains, one from group A1 (*Lab. acidophilus*:A) and one from group B1 (*Lab. gasseri*:B). Arrow indicates the surface layer protein over the cell wall in strain from group A1 (Ray, unpublished data).

grow in the presence of lactose; thus, they may not help overcome lactose-intolerance symptoms. Also, for such benefit, a strain with high β-galactosidase activity potential should be selected for use in commercial products. Also, the level of production of β-galactosidase by group A1 *Lab. acidophilus* strains vary greatly with growth conditions. A strain growing in a broth containing lactose produces considerably less β-galactosidase at 45°C and above, in the presence of bile salts, in the presence of higher than 0.5% lactose and in the presence of Ca^{2+}, Mg^{2+} and Mn^{2+} greater than 5 mM. During storage of cells even at 4°C the β-galactosidase activity decreased rapidly (Fisher *et al.*, 1985; Bhowmik *et al.*, 1987). It also is important to note that if one plans to use *Lab. gasseri* as probiotics, it cannot be effective to overcome lactose-intolerance.

Some fermented dairy products are available in the market that are produced by inoculating *Lab. acidophilus* with other lactic acid bacteria that have constitutive β-galactosidase system, such as *Lab. bulgaricus* + *Str. thermophilus* + *Bif. bifidus* (in some yogurt). We found that in milk *Lab. acidophilus* strains are poor competitors against *Lab. bulgaricus* and *Str. thermophilus*, even when a strain with high β-galactosidase potential and grown previously in lactose broth or milk to induce β-galactosidase is used. Under the incubation conditions of yogurt production, *Lab. acidophilus* with initial inoculation level of 10^6 cell/ml (similar to other two) were out numbered by *Lab. bulgaricus* and *Str. thermophilus*. A similar situation may exist when *Bifidobacterium* spp. are used in combination as favorable growth environment of *Bifidobacterium* is different from the other two LAB used to produce yogurt (Ray, unpublished data).

Among the *Lab. acidophilus* strains isolated from intestinal contents of humans, animals and birds, some are found to adhere to the intestinal epithelium while others do not. The adherence is species specific. Also, a strain can lose the adherence ability during subculturing in a bacteriological medium (Makinen *et al.*, 1983; Savage, 1987; Schneitz *et al.*, 1993; Muki & Arihara, 1994). We observed that among the *Lab. acidophilus* isolates from calf jejunum, ileum and colon, some were able to adhere to epithelium cell preparation from ileum of a 1 week old calf while other strains failed to do so (Fig. 3 A,B). The level of adherence differs with the isolates. Also, when some of these isolates, with strong adherence tendency with calf ileal epithelial cells, were mixed under similar test conditions with ileal epithelial cells of a 1 week old piglet, they either did not adhere or did so poorly (Ray, unpublished data).

The adherent strains of *Lab. acidophilus* have gained popularity in probiotic use over the non-adherent strains as it is assumed due to this property the adherent strains have the ability to establish and maintain a high population level in the intestine. Also, by remaining attached to intestinal epithelia, they can make the cells unavailable for the adherence and setting up infection by some enteric pathogens. However, for use in humans, it is important to use a strain that is adherent to human intestinal epithelial cells. Also, as the adherence trait can be lost during laboratory subculturing it will be important to develop means to stabilize the trait. One suggestion was to feed and reisolate the strain from a human; but the validity of the process has not been well studied (Speck, personal communication). The best approach will be to determine biochemical and genetic basis of the species-specific adherence trait. This can help develop methods to stabilize the trait in a suitable *Lab. acidophilus* strain. Available reports indicate that the lipoteichoic acid (LTA) and/or SLP of *Lab. acidophilus* may have important role in host-specific adherence. We observed that all A1 group strains we tested, irrespective of their ability to adhere to intestinal epithelium, have a 46 KDa protein. But that does not tell if those proteins differ functionally for species-specific attachment. In general, this protein(s) is hydrophobic, resistant to trypsin, pepsin, papain and protease IV, but not to protease XIV (pronase). The amounts present in cells differ with growth conditions. More proteins were extracted from cells growing in a broth containing Ca^{2+}, threonine or lactose, under anaerobic conditions and at the early stationary growth phase. In contrast, cells grown in the

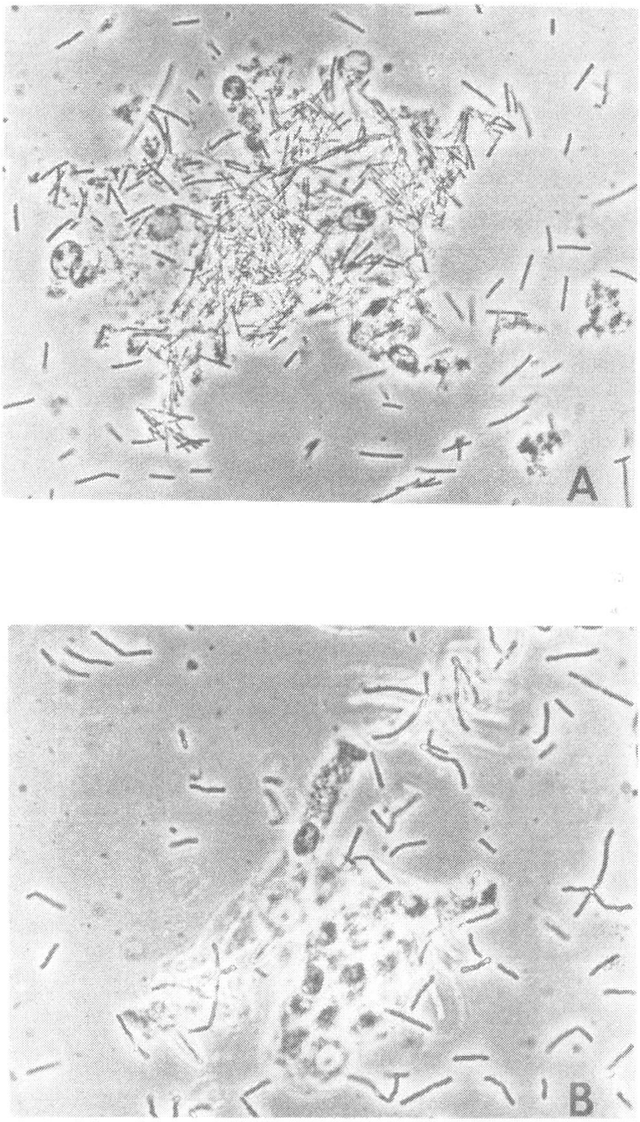

Fig. 3 Photograph showing adherence (A) and non-adherence (B) of two isolates of *Lab. acidophilus* to the ileal epithelial cells from a dairy calf. Both isolates were of calf origin (Ray, unpublished data).

presence of bile salts, with incubation temperature of 45°C and above and for longer periods produced less LSP (Bhowmik *et al.*, 1985; Johnson *et al.*, 1987). Several recent studies have suggested a cell surface component, probably a carbohydrate but not a protein and a proteinaceous extracellular compound are important in the host specific adherence of *Lab. acidophilus* strains to intestinal epithelial cells (Chauviere *et al.*, 1992; Coconnier *et al.*, 1992; Green & Klaenhammer, 1994). Other studies have reported that the strongly adherent strains have well defined SLP. In contrast, in the weekly adherent and non-adherent strains or strains that have lost the trait during subculturing, the SLP was either covered with a polymerized materials or absent (Schneitz *et al.*, 1993). A more recent study has shown that the SLP of *Lab. acidophilus* have lectin-like activity. Host intestinal epithelial cells have glycoproteins and glycolipids, the compositions of which could be species specific. The adherence of the adhering strains of *Lab. acidophilus* could be through the binding of cell surface lectins in bacterial cells to specific glycoproteins/glycolipids of host intestinal epithelial cells (Mukai & Arihara, 1994).

There are reports that several *Lab. acidophilus* strains produce proteinaceous antibacterial compounds, designated as bacteriocins (Ray & Daeschel, 1992). One such bacteriocin, designated as lactacin B, was found to be bactericidal to several other *Lactobacillus* spp. (Barefoot & Klaenhammer, 1983). Another bacteriocin, designated as lactacin F and produced by a separate strain, was also found to be bactericidal against several *Lactobacillus* spp. and a strain of *Enterococcus faecalis* (Muriana & Klaenhammer, 1991). Both lactacin B and lactacin F have been studied to determine their amino acid sequences and genetic determinants. Although there are other reports that *Lab. acidophilus* strains produce bacteriocins or bacteriocin-like compounds that are antibacterial against many Gram-positive and Gram-negative bacteria and even viruses, their identity is far from complete and thus are not discussed here.

In general, the available results tend to indicate that bacteriocins of *Lab. acidophilus* could have narrow spectrum of antibacterial property. Thus, it is difficult to assume that these bacteriocins (only those which have been well characterized) or these bacteriocin producing strains have any real impact on controlling the growth and viability of diverse types of bacteria found in the GI-tract of humans (also animals and birds). Also, it is very important to recognize that in general, the antibacterial activity of bacteriocins of lactic acid bacteria is destroyed by proteolytic enzymes of the GI-tract. Also, many of the GI-tract bacteria produce proteolytic enzymes. Thus, actual contributions of bacteriocins of lactic acid bacteria in maintaining health of the GI-tract of hosts is debatable.

It appears from the foregoing discussions that for effective use of *Lab. acidophilus* as probiotics in humans (as well as in animals and birds), a strain should have several desirable characteristics: (a) it should belong to homology group A1; (b) it should be a human isolate, preferably capable of adhering to the human intestinal epithelial cells; (c) it should produce relatively rapidly large amounts of lactic acid from fermentable carbohydrates; (d) it should be able to produce other antibacterial compounds effective against indigenous and transient undesirable intestinal bacteria; (e) if used to help lactose intolerance, it should have high levels of already induced ß-galactosidase (if possible, by genetic modifications such a strain can be changed to constitutive strain); (f) it should survive freezing, drying and other stresses used during processing and storage of the commercial products or techniques should be developed to achieve this goal. In addition, for benefit the viable cells of *Lab. acidophilus* should be consumed in high numbers over a period of time (two weeks or more). It is impossible to demand so much from a single strain. Maybe it is more practical to use several strains each with some desirable characteristics but not antagonistic to each other. Finally, an understanding of the biochemical and genetic control system may help in the future to design a strain that has many of the desired traits.

Benefits of intestinal lactobacilli: ***Lactobacillus reuteri***

Heterofermentative *Lab. reuteri* is closely related to *Lab. fermentum* and recently has been recognized to be a normal inhabitant of the GI-tract of humans and animals. It is considered to be a desirable species in maintaining health of the host intestine. It has not yet been studied well as in the case of *Lab acidophilus*. The available results indicate the strains produce lactic and acetic acids as well as an antimicrobial compound, designated as reuterine. Reuterine has a broad spectrum antimicrobial property, being able to inhibit growth and survival of many species/strains of Gram-positive and Gram-negative bacteria and some viruses (Ray & Daeschel, 1992; Speck *et al.*, 1993). Studies in pigs and chickens have shown the oral administration of live cells of *Lab. reuteri* resulted in the reduction of stress, controlling growth of some enteric pathogens and weight gain. Some products containing this species are available commercially. It is expected that future studies will reveal other benefits of this species.

Benefits of intestinal bifidobacteria

Many species/strains of *Bifidobacterium* have been isolated from the feces of humans and animals. In breast-fed babies bifidobacteria become predominant fecal flora (99%) within 3 to 7 d after birth. However, they are found in normal people throughout life in relatively high numbers. The predominant species in infants are *Bif. bifidus* and *Bif. infantis* while those in adults are *Bif. longum* and *Bif. adolescentis*. They, like lactobacilli, are Gram-positive, nonspore forming anaerobic rods. They predominantly inhabit at the proximal part of the colon, although are present in lower numbers in the small intestine.

Some oligosaccharides in human milk stimulate their growth which is thought to be the reason of their presence in very high numbers in the GI-tract of breast-fed babies, as opposed to formula-fed babies. They hydrolyze lactose and ferment glucose, galactose and fructose (and other metabolizable carbohydrates) to produce L(-) lactic acid and acetic acid in a ratio of 2:3, and can reduce the pH of the intestinal contents to 5.5 or below in the breast-fed babies (due to low buffering capacity of human milk as opposed to cows milk). At this low pH a large portion of the acetic acid produced remain in undissociate form which has greater antibacterial ability against the intestinal undesirable bacteria (Sandine, 1990; Hoover, 1993; Ishibashi & Shimamura, 1993).

Both adhering and non-adhering strains of *Bifidobacterium* species have been isolated from human feces. Limited information suggest that lipoteichoic acid of cell wall is associated with the attachment of a *Bifidobacterium* species to the human intestinal epithelial cells. Some studies have shown that selected *Bifidobacterium* strains are capable of controlling growth of pathogens and other undesirable bacteria in the intestinal tract. The beneficial effect was attributed not only to their ability to produce organic acids but also to activate the host immunity systems by the cell wall components. In addition, both whole cells and cell wall materials of bifidobacteria have been shown to reduce occurrence of tumors as well as to cause regression in tumor size in the colon. This was thought to be due to the ability of the cells and cell wall components to stimulate macrophages (Sandine, 199-, Ishibashi & Shimamura, 1993; Reddy & Rivenson, 1993).

While selecting *Bifidobacterium* species and strains for their beneficial effect as probiotics in humans, different criteria discussed previously for *Lab. acidophilus* may be necessary to consider. It is important to recognize that *Bifidobacterium* spp. predominantly inhabit in the colon. Thus, for some specific benefits, such as reducing lactose-intolerance, reducing serum cholesterol level, antagonizing enteric pathogens that infect small intestine, *Lab. acidophilus* and *Lab. reuteri* are probably better candidates than *Bifidobacterium* spp., if the latter is used alone. There is a trend now to use both groups in combination to get the most benefits.

DESIRABLE NON-INTESTINAL LAB

Non-intestinal LAB primarily include species and strains from several genera that include *Lactococcus, Leuconostoc, Pediococcus,* dairy *Streptococcus* and *Lactobacillus*. From these genera three subspecies of *Lactococcus lactis*, a few *Leuconostoc* species, *Streptococcus thermophilus*, and a few species of *Lactobacillus* (namely *Lab. bulgaricus, Lab. helveticus, Lab. lactis*) are used widely to produce a wide variety of fermented dairy products. Consumption of fermented dairy products, especially those containing viable cells of these species and strains, has been associated with many health benefits in humans (Table 1). As these are non-indigenous to the GI-tract, their beneficial effects are produced differently as compared to the indigenous species, discussed earlier. The cells of these LAB are quite sensitive to low pH, bile salts, lysozyme and other antibacterial factors present in the different sections of human GI-tract. Some studies have shown the viability of *Lab. bulgaricus* and *Str. thermophilus* was reduced in the duodenum by about 6 logs within 90 min following consumption. Other studies also have shown sensitivity of the cells of different LAB in this group to the GI-tract environment.

When consumed live through the fermented product, this group of LAB are considered transient. The benefits they produce are through some of their metabolites present in the fermented foods and specific cell components (especially after they die in the intestine) such as ß-galactosidase, lipoteichoic acid and cell wall. As these strains do not establish in the GI-tract, the cells through the fermented foods have to be consumed on a regular basis for the health benefits. Also, heating of the fermented products (such as heated yogurt) can reduce some of their benefits, such as a source of ß-galactosidase to reduce problems associated with lactose intolerance (Savino & Levitt, 1984; Robinson, 1991; Sanders, 1992).

EXPECTED BENEFITS OF PROBIOTICS

The objectives of this section are to discuss the overall results of the studies using lactic acid bacteria as probiotics to combat several intestinal ailments, evaluate the possible reasons of controversy between studies and suggest means that could reduce such controversy. Although only a few of the benefits listed in

Table 1 are discussed, the information provided here will be helpful in evaluating those that are not discussed.

To combat intestinal ailments

Consumption of fermented dairy products and live cells of LAB have been suggested to control different types of diarrhea (travellers, chronic, pediatric) in adults and children. The intestinal infections caused by transient pathogenic enteric bacteria, consumed through contaminated food and water, or uncontrolled growth of indigenous undesirable bacteria are associated with the diarrheal symptoms. Reduced normal resistance due to age (both old and young), antibiotic intake, pelvic irradiation, mental stress, sickness and abusive food habits can also act as predisposing causes. Some of the commercial products available to combat such diarrhea are: fermented milk products, such as yogurt containing live cells of *Lab. bulgaricus* and *Str. thermophilus* and their metabolites, similar products containing lives cells and metabolites of *Lab. acidophilus, Lab. reuteri*, and *Bifidobacterium* spp., pasteurized milk inoculated with live cells ($\geq 10^6$/ml) of the last three species, and tablets, capsules and powders containing one or more of those species (Table 2).

The results of many studies have shown that in general consumption of either the fermented dairy products or the live cells of indigenous desirable bacteria are capable of reducing the incidence of bacterial diarrhea in humans. However, results of some studies have suggested that such a benefit either was not observed or not significant.

It is quite apparent from the discussions presented before that the enteric pathogens are susceptible to one or more antimicrobial metabolites produced by the LAB generally used in such instances. One then needs to ask what could the reason(s) be for such differences in the results between the studies. While analyzing such studies, it becomes apparent the design and evaluation methods vary greatly between studies. In quite a few studies the effectiveness of the bacterial species/strains used for the purpose is not evaluated previously in separate studies. Many studies use commercial cultures with the information about their effectiveness provided by the supplies which may be claims without scientific basis. We have discussed before the variation in effectiveness among species and strains. Also, in many studies the levels of viable and injured cells of LAB present in a product prior to consumption were not determined. Again, it is important to recognize that to overcome intestinal ailments by consuming specific lactic acid bacteria one needs to consume $\geq 10^9$ physiologically normal viable cells/day over a period of 2 or more weeks. But processing and storage conditions of products can drastically reduce the population of normal viable cells of LAB. Other desirable characteristics, such as species identity of a strain, ability to adhere in human intestinal epithelial cells by a strain, ability to produce large amounts of lactic acid rapidly from the fermentable carbohydrates in the intestine and ability to produce other antibacterial compounds are very rarely determined in separate studies prior to their selection and use of a LAB to combat diarrhea. Similarly, in some commercial fermented products, such as yogurt, prepared with a combination of *Lab. bulgaricus, Str. thermophilus, Lab. acidophilus* and *Bifidobacterium* spp., no information is available on the viable population levels of each species prior to their use. It was mentioned before that strains of these four species do not grow in milk optimally under a single incubation condition and *Lab. acidophilus* has inducible ß-galactosidase system.

To reduce diarrheal infection in humans by using LAB and to obtain consistent results between studies it will be necessary to design the studies in such a way that they do not vary greatly. It will also be important to have authentic information about the suitability of the species/strains of LAB used for the purpose.

To combat lactose-intolerance

Lactose-intolerance or lactose-maldigestion in humans results from the inability of an individual to produce ß-galactosidase by the small intestinal epithelial cells due to a genetic defect. As a result, when an individual with lactose-intolerance consumes milk, the lactose is not hydrolyzed in the small intestine to galactose and glucose prior to their transport into the body for further metabolism. Lactose molecules, which cannot be absorbed from the small intestine, thus are moved to the large intestine where they are hydrolyzed by bacteria containing ß-galactosidase. Galactose and glucose are metabolized by homo- and heterofermentative bacteria to produce acids and large amounts of gas in the large intestine and the person develops diarrhea, flatulence and abdominal discomfort.

Consumption of some fermented dairy products, such as yogurt, due to their lower lactose content and being sources of ß-galactosidase from the starter cultures of LAB like *Lab. bulgaricus* and *Str. thermophilus*, have been suggested to reduce the symptoms of lactose intolerance. However, as these two species of LAB are killed in the digestive tract, benefits against lactose-intolerance from the consumption of such fermented dairy products or cells of these two species can only produce short-term relief. In contrast, consumption of live cells of *Lab. acidophilus* and *Lab. reuteri* either through fermented dairy products or through tablets, capsules, pasteurized milk can be expected to produce long-term benefits against this problem. It is important to recognize that *Lab. acidophilus* (from group A1) have inducible ß-galactosidase system, and the live cells in milk, tablet and capsule preparations have to be grown in lactose containing media so they can be the source of ß-galactosidase in the small intestine as soon as they are consumed. *Bifidobacterium* species are principally present in the large intestine in high numbers. Through fermented dairy products they may supply ß-galactosidase in the small intestine when consumed. Their effectiveness on a long-term basis as a source of ß-galactosidase in the small intestine may be much less than the *Lab. acidophilus* and *Lab. reuteri*.

The effectiveness of consuming fermented dairy products containing the lactic acid bacterial species listed above as well as milk containing live cells of those species on hydrolysis of lactose and metabolism of hydrolyzed galactose and glucose in the digestive tract of lactose-intolerant individuals has been studied measuring breath hydrogen concentrations. A high concentration of breath hydrogen indicates a greater hydrolysis and metabolism of lactose in the large intestine. The results of these studies by various researchers were contradictory.

In these studies many differences were present in the experimental designs. Also, effectiveness of species and strains of lactic acid bacteria with respect to their identity, desirable characteristics, viability and injury, and particularly ß-galactosidase producing potential was not identified in many studies. In some studies frozen or freeze-dried commercial preparation were used without identifying the viable cells present during a study or measuring the ß-galactosidase activity present in a product used in the study. Among lactic acid bacteria used as starter cultures, ß-galactosidase could be inducible or constitutive among species/strains and some strains may not carry this phenotype. Strains of LAB differ greatly in ß-galactosidase activity. In addition, ß-galactosidase activity can be destroyed during storage of cultures or products.

It is important for consistent results not only the experimental design in similar studies should be very similar, but also the species and strains of LAB used in these studies be similar in their characteristics particularly in their ability to supply ß-galactosidase on a short-term or a long-term basis. Finally, the analytical procedures used should produce acceptable and reproducible results.

To reduce serum cholesterol level

There are suggestions that consumption of fermented dairy products as well as live cells of *Lab. acidophilus* can reduce serum cholesterol levels which in turn can reduce the incidence of cardiac heart diseases. The scientific basis by which some fermented dairy products can lower serum cholesterol levels has not been properly tested or convincingly explained. But some studies have been conducted to explain the basis of reducing serum cholesterol levels by specific *Lab. acidophilus* strains. Two explanations for this suggested benefit are: (a) some strains of

Lab. acidophilus by their ability to metabolize cholesterol from foods in the small intestine make them unavailable for adsorption in the body; (b) also, some strains of *Lab. acidophilus* by their ability to deconjugate bile salts in the small intestine prevent their reabsorption in the body forcing liver to synthesize more bile salts from serum cholesterol. Both conditions lead to reduction in serum cholesterol level.

Some animal feeding trials with pigs have shown a reduction in serum cholesterol levels following feeding of live cells of specific strains of *Lab. acidophilus*. But other studies have shown contradictory results. Enough well designed studies have not been conducted with human cases with high serum cholesterol levels to validate that the *Lab. acidophilus* strains can really reduce the serum cholesterol level.

The two explanations suggested to validate the basis of the ability of specific *Lab. acidophilus* strains to reduce serum cholesterol levels have not been accepted by some researchers. It has been argued that the human body can synthesize enough cholesterol and food cholesterol may constitute only a minor portion of the total serum cholesterol. Also, deconjugation of bile salts can lead to diarrhea. Thus, before the ability of certain LAB to reduce serum cholesterol levels and cardiac heart disease is accepted it will be necessary to produce non-controversial data through well designed experiments. Special emphasis has to be given in the selection of species and strains of LAB suitable for the purpose.

To reduce colon cancer

Results of several animal studies have shown that feeding of either fermented dairy products or live cells of *Lab. acidophilus* and *Bifidobacterium* spp. or cell wall components of *Lab. bulgaricus, Str. thermophilus, Lab. acidophilus* and *Bifidobacterium* spp. resulted in the reduction of incidence of tumors and cancer in the colon. This benefit is produced through the control of growth of undesirable intestinal bacteria by the fermented dairy products and live cells of desirable indigenous bacteria. Many of these undesirable bacteria in the colon produce specific enzymes, such as ß-glucuronidase and azoreductase, that can convert food procarcinogens to carcinogens (such as NO_2 to nitrosoamines). Controlled growth of undesirable bacteria in turn produces less carcinogens and reduces incidence of colon cancer. These beneficial LAB also increase the bowel motion which also helps to reduce concentrations of procarcinogens and carcinogens in the colon. Cell wall components of some LAB have been demonstrated to stimulate phagocytosis of the macrophages which in turn suppress formation of tumors and cancers in the colon.

However, like in other studies, these studies also have not shown that LAB are capable of reducing incidence of colon cancer significantly. A recent study has even suggested that *Bifidobacterium* spp. can be related to the incidence of colon cancer in humans (Moore & Moore, 1995). To reduce this controversy it will be necessary to conduct well designed studies in which the importance of specific characteristics of species and strains of LAB are taken into consideration and results are analyzed by individuals who have expertise in the area of colon cancer.

CONCLUDING REMARKS

From the discussion presented in this article, based on experimental evidence, there are reasons to expect that some species and strains of LAB may have health benefit effects in humans (also in animals and birds). The inconsistency of the results could be due to the methods used in the studies. Unless that is corrected it may be improper to disqualify the benefits as myths. Several possible reasons for the inconsistency in results among studies are listed here with suggestions to overcome them.

a. Collaboration of research expertise

Often studies to determine health benefits of LAB are conducted by a supervisor who could have expertise in one area but not in another that a study might require for proper designing of experiments and analysis and interpretation of results. A well designed feeding study may include expertise from several disciplines, namely, food microbiologists, gastroenterologists, immunologists, physiologists, nutritionists, and others depending upon the nature of a study. In the absence of such a collaborative effort an experimental design could be incorrect and results could be overlooked and misinterpreted.

b. Live microbial cells are different from medicinal dose

Because the microbial cell preparations often come as tablets and capsules, they will not produce expected benefits like an aspirin tablet when consumed. These preparations can contain LAB that differ widely in their response against an intestinal ailment. They can also have low levels of physiologically normal viable cells and high levels of injured and dead cells depending upon the methods used in the preparation and preservation. It is the normal viable cells of indigenous beneficial species and strains that produce a benefit in the intestinal tract. Also, viable cells of non-indigenous LAB can only produce temporary benefits while the indigenous species and strains can produce long-term benefits.

c. Importance of bacterial species and strains

The LAB differ greatly in their ability to produce a specific benefit in the intestine of humans. The species and strains used in a study for a specific benefit need to be identified through well designed experiments. Use of commercial strains with very little authentic information about the characteristics of a strain will only give inconsistent results. If a *Lab. acidophilus* strain is needed in a study, it is important to select one from group A1. Also, for some studies such a strain should be of human origin. The ability of a strain of *Lab. acidophilus* and *Bifidobacterium* to adhere to intestinal epithelial cells of humans and the ability of a *Lab. reuteri* strain to produce reuterine have to be examined if such strains are included in a study. Also, when different species and strains are used in combinations, they should not have antagonistic action among themselves.

REFERENCES

Barefoot SF, Klaenhammer TR (1983) Detection and activity of lactacin B, a bacteriocin produced by *Lactobacillus acidophilus*. Appl Environ Microbiol 45:1808-1815.

Bhowmik T, Johnson MC, Ray B (1985) Isolation and partial characterization of surface protein of *Lactobacillus acidophilus*. International J Food Microbiol 2:311-321.

Bhowmik T, Johnson MC, Ray B (1987) Factors influencing synthesis and activity of ß-galactosidase in *Lactobacillus acidophilus*. J Industrial Microbiol 2:1-7.

Brennan M, Wanismail B, Ray B (1983) Prevalence of viable *Lactobacillus acidophilus* in dried commercial products. J. Food Prot 46:887-892.

Brennan M, Wanismail B, Johnson MC, Ray B (1986) Cellular damage in dried *Lactobacillus acidophilus*. J Food Prot 49:47-53.

Chauviere G, Coconnier M-H, Kerneis S, Fourniat J, Servin AL (1992) Adhesion of human *Lactobacillus acidophilus* strain LB to human enterocyte-like Caco-2 cells. J Gen Microbiol 138:1689-1696.

Coconnier M-H, Klaenhammer TR, Kerneis S, Bernet M-F, Servin AL (1992) Protein-mediated adhesion of *Lactobacillus acidophilus* BG2FO4 on human enterocyte and mucus-secreting cell lines in culture. Appl Environ Microbiol 58:2034-2039.

Drasar, BS, Hill, MJ (1974) Human intestinal flora. Academic Press, New York.

Drasar, BS, Barrow, PA (1985) Intestinal microbiology. Am Soc Microbiol, Washington, DC.

Fisher K, Johnson MC, Ray B (1985) Lactose hydrolyzing enzymes in *Lactobacillus acidophilus* strains. Food Microbiol 2:23-29.

Fuller R (ed) (1992) Probiotics. Chapman & Hill Inc, New York

Gilliland SE (1990) Health and nutritional benefits from lactic acid bacteria. FEMS Microbiol Rev 87:175-188.

Green JD, Klaenhammer TR (1994) Factors involved in adherence of lactobacilli to human Caco-2 cells. Appl Environ Microbiol 60:4487-4494.

Hentges DH (ed) (1983) Human intestinal microflora in health and diseases. Academic Press, London.

Hoover DG (1993) Bifidobacteria: activity and potential benefits. Food Technol 47(6):120-124.

Ishibashi N, Shimamura S (1993) Bifidobacteria: research and development in Japan. Food Technol 47(6):126-135.

Johnson JL, Phelps CF, Cummins CS, Lordan J, Gasser F (1980) Taxonomy of the *Lactobacillus acidophilus* group. International J Sys Bacteriol 30:53-68.

Johnson M, Ray B, Speck ML (1984) Freeze-injury in cell wall and its repair in *Lactobacillus acidophilus*. Cryo-Letters 5:171-176.

Johnson MC, Ray B, Bhowmik T (1987) Selection of *Lactobacillus acidophilus* strains for use in "acidophilus products". Antonie van Leewenhoek 53:215-217.

Makinen AM, Mannienen M, Gyllenberg H (1983) The adherence of lactic acid bacteria to columnar epithelial cells of pigs and calves. J Appl Bacteriol 55:241-245.

Marteau P, Rambaud J-C (1993) Potential of using lactic acid bacteria for therapy and immunomodulation in man. FEMS Microbiol Rev 12:207-220.
Metchinkoff, E (1907) The prolongation of life. Optimistic studies. William Heinemann, London.
Mitsuoka T (1989) Microbes in the intestine. Yakult Honsha Co. Ltd, Tokyo, Japan.
Montes RG, Bayless TM, Saavedra JM, Perman JA (1995) Effect of milk inoculated with *Lactobacillus acidophilus* or a yogurt starter culture in lactose-maldigesting children. J Dairy Sci 78:1657-1664.
Moore WEC, Moore LH (1995) Intestinal floras of populations that have a high risk of colon cancer. Appl Environ Microbiol 61:3202-3207.
Mukai T, Arihara K (1994) Presence of intestinal lectin-binding glycoproteins on the cell surface of *Lactobacillus acidophilus*. Biosci Biotech Biochem 58:1851-1854.
Muriana PM, Klaenhammer TR (1991) Purification and characterization of lactacin F, a bacteriocin produced by *Lactobacillus acidophilus* 11088. Appl Environ Microbiol 57:114-121.
Pasteur, L, Joubert, JF (1877) Charbon et septicemic. C.R. Soc. Biol. Paris 85:101-115.
Pot B, Hertel C, Ludwig W, Descheemaeker P, Kersters K, Schleifer K-H (1993) Identification and classification of *Lactobacillus acidophilus, L. gasseri* and *L. johnsonii* strains by SDS-PAGE and r-RNA-targeted oligonucleotide probe hybridization. J Gen Microbiol 139:513-517.
Ray B, Johnson MC (1986) Freeze-dried injury of surface layer proteins and its protection in *Lactobacillus acidophilus*. Cryo-Letters 7:210-217.
Ray B, Daeschel MA (1994) Bacteriocins of starter culture bacteria. In: Dillon VM, Board RG (ed) Natural Antimicrobial Systems and Food Preservation. CAB International, UK, p 133.
Reddy BS, Rivenson A (1993) Inhibitory effect of *Bifidobacterium longum* on colon, mammary and liver carcinogenesis induced by 2-amino-3 methylimidazole [4,5-f] quinoline, a food mutagen. Cancer Res 53:3914-3918.
Robinson RK (ed) (1991) Therapeutic properties of fermented milks. Elsevier Appl Sci Inc, New York.
Sanders ME (1993) Healthful attributes of bacteria in yogurt. Contemporary Nutrition 18(5). General Mills Nutrition Department, P.O. Box 1113, Minneapolis, MN 55440.
Sandine WE (1990) Roles of bifidobacteria and lactobacilli in human health. Contemporary Nutrition 15(1). General Mills Nutrition Department, General Mills Inc., Stacy, MN 55079, USA.
Savage DC, Fletcher M (ed) (1985) Bacterial adhesion. Plenum Press, New York.
Savage DC (1987) Factors influencing biocontrol of bacterial pathogens in the intestine. Food Technol 41(7):82-87.

Savaiano DA, Levitt MD (1984) Nutritional and therapeutic aspects of fermented dairy products. Contemporary Nutrition 9(6). General Mills Nutrition Department, P.O. Box 1113, Minneapolis, MN 55440.

Schneitz C, Nuotio L, Lounatma K (1993) Adhesion of *Lactobacillus acidophilus* to avian intestinal epithelial cells mediated by crystalline bacterial cell surface layer (S-layer). J Appl Bacteriol 74:290-294.

Speck ML (1978) Acidophilus food products. Development in Industrial Microbiology 19:95-101.

Speck ML, Dobrogosz WJ, Casas IA (1993) *Lactobacillus reuteri* in food supplement. Food Technol 47(6):90-94.

Wood BJB (ed) (1992) The lactic acid bacteria in health and diseases. Vol 1. Elsevier Appl Sci Inc, New York

Use of bacteriocin producing starters advantageously in dairy industry.

Jean A. Richard
INRA, Dairy Research Unit
F78352 Jouy-en-Josas Cedex
France

Firstly, the technological interest drawn from the determination of MICs (Minimum Inhibitory Concentrations) is discussed along with results demonstrating the efficacy of the bacteriocins in different food products. Then, the results of a reference study on the anti-listerial activity of low concentrations of nisin are presented, using a culture medium as model and a non-pathogenic test organism, *Listeria innocua*. Addition of nisin to the medium caused a rapid drop in cell viability followed by regrowth of the survivors at the same rate as the control culture. The total decrease in population was used to study the combined effect of pH and temperature on the anti-listerial activity of nisin. It was also observed that survivors to nisin had increased resistance not only to this bacteriocin but towards two other bacteriocins quite different in their molecular structure.
Finally, inhibition of *Listeria monocytogenes* in Camembert cheese made using a nisin-positive starter was compared to that in a control made with a nisin-negative starter. It is shown that unlike the two other bacteriocins under study, inhibition by nisin is limited to a narrow range of milk pH. Inhibition by nisin can be considerably improved by both adding small amounts of this compound in cheese milk when pH is most appropriate for bactericidal activity. Also, heating milk at sublethal temperatures to listeria considerably increase the bactericidal activity of nisin. Thus, use of these improvements may lead to complete eradication of listeria in soft surface-ripened cheeses if raw milk is produced under good hygienic conditions.

1. Introduction

Nisin is an antibacterial polypeptide of 34 amino-acid residues produced by some strains of *Lactococcus lactis* subsp. *lactis*. The molecule contains dehydro amino acids (dehydroalanine and dehydrobutyrine) and thioester linkages. It has recently been postulated (Liu & Hansen, 1990) that the dehydro residues could play an important role in the mechanism of nisin anti-bacterial action by reacting with one or more nucleophiles (e.g. sulphidryl residues) in the membrane of sensitive bacteria. The overall result is the formation of large pores in the cytoplasmic membrane which allow the efflux of ions and small intracellular molecules (potassium, amino acids, ATP...), finally resulting in cell death, and possibly lysis (Henning *et al.*, 1986; Sahl *et al.*, 1987; Gao *et al.*, 1991; Garcerà *et al.*, 1993). Nisin has long been used as a food preservative (Hurst, 1981; Delves-Broughton, 1990). Despite its activity against most Gram-positive food-borne pathogens or

NATO ASI Series, Vol. H 98
Lactic Acid Bacteria:
Current Advances in Metabolism, Genetics and Applications
Edited by T. Faruk Bozoğlu and Bibek Ray

food spoilage organisms, the only general application of nisin in the dairy industry remains the production of processed cheese and cheese spreads (Delves-Broughton, 1990). Nisin (200-500 IU/g *i.e.* 5-12.5 ppm of pure nisin A) is added to these products to avoid outgrowth of clostridia.

Many other lactic acid bacteria have been found to produce bacteriocins but so far, those produced by some strain of the genus *Pediococcus* have been studied in detail and provide good inhibitory properties to be used in foods. Pediocin AcH and pediocin PA1 are the same molecule which is produced by two different strains of *Pediococcus acidilactici* (Bhunia *et al.*, 1988, Jack *et al.*; 1995). It is a basic polypeptide (pHi 9.6) consisting of 44 amino acids with two disulfides bonds. Properties of pediocins in general have been reviewed by Ray & Daeschel (1992). In particular they display antilisterial activity which has been demonstrated by adding this bacteriocins in different foods, including dairy products. As pediocin AcH was found in our laboratory to have no activity on most lactococci, its inhibitory activity on listeria in soft surface-ripened cheeses warranted to be investigated.

Enterococcin EFS2 is a new bacteriocin of 67 amino acid residues produced by a strain of *Enterococcus faecalis* which has been isolated in our laboratory from the surface of a traditional French cheese (Maisnier-Patin & Richard, 1995).

These three bacteriocins display a good activity against *Listeria monocytogenes*, a potentially pathogenic bacteria present in raw milk which has the ability to withstand the acidic conditions prevailing in cheese during manufacture. Moreover, these bacteria can resume growth in cheeses exhibiting pH raise during ripening, as is the case for Camembert cheese (Ryser & Marth, 1987; Maisnier-Patin *et al.*, 1992). The same behaviour for *L. monocytogenes* is expected to occur in any soft and semi-hard surface-ripened cheeses.

The results of a study of making Camembert cheese free from listeria thanks to the use of nisin and a nisin-producing starter will be presented, along with some improvements of the inhibitory system. Before, it is necessary to present some traits of the anti-listerial activity of nisin, using a laboratory model-system which allows kinetics of viability loss and regrowth of this organism to be observed under controlled experimental conditions, and the properties of the surviving bacteria to be studied.

2. Anti-listerial activity of nisin

2.1. Relevant properties of nisin for its use in cheese technology

Nisin, a bacteriocin produced by some strains of *Lactococcus lactis* subsp. *lactis* has a general anti-microbial activity against Gram-positive bacteria, including non producing lactococci and other genera of lactic acid bacteria. It has been demonstrated that Gram-negative as well as nisin-resistant Gram-positive bacteria injured by sublethal heat treatment become sensitive to this bacteriocin (Kalchayanand *et al.*, 1992). This phenomenon is due to cell wall damage of these bacteria, which allow nisin to reach the cytoplasmic membrane, the ultimate target of this compound.

It has long been recognized that nisin activity and solubility are pH dependent, both increasing in acidic conditions. As an example, Liu and Hansen (1990) observed that nisin solubility dropped from 1.5 mg/ml (*i.e.* 60,000 IU/ml for nisin A) at pH 6.0 to 0.25 mg/ml (*i.e* 10,000 IU/ml) at pH 8.5. However, solubility should not be of concern in the dairy industry, considering the low levels of nisin used (usually less than 1,000 IU/ml) and the low pH of most products in which this substance is added, or produced. Interestingly, there are two natural variants of this bacteriocin, nisin A and nisin Z which differ by only one amino acid. It is claimed that this minor change gives nisin Z a better diffusion property in solid media than nisin A (Hugenholtz and de Veer, 1991) but this interesting feature has not yet been validated for dairy products.

2.2 Need for a model-system to understand nisin and other bacteriocin activity in food

Technological interest of MIC determinations

Although nisin has been reported by different authors to have a bactericidal effect against *Listeria monocytogenes,* large discrepancies are commonly found about the Minimal Inhibitory Concentrations (MICs). This could be due to use of different techniques *e.g.* determination of inhibition zone in agar diffusion tests or inhibition of growth in liquid cultures. For instance, Benkerroum and Sandine (1988) found MICs ranging from 740 IU/ml for the most vulnerable strain to ca. 1.2×10^5 IU/ml for the most resistant one (strain V7) at pH 7.3 in TSA (Tryptic Soy Agar, Difco). But in MRS at pH 6.8 these figures were reduced to 1.9 IU/ml and 3.4×10^3 IU/ml, respectively. Sodium chloride (at least 2.5%) or pH reduction from 6.5 to 5.5 were found to significantly increase the effectiveness of nisin. In the study by Hugenholtz and de Veer (1991) MICs of 200-2000 IU/l (corresponding to 8-80 IU/ml of nisin A) were observed, but the experimental conditions were not specified, particularly concerning the pH of the medium. On the other hand, Ferreira and Lund (1991) observed that the majority of the strains tested had MICs of 100-400 IU/ml at pH 5.5 and 20°C. Besides differences in suceptibility between strains these examples emphasize the major role of the media in determining the MICs for a particular bacteriocin.

Experiments in food- systems

Thus, only experiments carried out in real food systems are expected to provide reliable results on the inhibitory activity of a bacteriocin. For conveniences, the bacteriocins are used at concentration under the MICs, and the food is inoculated with high levels of the test organisms. Loss of viability are observed which are generally followed by regrowth of the survivors, as illustrated in Table 1.

Although experiments with solid food-systems provide reliable results, it is difficult to evaluate the effect of factors as pH, water activity, food composition, addition of particular ingredients... on cell destruction, since good mixture of additives is not easy. Similarly, it is nearly impossible to determine whether the survivors are resistant or not to the bacteriocin under study because separation of cells from the solid phase is difficult and potentially harmfull for injured bacteria. It is why we found necessary to develop a reference liquid system more easy to handle than solid food.

Table 1: Effect of pediocin PA.1 (100 AU per gramme) on *Listeria monocytogenes* (MIC : 55 AU/ml) in different dairy products stored at 4°C (adapted from Pucci *et al.*, 1988)

Food system	Incubation time (days)	Number of listeria (log_{10}CFU/g)	
		Control	With pediocin
Half-and-half cream	0	2.5	< 2.0
(18 % fat, pH 6.6)	1	4.2	<2.0
	7	6.7	3.8
	14	8.3	6.0
Cheddar Soup	0	2.8	<2.0
(1%NaCl, pH 6.0)	1	3.4	<2.0
	7	7.2	2.0
	14	8.3	4.5

Development of a reference model-system

In the developement of such a system, it is necessary to take into consideration two main principles : (i) when inoculated the test organism must be in a stationary phase of growth because cells are less sensitive to a bacteriocin than in the exponential phase of growth. Thus, the risk of over-estimating the inhibitory effect of the bacteriocin is minimized. In addition it is assumed that contaminants are generally in stationary phase of growth in their source, but this assumption remains to be validated,

(ii) the medium must allow the test organism to grow or at least to be energized, as energization is claimed to be a prerequisite for sensitivity to certain bacteriocins like nisin (Bruno and Montville, 1993). In addition, survivors should be able to resume growth, as it is the case in many foods.

Morever, it is essential that the kinetics of cell destruction be studied to determine the maximum bactericidal effect of the bacteriocin under study. Otherwise, this effect would be under-estimated either by incomplete cell destruction or regrowth of the survivors if sampling is too early or too late, respectively as is illustrated in Figure 1.

One of the earlier system we used consisted of the rich laboratory culture medium TSBYE (Tryptic Soy Broth, Difco supplemented with 0.6% Yeast Extract) adjusted to different pH values using concentrate HCl or NaOH, and in which chlorhydric solution of purified nisin (Aplin & Barrett Ltd, England) was added. A strain of the non-pathogenic species *Listeria innocua* (hereafter referred to as Lin11) of cheese origin was chosen as test organism. This strain was selected in a preliminary study as representative of the listeria for resistance to nisin and acidic (lactic acid) conditions encountered in Camembert cheese. Before use, the strain was cultivated overnight at

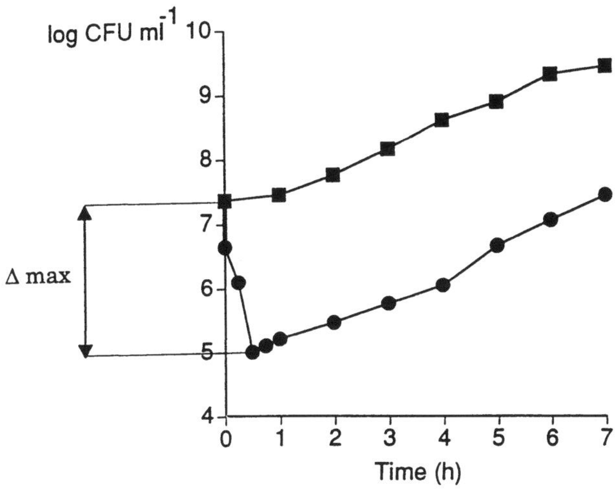

Figure 1 : Kinetics of destruction of *Listeria innocua* (strain 11) by 20 IU/ml of nisin and regrowth in TSBYE (Tryptic Soy Broth + 0.6% Yeast extract) at 30°C and pH 7.0
square and round symbols: control and experiment, respectively

30°C in TSBYE. Unless otherwise stated, fresh TSBYE adjusted to different pH values was inoculated with 1% of the culture giving ca. 10^7 CFU/ml. Appropriate dilution in 0.02N HCl of pure nisin was added to the medium and the culture was incubated at different temperatures. The surviving bacteria were enumerated on TSAYE (TSBYE to which 1.2 % agar was added) after dilution in TSBYE if necessary at intervals selected to detect the maximum population decrease. This is expressed as $\Delta max = \log_{10} N_o/N$, with N_o and N the numbers of colony forming units per ml at time of nisin addition and that of maximum cell destruction is observed, respectively. It is worth mentioning that Δmax is related to the « survivor ratio » R_t described by Mayr-Harting *et al.*, 1972), with $R_t=N_t/No$. Thus, $\Delta max = - \log R_t$. The time at which Δmax occured, as well as the growth rate of the survivors were also recorded.

Figure 1 gives an example of the results obtained with 20 IU/ml nisin at pH 7.0 and 30°C. Addition of nisin is followed by a rapid drop in population leading to a Δmax of ca. 2.3 log units in approximately half an hour. The surviving cells then resume growth at the same rate as the control. The same curve pattern was observed regardless of the nisin concentration (up to 200 IU/ml), the pH of the medium or the temperature of incubation.

Figure 2 illustrates the mean values of Δmax for 3 repetitions with 20 IU/ml at 4 different pHs and 3 temperatures of incubation. At 30°C, the mean Δmax value at pH 6.0 was the same as at pH 6.5 and a little lower but not significantly different from those observed at pH 5.0 and 5.5. These results are consistent with reports in the literature that the optimal pH for nisin acivity is in the range of 5.0-6.5, regardless of the test organism. At pH 4.5, the mean Δmax value is

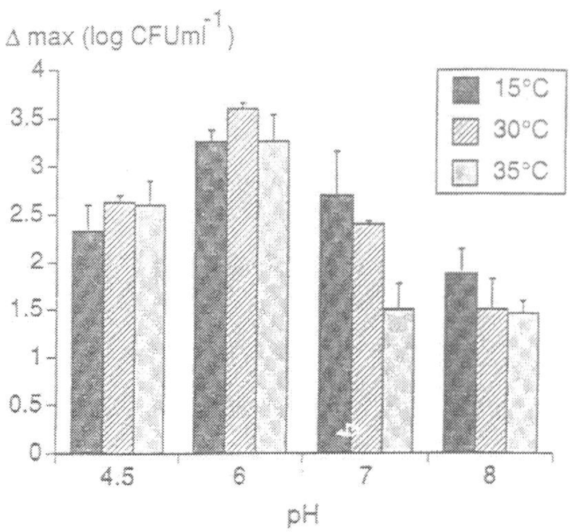

Figure 2: Combined effect of pH and temperature on the maximal reduction in population (Δmax) of *Listeria innocua* strain 11, by 20 IU/ml of nisin added in TSBYE (Tryptic Soy Broth + 0.6% Yeast extract)

approximately 1 log less than at pH 6, regardless the temperature. This means that nisin kills 10 times less listeria than at pH 6. As nisin is stable at low pH, this loss of activity can be attributed to change in physiology of the test bacterium: low external pH causes internal pH decrease which in turn results in lower energetic activity (Rhur & Sahl, 1985; Sahl *et al.*, 1987; Gao *et al.*, 1991; Garcera *et al.*, 1993). The loss of bactericidal activity of nisin at pH 7.0 and 8.0 is worse, with more effect of temperature. This may be due to less stability of nisin (Liu & Hansen, 1990). It is noteworthy that Δmax was not dependent on the temperature of incubation, except at pH 7.0. However, it took longer to reach Δmax at 15°C than at 35°C, with no significant difference between 30 and 35°C.

At 15°C, nisin activity at pH 7 is the same as at pH 6.0, whereas it sharply decreased at higher temperatures. This could be explained by a degradation process (decrease in solubility and/or aggregate formation) which is more temperature-dependent than loss of bactericidal activity of nisin.

Figure 3 shows the influence of the inoculum size on Δmax. If we omit the result obtained with the lowest population which needed the use of a different enumeration technique (membrane filtration), it is clear that Δmax does not depend on the size of the population. This should be not surprising if we bear in mind that Δmax is the log of the proportion of the surviving bacteria in the total population. This proportion (ratio) does not depend on the size of the population.

More interesting is the fact that the survivors were found resistant not only to nisin at the concentration under study, but at 4 times higher concentrations. In addition, the survivors to 20 IU/lm of nisin displayed higher resistance to pediocin AcH and enterococcin EFS2 than the original population (results not shown).

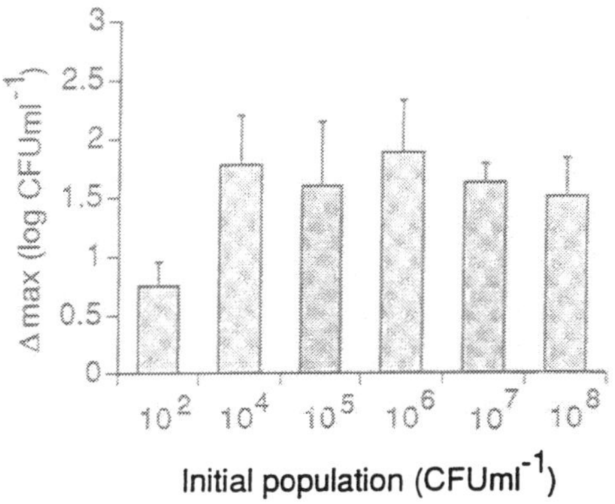

Figure 3: Effect of population size on destruction of *Listeria innocua* strain 11 by 20 IU/ml of nisin at 30°C in TSBYE and pH 7.0

3. Inhibition of *Listeria monocytogenes* by nisin during Camembert cheese making

3.1 Use a nisin-producing starter

Camembert cheeses were made in our pilot plant on three different occasions with pasteurized milk (72°C, 15 s) artificially contaminated with 10^5, 10^3, and 10^1 CFU/ml of *Listeria monocytogenes* strain V7 in the stationary phase (overnight culture at 30°C in skim milk). This strain was chosen as relatively resistant to acidic conditions prevailing in this kind of cheese (Ryser & Marth, 1987). A nisin-producing (Nis+) starter composed of protease-positive cells and protease-negative variant of a selected strain of *Lactococcus lactis* subsp. *lactis*, was used in comparison with a single strain Nis- starter. In each starter the protease-positive/protease-negative ratio was adjusted to obtain similar milk acidification rates. The surviving listeria were enumerated on Oxford medium. It has been shown during the first part of this study that this selective medium gives the same results as two non-selective ones (Maisnier-Patin *et al.*, 1992). Nisin concentration in milk and cheese was determined by the method of Fowler *et al.* (1975) which concerns total nisin (bound as well as free nisin), since extraction is performed at pH 2.

Figure 4 illustrates the change in lactococcal populations and nisin production during cheese making and ripening. As expected from Hurst (1966), nisin production started only at the beginning of the second half of the cell growth (ca. after 6 hours) and paralleled the lactococal growth. Maximum nisin production of ca. 700 IU/g was obtained in curd at 9 h. Then, nisin concentration decreased slowly in the core of the cheese throughout the 6-week period of ripening, whereas it dramatically dropped in the crust of cheese during the first week. The decrease in the crust may be attributed to the strong proteolytic activity exhibited by the mold, *Penicillium caseoculum.*

Figure 5 illustrates the fate of the test organism (*L. monocytogenes*, strain V7) in cheese. In presence of the Nis+ starter, the number of listeria decreased dramatically from 6 h to 9 h,

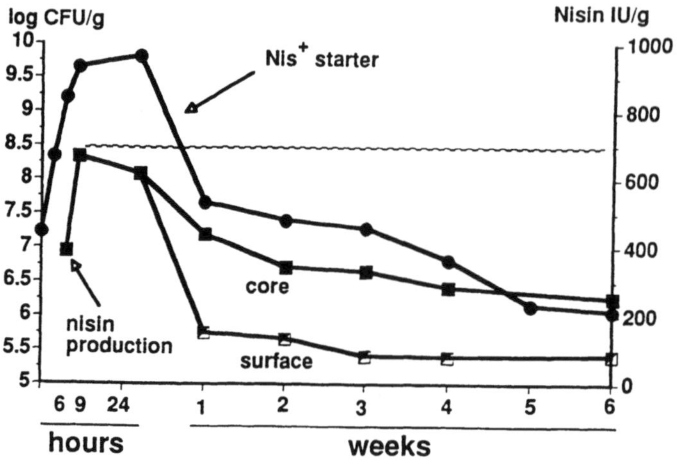

Figure 4: Change in *Lactococcus lactis* subsp. *lactis* CNRZ 150 (round symbols) and concentration of nisin (square symbols) during manufacture and ripening of Camembert. Solid and semi-open symbols indicate nisin determinations in core and crust of cheeses, respectively.

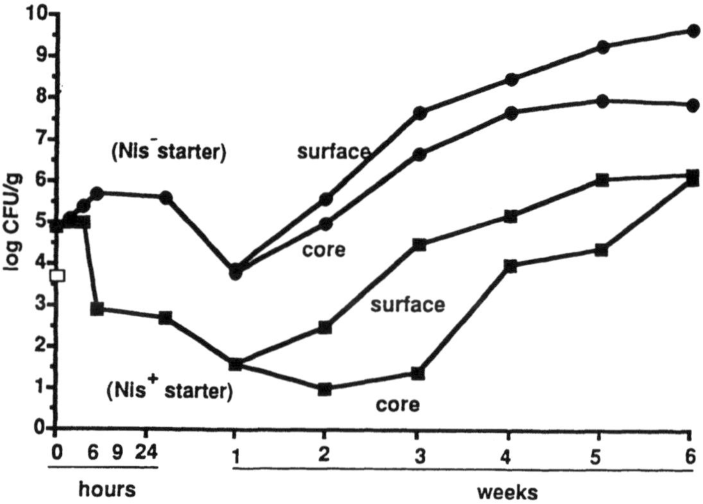

Figure 5: Behaviour of *Listeria monocytogenes* strain V7, during making and ripening of Camembert cheese made with a nisin-producing (square symbols) and a nisin-negative (round symbols) starter.

whereas it increased a little in the control cheese. Afterwards, the listeria exhibited a similar fate in both kinds of cheeses, with regrowth of the surviving cells sooner in the crust than in the core. Regrowth is obviously related to pH raise, but it occurred in the core of the cheese made with the Nis+ starter despite the presence of at least 200 IU/g of nisin. In cheeses made from milk inoculated with 10^3 listeria/ml and the Nis+ starter, counts of *L. monocytogenes* during the manufacture were close to the predicted numbers on the basis of the experiments with milk containing 10^5 listeria/ml but largely below the predicted numbers at the end of the ripening period: less than 10 CFU/g in both core and crust instead of the predicted 10^4 and 10^3, respectively. This suggests that nisin is more efficient with low than high numbers of listeria, as postulated by Mohamed *et al.* (1984) and Monticello and O'Connor (1990) but this is in contradiction to the results of the reference study presented above. With 10^1 listeria/ml, counts were under the limit of detection in 25g after enrichment.

Table 2 summarizes the results of 3 repetitions of experiments carried out with milk inoculated with 10^5 listeria/ml. They confirm that the inhibitory activity of nisin was restricted to the first 24 hours of cheese-life, with a mean difference of 2.4 log unit cfu/g between the Nis+ and the Nis- starters in cheese aged 24 hours. During the following week, the mean decrease of about 1.65 log unit cfu/g in the core and 1.75 in the crust can be attributed to acidic conditions prevailing in the two kinds of

Table 2: Fate of *Listeria monocytogenes* strain V7 in Camembert cheese made with a nisin-producing (Nis+) and a nisin-negative (Nis-) starter (adapted from Maisnier-Patin *et al.*, 1992)

Starter	Trial	Changes in counts (log_{10}CFU/g)		
		0-24 h	24hrs-1 week	
			core	crust
Nis^-	1	+0.7	-1.8	-1.7
	2	+0.9	-1.5	-1.8
	average A	+0.8	-1.65	-1.75
Nis^+	1	-2.2	-1.1	-1.1
	2	-1.1	-1.8	-1.9
	3	-1.4	-2.0	-2.3
	average B	-1.6	-1.63	-1.76
Mean difference B-A		-2.4	NS	NS

cheeses. Although in the reference liquid system pH and temperature provided fairly good conditions for bactericidal activity of nisin, in cheese aged 1-7 days, no further decrease due to residual nisin (between 400 and 600 IU/g in the interior) was observed. This is in agreement with previous observation of Ferreira and Lund (1991) who pointed out that nisin had a lower listericidal activity in cottage cheese with pH between 4.6 and 5.0 than in the culture medium, despite the fact that the nisin retained its activity as measured by assay against *Micrococcus luteus* (i.e. with chlorhydric acid extraction). This surprising loss of activity of nisin in cheese warranted further investigation.

3.2 Factors affecting nisin activity in cheese

The main factors known to counteract the inhibitory activity of nisin are listed in Table 3. Besides possible degradation by proteolytic enzymes, the main factor expected to cause loss of nisin activity in cheese was fat. This because several studies have shown that nisin activity is diminished in food containing fat (Jones, 1974; Jung *et al.*, 1992). As these studies deal mostly with fluid milk (pH 6.5-6.8), we decided to carry out some experiments with milk at pH 4.5, ensuring strict control of all parameters, particularly with respect to heat treatment of milk and physical state of fat.

Table 3: Factors assumed to be responsible for the loss of activity of nisin in Camembert cheese during the first week of ripening

1. Inappropriate conditions for pH and temperature for nisin activity

2. Adsorption of nisin on cheese components:
   - fat
   - proteins
   - lactic acid bacteria

3. Reduced nisin diffusion in the curd matrix

4. Resistance of target bacteria:
   - loss of sensitivity due to inappropriate physiological conditions (too low bioenergetics)
   - occurrence of naturally resistant cells in the population
   - acquired resistance in presence of nisin

5. Combination of the above factors

Effect of fat

Raw milk of good bacteriological quality (i.e. less than 50,000 cfu/ml) from an experimental farm located in our area was skimmed and cream reincorporated to have 0, 32 and 160 g/l, the latter concentration corresponding to the fat content of a 24-hr-old soft cheese. Milk was subjected to two different heat treatments : (i) rapid heating at 72°C in a boiling water bath followed by rapid cooling at O°C, to mimic HTST pasteurization and (ii) steam heating for 10 min, as done by Jung *et al.* (1992), followed by cooling at O°C. Milk pH was then adjusted to 4.5 by lactic acid and 10^5 cfu/ml of Lin 11 (stationary phase after overnight culture at 30°C in TSBYE) were added to milk. After 30 min, 800 IU/ml of pure nisin were added to the acidified milk and the mixture was vigourously mixed. The samples with and without nisin were stored at 15°C for 4 hours before enumerating the surviving bacteria. It has been shown during this experiment that Δmax occurred before 4 hours, with no further change in population in the following 4 hours. Thus, the drop in population in 4 hrs at pH 4.5 and 15°C can be considered as a good estimate of Δmax.

Table 3 shows the change in population observed in two repetitions of the experiment. It is clear that milk fat does not significantly affect nisin activity at pH 4.5. More remarkable still, is the fact that nisin loses most of its effectiveness when added to milk subjected to the HTST-like heat treatment. This could be related to protein changes that are expected to occur during heating: the less drastic the heat treatment, the more nisin bound to proteins.

Additional experiments performed with heat treament (ii) confirmed that fat has less effect on nisin activity than expected from the study by Jung *et al.* (1992). The discrepancy between the results of our study and that of these authors could be explained by the physical condition of fat (*i.e.* fat structure), if we agree that only the hydrophobic free butter fat or not soluble free fatty acids are able to bind nisin. It is generally accepted that in fresh raw milk of good bacteriological quality, fat is mostly in intact globules whereas in market milk (and more likely in half-and-half), more free fat and free fatty acid are present, due to the intense mechanical effect of pumping and fast changes in temperature on the fat globule membranes.

Effect of pH

The effect the pH has on the inhibitory activity of 50 IU/ml of nisin after 4 hours at 15°C in TSBYE and 3.2% fat milk from our experimental farm heated according to (i) is illustrated on Table 4. At pH 6.7 (the regular milk pH), a difference of 0.5 unit in Δmax is observed between milk and TSBYE. This means that 50 IU/ml of nisin in milk kills 3 times less bacteria than in TSBYE. At pH lower than 6, there is a sharp decrease in nisin activity in milk, with nearly complete loss at pH 4.5 whereas the corresponding loss in TSBYE is less dramatic. The reasons why nisin activity is lost more rapidly in milk than in TSBYE as the pH decreases is not yet understood. We can deduce from Table 4 that during cheese making, nisin has the same activity as 50 IU/ml in milk at pH 6.7 despite the production of at least 700 IU/g of curd.

Table 4: Influence of milk fat on the activity of nisin (800 IU/ml) on *Listeria innocua* strain Lin11 in milk at pH 4.5 and 15°C

Fat content	Viability loss (log_{10}CFU/ml)	
(%)	milk heated at 75°C for 15 s	milk heated at 100°C for 10 min
0	– 0.26 ± 0.05	– 1.10 ± 1.01
3.2	– 0.38 ± 0.78	– 1.47 ± 0.23
16	– 0.72 ± 1.33	– 1.46 ± 1.05

Miscellaneous factors

In the course of this study, we observed that nisin activity was only slightly reduced by adsorption to producing cells (10^9 CFU/ml) and that the diffusion of nisin in acid clotted milk was not an important factor, since inhibition was not enhanced by addition of up to 1% of an emulsifier (Tween 20). Contrary to what we observed with the reference liquid model, the surviving listeria in cheese were not resistant to nisin. This probably because the technique used to separate the surviving cells from curd (dispersion of cheese in 2% sodium citrate followed by centrifugation) increased their sensitivity to nisin.

3.3 Improvement of the anti-listerial activity of nisin in soft cheese technology

Combined effect of added nisin and use of a nisin-producing starter

The above results have demonstrated that the higher the milk pH, the more effective nisin is. This strongly suggested that adding nisin to milk before it is produced by a Nis+ starter should improve its inhibitory effect, provided the producing strain itself is not affected by added nisin.

We first observed that addition of nisin to milk up to 500 IU/ml of had no effect on the acidification rate of the Nis+ starter when it was in the early stationary phase of growth (corresponding pH of ca. 4.6). Use of starters at pH values higher than 5.3 (in exponental phase of growth) resulted in delay in acidification (*e.g.* the time to reach pH 5.0) of at least 20 min. These observations that a nisin-producing strain is resistant to nisin when it is in the early stationary phase are quite in agreement with those of Hurst and Kruse (1972).

The combined effect of added nisin (up to 500 UI/ml) and nisin produced by the starter was then investigated, using *L. innocua* Lin11 in stationary phase on TSBYE at 30°C at 10^5 cfu/ml in milk. The mean results of two repetitions are presented in Table 5. It is clear that the addition of a small amount of nisin to milk at the time of inoculation by a Nis+ starter can considerably improve the inhibitory effect of this bacteriocin. For instance, addition of 50 UI/ml led to a 4-fold decrease in population in 8 hours, as compared to the control without nisin.

Table 5: Influence of pH on the activity of nisin (50 IU/ml) on *Listeria innocua* strain Lin11 in TSBYE (Tryptic Soy Broth plus 0.6 %Yeast Extract) and in 3.2 % fat milk after 4 hours of incubation milk at 15°C

pH	Changes in counts (log_{10}CFU/ml)	
	in TSBYE (n=2)	in milk (n=4)
6.7	2.6	2.1
6.0	2.5	1.9
5.0	2.1	0.6
4.5	2.0	0.2

Combined effect of added nisin and sublethal heat treatment

Two strains of *L. monocytogenes* (V7 and Scott A), were used in this study. Heat inactivation was performed in capillary tubes as described by El-Shenawy *et al.* (1989)]. Nisin was added to the inoculated milk just before filling the tubes. Then, they were kept at room temperature for 5-6 min before immersion in a water bath set at 54, 56, 58, 60 and 65°C. After heating for specified times, tubes were rapidly removed from the water bath and promptly cooled in an ice-water mixture. The surviving bacteria were enumerated on TSAYE after 5 days of incubation at 30°C in aerobic conditions. A preliminary study showed that in our experimental conditions, the anaerobic incubation proposed by Knabel *et al.* (1990) had no significant advantage on the recovery of listeria.

The kinetics of destruction of strain V7 heated in milk at 60°C in presence of nisin is depicted in Figure 6. Similar curves were obtained at all temperatures for both strains V7 and Scott A. Survivor curves exhibited a shoulder preceding an accelerated death rate. The presence of nisin reduced or eliminated the initial lag in killing and increased the killing rate *of L. monocytogenes.*

Predicted times to decrease 10^3 and 10^{6-}fold the number of the listeria were derived from the curves. Figure 7 gives an example of the results for strain V7. At 54°C, addition of 25 IU/ml nisin to milk considerably reduced the time needed to achieve a 3 or 6 log decrease in the milk cell content. Doubling the nisin concentration of milk did not appreciably affect these two times. At higher temperatures, the difference between the control and milk with nisin tended to decrease. For instance, at 60°C, the times to achieve a 3 log reduction in milk containing 0, 25 and 50 IU/ml was 4.2, 3.4 and 1.8 min, respectively.

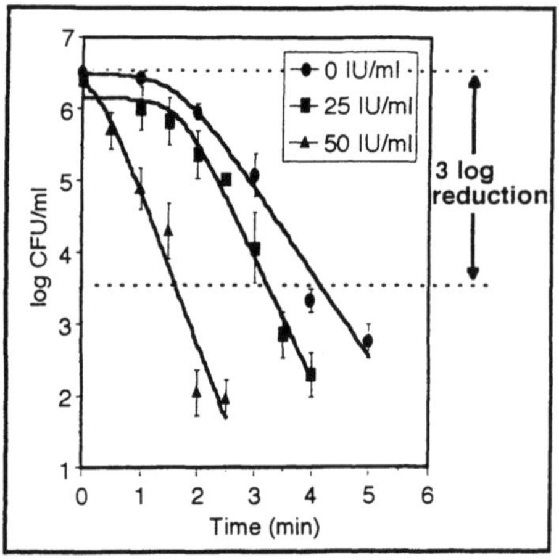

Figure 6: Combined effect of adding nisin to milk and heating at 60°C on the survival of *Listeria monocytogenes* (strain V7).
round symbols: control
square symbols: addition of 25 IU/ml
triangular symbols: addition of 50 IU/ml

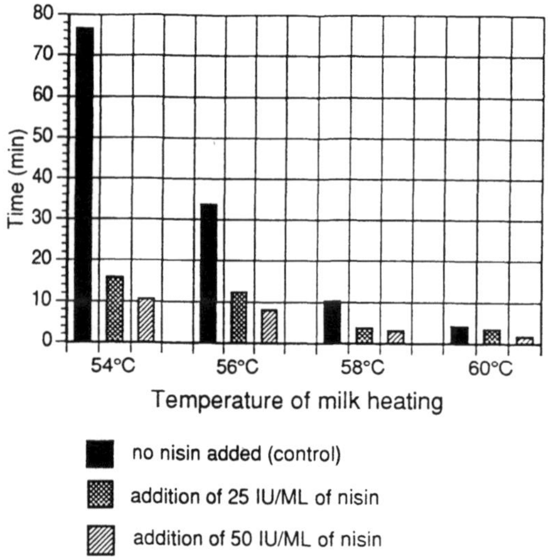

Figure 7: Predicted time to decrease 10^3-fold the population of *Listeria monocytogenes* (strain V7) by heating milk with or without added nisin.

Possible eradication of listeria from soft cheeses by combining addition of nisin to cheese milk, moderate heat treatment, and use of a nisin-producing starter

Our results enable us to compute the probabilities for the presence of one listeria in 100 g of cheese aged one week as a function of milk contamination and use of a nisin-based strategy to control this organism. According to the results of different surveys (Beckers *et* al., 1987; Lovet *et al.*, 1987; Fenlon & Wilson, 1989; Ryser & Marth, 1991; Rohrbach *et al.*, 1992; Sanaa *et al.*, 1993; Peeler & Bunning, 1994) listeria are detected in less than 10% samples of raw milk, with counts generally lower than one per ml in positive samples. From these results we can expect listerial counts between 1 and 100 listeria per litre of commingled milk. Table 6 shows the reduction in cell viability in cheeses aged one week, derived from our cheese experiments. To simplify the calculation, we assumed that one litre of milk yields 100 g of cheese aged 24 hours. During the first week in cheese made with the Nis- starter a decrease averaging 1.7 log unit due to acidic conditions was noted (see Table 2). Use of the Nis+starter caused a viability loss of 2.4 log units during cheese making followed by a decrease due to acidic conditions of 1.7 log unit during the first week. This gives a total loss of 4.1 log units in one week. Addition of 1 mg/l of nisin (about 50 IU/ml) in milk should cause an extra destruction of one log, and heating milk at chosen times-temperatures to increase this destruction by 3 log units is easy. As shown in Table 6, making cheese from milk with 100 listeria per litre and use of a regular (Nis-) starter, would result in presence of one listeria per 100 g of cheese aged one week. Use of a nisin-producing starter in combination with addition of minute amount of nisin to milk would diminish the probability by 1,000 without increasing the cost of production more than 1%. Heating milk at moderate temperature would add an extra decrease by 1,000 while preserving the technological propertie of cheese milk. If milk is produced under reinforced hygienic conditions so that it contains less than 1 listeria per litre, then the nisin strategy herein described would almost eradicate *L. monocytogenes* from this kind of cheeses without impairing significantly their taste as compared to that of similar cheeses made from raw milk.

Table 6 : Probability to find 1 listeria in 100 g of cheese aged one week, as a function of milk content and technology

Starter used	Nisin added (mg/l)	Viability loss ($\log_{10}$CFU/g)	Number of listeria per litre of milk: 1	10	100
Regular (nisin-negative)	0	1.7	2.10^{-2}	2.10^{-1}	1
Nisin-producer	0	4.1	10^{-4}	10^{-3}	10^{-2}
• without heating milk	1	5.1	10^{-5}	10^{-4}	10^{-3}
• with moderate heating	1	8.1	10^{-8}	10^{-7}	10^{-6}

4. Conclusions

The simplest way compatible with technology, and economically acceptable, for improving nisin effectiveness during cheese manufacture is adding minute amounts of nisin to milk, heating milk at moderate temperature, and using a nisin-producing starter. Thus, a decrease up to 8 log cycles in listerial population could be achieved. Unfortunately, nisin presents two main drawbacks. First, it is active on non nisin-producing lactococci and other lactic acid bacteria (*e.g. Leuconostoc* spp. and lactobacilli) which are present in the complex starters used by most cheese makers in France. It is possible, and generally accepted, to transfer the genes coding for immunity towards nisin in lactococci (Hugenholz & de Veers, 1991), but such a transfer is not permitted in bacteria not belonging to the same species. In addition, for a reason not yet elucidated, nisin activity is lost in milk at low pH, so that the curd is no longer protected from possible re-contamination by listeria during cheese making. It is worth noting that the same remark stands for cheese made from pasteurized milk. Therefore, there is a need for a bacteriocin which could be active in milk at low pH, and not active against the bacteria constituting the complex starters. It is the case for pediocin AcH. Ideally the bacteriocin should be produced during the cheese manufacture and retain an inhibitory activity throughout the life of surface-ripened soft cheeses. To our knowledge, such a bacteriocinogenic strain has not yet be discovered, if any it exists. Genetic modification of existing bacteriocinogenic strains and protein engineering could provide the solution to the problem.

On the other hand, routine use of a bacteriocin under bad hygienic conditions of manufacture could cause the build-up of resistant bacteria in the industrial premises. It has been shown by Harris *et al.* (1991) that nisin resistant cells are present in large populations of *L. monocytogenes.* Our own results (not yet published) suggest that it is the case for any bacterium and any bacteriocin. It is why we must emphasize that use of a bacteriocin must be considered only as a complement to good manufacturing practices and not at all as a substitute.

References

Beckers H. J., Soentoro P. S. S., Delfgou-van Asch E.H.M. (1987) The occurrence of *Listeria monocytogenes* in soft cheeses and raw milk and its resistance to heat. Int. J. Food Microbiol., 4, 249-256.

Benkerroum N. and Sandine W.E. (1988) Inhibitory action of nisin against *Listeria monocytogenes.* J. Dairy Sci., 71, 3237-3245.

Bruno M.C.E. and Montville T. J. (1993) Common mechanistic action of bacteriocins from lactic acid bacteria. Appl. Environ Microbiol., 59, 3003-3010.

Bhunia A.K., Johnson M.C. and Ray B. (1988) Purification, characterization and antimicrobial spectrum of a bacteriocin produced by *Pediococcus acidilactici.* J. Appl. Bacteriol., 65, 261-268.

Delves-Broughton, J. (1990) Nisin and its use as a food preservative. Food Technol., 44, 100-117.

El-Shanawy M. A., Yousef A.E. and Marth E. H. (1989) Thermal inactivation and injury of *Listeria monocytogenes* in reconstituted non fat dry milk. Michwissenschaft, 44, 741-745.

Fenlon D.R. and Wilson J. (1989) The incidence of *Listeria monocytogenes* in raw milk from farm bulk tanks in North-East Scotland. J. Appl. Bacteriol, 66, 191-196.

Ferreira, M.A.S.S. and Lund, B. M. (1991) The effect of nisin on *Listeria monocytogenes*. J. Appl. Bacteriol. 71, p. xxiii.

Fowler G.C., Jarvis B. and Tramer J. (1975) The assay of nisin in foods. *in* "Some Methods for Microbiological Assay". R.G. Board & D.W. Lovelock eds., Academic Press, London.

Gao F. H., Abee T. and Konings W. N. (1991) Mechanism of action of the peptide antibiotic nisin in liposomes and cytochrome c oxidase-containing proteoliposomes. Appl. Environ. Microbiol. 57, 2164-2170.

Garcerà M. J., Elferink M. G. L., Driessen A. J. M. and Konings W. N. (1993) In vitro pore-forming activity of the lantibiotic nisin. Role of the protomotive force and lipid composition. Eur. J. Biochem. 212, 417-422.

Harris L. J., Fleming H. P. and Klaenhammer T. D. (1991) Sensitivity and resistance of *Listeria monocytogenes* ATCC 19115, Scott A, and UAL5500 to nisin. J. Food Prot. 54, 836-840.

Henning S., Metz R. and Hammes W. P. (1986) Studies on the mode of action of nisin. Intern. J. Food Microbiol., 3, 121-134.

Hugenholtz J. and de Verr G.J.C.M. (1991) Application of nisin A and nisin Z in dairy technology. Nisin and Novel lantibiotics, G. Jung & H.-G. Sahl ed., ESCOM, Leiden, The Netherlands.

Hurst A. (1966) Biosynthesis of the antibiotic nisin and other basic peptides by *Streptococcus lactis* grown in batch culture. J. Gen. Microbiol.,45, 503-513.

Hurst A. (1981) Nisin. Adv. Appl. Microbiol., 27, 85-123.

Hurst A. and Kruse H. (1972), cited by Hurst (1981)

Jones L. W. (1974) Effect of butter fat on inhibition of *Staphylococcus aureus* by nisin. Can. J. Microbiol., 20, 1257-1260.

Jung D. S, Bodyfelt F. W. and Daeshel M. A. (1992) Influence of fat and emulsifiers on the efficacy of nisin in inhibiting *Listeria monocytogenes* in fluid milk. J. Dairy Sci. 75, 387-393.

Jack R.W., Tagg J.R. and Ray B. (1995) Bacteriocin of Gram-positive bacteria. Microbiol. Rev., 59, 171-200.

Kalchayanand N., Hanlin M.B. and Ray B. (1992) Sublethal injury makes Gram-negative and resistant Gram-positive bacteria sensitive to the bacteriocins, pediocin AcH and nisin. Letters Appl. Microbiol.,15, 239-243.

Knabel S.J., Walker H. W., Hartman P.A. and Mendoca A.F. (1990) Effect of growth temperature and strictly anaerobic recovery on the survival of *Listeria monocytogenes* during pasteurization. Appl. Environ. Microbiol., 56, 370-376.

Liu W. and Hansen J. N. (1990) Some chemical and physical properties of nisin, a small-protein antibiotic produced by *Lactococcus lactis*. Appl. Environ. Microbiol., 56, 2551-2558.

Lovett J., Francis D.W. and Hunt J.M. (1987) *Listeria monocytogenes* in raw milk: Detection, incidence and pathogenecity. J. Food Prot., 50, 188-192.

Maisnier-Patin S. and Richard J. (1995) Purification, partial characterization and mode of action of enterococcin EFS2, an antilisterial bacteriocin produced by a strain of *Enterococcus faecalis* isolated from a cheese. Int. J. Food Microbiol. (in press)

Maisnier-Patin S., Deschamps N., Tatini S.R., Richard J. (1992) Inhibition of *Listeria monocyto-genes* in Camembert cheese made with a nisin-producing starter. Lait, 72, 249-263.

Maisnier-Patin S., Tatini S.R. and Richard J. (1995) Combined effect of nisin and moderate heat on destruction of *Listeria monocytogenes* in milk. Lait, 75, 81-91.

Mayr-Harting A., Hedges A.J. and Berkeley R.C. W. (1972) Methods for studying bacteriocins. *in* "Methods in Microbiology", vol. 7A, J.R. Norris & D.W. Ribons eds., Academic Press, London.

Mohamed G.E.E, Seaman A. and Woodbine M. (1984) Food antibiotic nisin: comparative effects on *Erysipelothrix* and *Listeria*. *in* "Antimicrobials and Antibiosis in Agriculture" Woodbine ed. Butterworth, London.

Monticello D.J. and O'Connor D. (1990) Lysis of *Listeria monocytogenes* by nisin. *in* "Foodborne Listeriosis" A.J. Miller, J.L. Smith & G.A. Somkuti eds., Academic Press, London.

Molitor E. and Shal H.-G. (1991) Application of nisin a literature survey. Nisin and Novel lantibiotics, G. Jung & H.-G. Sahl ed., ESCOM, Leiden, The Netherlands.

Peeler J.T. and Bunning V.K. (1994) Hazard assessment of *Listeria monocytogenes* in the processing of bovine milk. J. Food Prot., 57, 689-697.

Pucci M.J., Vedamuthu E.R., Kunka B.S. and Vanderbergh P.A. (1988) Inhibition of *Listeria monocyto-genes* by using bacteriocin PA-1 produced by *Pediococcus acidilactici* PAC1.0. Appl. Environ. Microbiol., 54, 2349-2353.

Ray B. and Daeschel M. (1992) Food Biopreservatives of Microbial Origin. CRC Press, Inc., 2000 Corporate Blvd., N. W., Boca Raton, Florida 33431, USA.

Rohrbach B.W., Draughon F.A., Davidson P.M. and Oliver S.P. (1992) Prevalence of *Listeria monocytogenes, Campylobacter jejuni, Yersinia enterolitica,* and *Salmonella* in bulk tank milk: risk factors and risk of human exposure. J. Food Prot., 55, 93-97.

Rhur E. and Sahl H.-G. (1985) Mode of action of the peptide antibiotic nisin and influence of the membrane potential of whole cells and on cytoplasmic and artificial membrane vesicles. Antimicrob. Agents Chemother., 27, 841-847.

Ryser E.T. and Marth E.H. (1987) Fate of *Listeria monocytogenes* during the manufacture and ripening of Camembert cheese. J. Food Prot., 50, 372-378.

Ryser E.T. and Marth E. H. (1991) *Listeria*, listeriosis, and food safety. Marcel Dekker, Inc. New York, USA.

Sanaa M., Audurier A., Bind J. L., Millet A., Menard J. L., Poutrel B. and Serieys F. (1993) Origine et prévention de la contamination du lait de vache par *Listeria monocytogenes*. Flash sur les microorganismes pathogènes dans les aliments. Colloque de la Société Française de Microbiologie, Paris, 28-29 avril 1993.

Sahl H.-G., Kordel M. and Benz R. (1987) Voltage-dependent depolarization of bacterial membranes and artificial lipid bilayers by the peptide antibiotic nisin. Arch. Microbiol., 149, 120-124.

CHARACTERISTICS AND APPLICATIONS OF PEDIOCIN(S) OF *PEDIOCOCCUS ACIDILACTICI:* PEDIOCIN PA-1/AcH.

Bibek Ray
Animal Science Department
University of Wyoming
Box 3684
Laramie, WY 82071, USA

SUMMARY

Many food-grade lactic acid bacteria produce a group of peptides, classified as bacteriocins, that are bactericidal to many gram-producer bacteria. On the basis of the presence of a modified amino acid, lanthionine, they are grouped as lantibiotics and nonlanthionine bacteriocins. Due to the stability of the antibacterial action at high heat and in the environment of many foods, there is an interest in using bacteriocins of lactic acid bacteria as food biopreservatives.

However, many of the bacteriocins of lactic acid bacteria are bactericidal only to a few strains of gram-positive bacteria. Among those with antibacterial action against different spoilage and pathogenic gram-positive bacteria, the lantibiotic nisin, and nonlanthionine bacteriocins, pediocin PA-1 and pediocin AcH of *Pediococcus acidilactici* strains have been thoroughly studied. As pediocin PA-1 and pediocin AcH are identical peptides, a name pediocin PA-1/AcH has been used in this article.

Nisin has been studied for over 50 years and at present, approved for used in some foods. It has been discussed in other chapters. In this chapter the results of studies on various aspects of pediocin PA-1/AcH are discussed. These include: antibacterial spectrum, mode of antibacterial action, bactericidal effect against sublethally injured bacteria, factors influencing bactericidal effect, safety and antigenic property, immunity and resistance, characteristics of the peptide, genetic control in production, optimum production and purification, and effectiveness in food systems.

NATO ASI Series, Vol. H 98
Lactic Acid Bacteria:
Current Advances in Metabolism, Genetics and Applications
Edited by T. Faruk Bozoğlu and Bibek Ray

INTRODUCTION

Many food-grade lactic acid bacterial strains from the genera *Lactococcus, Streptococcus, Leuconostoc, Pediococcus, Lactobacillus* and *Carnobacterium* produce many small antibacterial proteins, grouped together with bacteriocins, that are bactericidal to other gram-positive bacteria, some of which are associated with food spoilage and foodborne diseases (Ray & Daeschel, 1992; Klaenhammer, 1993; Hoover & Steenson, 1993; DeVuyst & Vandamme, 1994; Jack *et al.*, 1995). These and other similar types of bactericidal proteins from different gram-positive bacteria have been broadly divided into two groups based on single chemical characteristics: presence or absence of a modified amino acid, lanthionine (Ala-S-Ala). Those with lanthionine are designated as lantibiotics while the others are designated as nonlatibiotics, or more appropriately nonlanthionine bacteriocins (Fig. 1). However, presence or absence of lanthionine(s) in such a molecule does not have any relationship with the antibacterial spectrum of bacteriocin (Jung & Sahl, 1991; Jack *et al.*, 1994; Sahl *et al.*, 1995). Nisin, a lantibiotic produced by several strains of *Lactococcus lactis* and discovered over 50 years ago, is the most thoroughly studied bacteriocin of lactic acid bacteria. Because of its bactericidal action against many gram-positive spoilage and pathogenic bacteria as well as for the availability of necessary information about its properties, chemical composition, genetic characteristics, nontoxicity, stability in food systems and other, nisin has been approved for use in several foods as a biopreservative in many countries (Jung & Sahl, 1991; Ray & Daeschel, 1992; Hoover & Steenson, 1993; DeVuyst & Vandamme, 1994). Information about nisin has been provided separately in this volume and thus will not be discussed in this chapter. The objectives of this presentation is to discuss the available information about another well studied bacteriocin of *Pediococcus acidilactici* strains, designated as pediocin PA-1 and pediocin AcH, respectively, by the research groups directly or indirectly associated at present with the Unilever Research Laboratory at the Netherlands and those working mainly at the Food Microbiology Laboratory at the University of Wyoming, USA.

<u>Nisin A(34 amino acids)</u>

Ile-Dhb-Ala-Ile-Dha-Leu-Ala-Abu-Pro-Gly-Ala-Lys-Abu-Gly-Ala-Leu-Met-

Gly-Ala-Asn-Met-Lys-Abu-Ala-Abu-Ala-His-Ala-Ser-Ile-His-Val-Dha-Lys

<u>Pediocin AcH or PA-1</u><u>(44 amino acids)</u>

Lys-Tyr-Tyr-Gly-Asn-Gly-Val-Thr-Cys-Gly-Lys-His-Ser-Cys-Ser-Val-Asp-Trp-Gly-Lys-Ala-Thr-Thr-

Cys-Ile-Ile-Asn-Asn-Gly-Ala-Met-Ala-Trp-Ala-Thr-Gly-Gly-His-Gln-Gly-Asn-His-Lys-Cys

<u>Leucocin A(37 amino acids)</u>

Lys-Tyr-Tyr-Gly-Asn-Gly-Val-His-Cys-Thr-Lys-Ser-Gly-Cys-Ser-Val-Asn-Trp-Gly-Glu-

Ala-Phe-Ser-Ala-Gly-Val-His-Arg-Leu-Ala-Asn-Gly-Gly-Asn-Gly-Phe-Trp

<u>Sakacin A(41 amino acids)</u>

Ala-Arg-Ser-Tyr-Gly-Asn-Gly-Val-Tyr-Cys-Asn-Asn-Lys-Lys-Cys-Trp-Val-Asn-Arg-Gly-

Glu-Ala-Thr-Gln-Ser-Ile-Ile-Gly-Gly-Met-Ile-Ser-Gly-Trp-Ala-Ser-Gly-Leu-Ala-Gly-Met

Fig. 1. Amino acid sequences of a lanthionine (Ala-S-Ala) bacteriocin, nisin A from *Lactococcus lactis* and three nonlanthionine bacteriocins, pediocin PA-1/AcH (from *Pediococcus acidilactici*), leucocin A (from *Leuconostoc gelidum*) and sakacin A (from *Lactobacillus sake*), from lactic acid bacteria. All three nonlanthionine bacteriocins probably have disulfide (-S-S-) bonds.

Use of two designations for the same bacteriocin has created some confusion. To avoid that in this article pediocin PA-1/AcH will be used. Also, a brief historical background will be provided to present the parallel sequence of events in the study of this bacteriocin by the two groups. One major reason of getting in such a state of confusion is due to the absence of any scientific system or legitimate basis for naming a bacteriocin. The literature is almost inundated with publications where "a novel bacteriocin" has been designated with a name just showing a bacterial culture or culture broth of lactic acid bacteria has antibacterial property against one or a few gram-positive bacterial strains and the activity is lost following a treatment of the test material(s) with one or more proteolytic enzymes to show that the antibacterial effect is most likely associated with proteinaceous component(s). This practice of premature naming, before knowing the amino acid sequence of bacteriocins, has resulted in assigning two or more names to the same bacteriocin. In a recent review several such examples have been listed and a rational approach to overcome this problem has been suggested (Jack *et al.*, 1995).

Prior to discussion on the available information on the characteristics of pediocin PA-1/AcH, it might be rational to briefly discuss the basis of sudden burst of research activities in bacteriocins of lactic acid bacteria (LAB) worldwide in the last few years. A detailed discussion can be found in several recent reviews and books (Jung & Sahl, 1991; Ray & Daeschel, 1992; Klaenhammer, 1993; Hoover & Steenson, 1993; DeVuyst & Vandamme, 1994; Jack *et al.*, 1995). Since the 1970's the philosophy of humans for the quest for life, especially in the developed countries, has changed from "How long shall I live" to "How well will I live". Relative abundance of raw food supply, technological innovations along with the changes in socio-economic pattern and life-style have changed the food consumption pattern towards a dependency on semi- or fully prepared foods that are not only convenient but also include a great variety. A great majority of these foods have many types of additives and preservatives, some of which are of nonfood origin. In addition, many are processed and preserved under conditions that can adversely affect and alter their normal acceptance and nutritional qualities. Moreover, some of the additives used before were later found to be harmful to the human body. This has raised questions on consumers' minds about the healthfulness of the "harshly processed" and "harshly preserved" foods. Food is something over which we as consumers emotionally think that we have complete control and we need it to be "absolutely" safe. However, satisfactory answers to the questions about some of the food additives are not available. These include: their effect on the human body from lifelong use, effect due to interactions of hundreds of additives from different foods inside the body, their effect on children, among the unborn and in the next generations. These concerns, fueled by the zealous lobbyists and media talk show hosts, and self-imposed experts have created a mistrust against the safety of many foods. The regulatory agencies, the scientists and the food processing companies have done very little, especially in the past, to inform and educate about the safety of the food additives or processed food to build consumer confidence. There are consumers who are now interested in foods that are "less processed", "less preserved" and close to "natural". However, at the same time they want foods that are convenient and tasty. This demand has created a new market and the food processors are scrambling to capture that niche by making such foods available. Many of these are ready-to-eat and heat-n-eat (in microwave) and may not contain effective preservatives. Most have over 21 day shelf-lives and some have 4 months or more. Vacuum-packaging (or gas flushing) and refrigeration at lower temperature (1 to 4°C) have been largely used to accomplish such a long shelf-life. Some of these products are designed as "new generation refrigerated foods", and "Sous vide foods" and include many categories containing large varieties of raw materials from animal and plant sources. Many of these foods, due to low heat processing harbor spoilage and pathogenic bacteria and bacterial spores. Also, many are handled following heating and prior to packaging during which various spoilage and pathogenic bacteria can get in food as post-heat contaminants. Some of these bacteria (and bacterial spores) can multiply (spores can germinate and outgrow) at refrigerated temperature under anaerobic conditions (psychrotrophic anaerobes or facultative anaerobes), which include many spoilage and some pathogenic bacteria. If these products are temperature abused (≥10°C)even for 2 to 4 h between the time of processing and storage and the time of consumption, they can grow very rapidly. Thus, during long shelf-life the population of these bacteria, even from a low initial number, can reach to a level to cause spoilage of the products or make them unsafe for consumption (Ray *et al.*, 1992, 1995).

To overcome this problem the regulatory agencies have advocated the selection of good quality raw materials, use of hazard analysis critical control point (HACCP) at all stages of handling, use of highest acceptable temperature for processing, refrigeration at lowest available temperature with the least temperature fluctuation during storage, transport and handling until consumed, and use of a combination of hurdles to kill, prevent or reduce growth of undesirable microorganisms from the following - low pH, low A_w, vacuum-packaging and acceptable preservatives (Anonymous, 1988). Due to current consumer apathy towards "harsh chemical preservatives", especially those that are of "non-food origin", the "acceptable preservatives" not only have to have the antimicrobial effectiveness of these "harsh preservatives" but also have to satisfy the consumers preference of being "natural", regulatory requirements of being safe, and food processors approval of being acceptable. Some of the characteristics this hypothetical preservative(s) needs to have are: (a) effective against different microorganisms important for spoilage and health hazard in food; (b) preferably kill pathogens and kill or prevent or reduce growth of spoilage microorganisms; (c) be able to prevent germination or kill outgrowing cells following germination of spores, especially of pathogens; (d) effective in small concentrations; (e) applicable in different food systems without altering their acceptance and nutritional qualities; (f) remain stable and effective during the shelf-life of a food; (g) produced easily and economically preferably from an acceptable source ("natural"?); (h) availability of necessary information about chemical and biological characteristics for regulatory approval; and (i) least objectionable to the regulatory agencies, food industries, and consumers.

Several metabolites of food-grade lactic acid bacteria, used in food fermentation, are known to have antimicrobial properties. They provide safety and shelf-stability of the fermented foods. It was assumed that since cells and metabolites of these bacteria have been consumed through different fermented foods for thousands of years without any health hazard, use of these antimicrobial metabolites of lactic acid bacteria will be least objectionable by the regulatory agencies. In addition, health conscious consumers view fermented foods and some lactic acid bacteria as natural and healthy. Some of the metabolites of lactic acid bacteria and other starter culture bacteria have already been permitted for use in foods as food additives. Examples are lactic acid, diacetyl, propionic acid and acetic acid. Also the bacteriocin, nisin, of *Lactococcus lactis*, has been approved as an antibacterial biopreservative for use in some foods, particularly some dairy foods in many countries. Information available by the 1980's revealed that nisin was effective normally against some gram-positive bacteria among the many spoilage and pathogenic bacteria in foods. By the early 1980's the existence of a few more bacteriocins produced by several strains of lactic acid bacteria was recognized. From the mid 1980's a great deal of interest developed worldwide to search for bacteriocins of lactic acid bacteria and other starter culture bacteria, preferably with wide antibacterial spectrum. Between 1988 and 1994 many research articles were published showing different species and strains from the genera *Lactococcus, Leuconostoc, Pediococcus, Streptococcus, Lactobacillus, Carnobacterium* and *Propionibacterium* produced bacteriocins. Most of them are small, cationic, heat resistant, bactericidal proteins and effective normally against some gram-positive bacteria (Table 1). Except for a few, most have limited antibacterial spectrum against gram-positive bacteria. Besides nisin, several bacteriocins from *Pediococcus* species and strains

Table 1. Common characteristics of bacteriocins of lactic acid bacteria[a]

Cationic peptides containing 34 to 57 amino acids
Bactericidal action against sensitive gram-positive bacteria
Cell death by destabilizing membrane functions
Bactericidal actions generally resistant to high heat, wide pH range, and food environment
Bactericidal action is lost by proteolytic enzymes
Bactericidal spectrum is narrow to wide
A gram-positive strain sensitive to a bacteriocin, has cells resistant to it
Cells of a gram-positive strain can be sensitive to one bacteriocin but resistant to another
Spores are resistant; outgrowing cells of germinated spores are sensitive to a bacteriocin to which the strain is sensitive
Gram-negative bacteria are resistant; sublethally injured cells can be sensitive
Translated as an inactive prepeptide, but undergoes processing which include removal of N-terminal leader peptide
Producer cells are genetically immune to own bacteriocin
A sensitive strain can become temporarily resistant while growing in the presence of a bacteriocin
Bacteriocin molecules are adsorbed on the cell surface of gram-positive bacteria including producer cells

[a]Source: Jack *et al.*, 1995.

were found to have relatively wide bactericidal property against many spoilage and pathogenic gram-positive bacteria (Ray & Daeschel, 1992; Klaenhammer, 1993). Among the vast numbers reported only a few bacteriocins were studied for their chemical identity, genetic basis, optimum production, toxicity, storage stability and effectiveness in food systems (Jack *et al.*, 1995). In that respect, besides nisin, pediocin PA-1/AcH of *Ped. acidilactici* have been studied, even though by separate groups, most thoroughly.

PEDIOCIN(S) OF PEDIOCOCCUS ACIDILACTICI

History

From the results published by the two groups between 1987 and 1991 there was reason to assume that pediocin PA-1 and pediocin AcH are two different bacteriocins from two strains of *Ped. acidilactici*. Only in 1992 two publications by the groups, first for pediocin PA-1 and then for pediocin AcH revealed that the two molecules have identical amino acid sequence. Both groups or individuals working with strains from the two groups have been using their specific name, indicating an unfortunate situation that has developed in naming a bacteriocin prior to identification of amino acid sequence as well as in the absence of scientific basis of naming a bacteriocin.

Historically, an abstract submitted by the University of Wyoming group in December, 1986 for the 1987 Annual Meeting of the Institute of Food Technologists, first revealed the identification of bacteriocins in several strains of *Ped. acidilactici*, isolated from different fermented sausages (Bhunia *et al.*, 1987a). A research note by the same group, submitted in May, 1987 and published in the same year in J. Industrial Microbiology revealed that the bacteriocin from *Ped. acidilactici* H had a mol. wt of about 2.7 KDa and this antibacterial protein could be identified on the SDS-PAGE gel by a simple bioassay method (Bhunia *et al.*, 1987b). As this group studied several strains of bacteriocin producing strains of *Ped. acidilactici* and there was no set rule to follow naming a bacteriocin, it was decided not to name this bacteriocin from strain H until more characteristics were known. In the same year a research article, submitted on May 4 to Applied and Environmental Microbiology and published in October by Microlife Techniques (now known as Quest International, a subsidiary of Unilever Research Laboratories), revealed that *Ped. acidilactici* strain PAC 1.0 (originally designated as NRRL-B-5627 by the National Regional Research Laboratories of USDA, Peoria, Illinois; also later designated as NRRL-B-18050) produced a bacteriocin, designated as pediocin PA-1 that had a mol. wt of 16.2 KDa and was encoded in a 6.2 MDa plasmid with the genetic determinant(s) for the host immunity linked to chromosome (Gonzalez & Kunka, 1987; Gonzalez, 1989). In 1988, from the characteristics and antibacterial spectrum of the purified bacteriocin from *Ped. acidilactici* H and from the previously published results, it was assumed that the bacteriocin from strain H as different from the pediocin PA-1. Following a suggestion of Tagg *et al*. (1976), this bacteriocin was designated as pediocin AcH using a combination of genus, species and strain (Bhunia *et al.,* 1988). Later studies showed that the pediocin AcH phenotype is linked to a 7.4 MDa plasmid and the host immunity to pediocin AcH is also linked to this plasmid (Ray *et al.,* 1989). Between 1987 and the middle of 1992 many articles on pediocin PA-1 and pediocin AcH were published assuming that the two bacteriocins were different (Ray & Daeschel, 1992).

Other publications, following the recognition that both pediocin PA-1 and pediocin AcH have identical amino acid sequence, subsequently revealed that the structural gene encoding the prepediocin molecule and the genes involved in host immunity, transmembrane-translocation and processing of prepediocin are also identical for both pediocin PA-1 and pediocin AcH. A total of four identical open reading frames (ORF) with a common promoter and a common terminator are involved in these activities (Marugg *et al.,* 1992; Motlagh *et al.,* 1994; Bukhtiyarova *et al.,* 1994). However, the two groups designated the four ORF and the proteins they encode differently: *ped* A, B, C, D and Ped A, B, C D for pediocin PA-1 and *pap* A, B, C, D and Pap A, B, C, D (for Pediocin AcH Production) for pediocin AcH.

Since the first publication in 1987, many other isolates/strains of *Ped. acidilactici* have been reported to produce pediocin, some of which have been designated differently (Table 2). However, none of these, except for pediocin PA-1 and AcH have information for both amino acid sequence and nucleotide sequence. From the available information about these pediocins, such as antibacterial spectrum, size of the plasmid encoding the trait, and some chemical and biological characteristics, there are reasons to assume that probably all of them are identical to pediocin PA-1 and Ach. It is worth noting that analysis of the amino acid sequences from three other *Ped. acidilactici* strains also revealed the same amino acid sequence as pediocin PA-1 and pediocin AcH. These include a commercial strain of *Ped. acidilactici* from Chr. Hansen Laboratories, Denmark (Lozano *et al.*, 1992), a strain UL5 (Daba *et al.*, 1994) and a strain RS_2, isolated from a fermented meat product (Bhunia *et al.*, 1993). At first partial amino acid sequence data suggested that the strain RS_2 has two different amino acids at positions 13 and 14 as compared to pediocin AcH (His-His in place of Ser-Cys). However, later studies on nucleotide sequence of the ORF 1 revealed that it was identical in both strains indicating pediocin AcH and pediocin from strain RS_2 are the same (Yang, 1994).

Recently, however, a report showed that *Ped. acidilactici* L50, isolated from a Spanish sausage produces a bacteriocin that has an amino acid sequence different from pediocin PA-1/AcH. It has been designated as pediocin L50. Although complete amino acid sequence of pediocin L50 is not yet known, it appears to have a higher molecular mass and the N-terminus is blocked. It also has a different bactericidal spectrum against gram-positive bacteria than pediocin PA-1 or AcH. The genetic determinant for pediocin L50 is not yet known. Only after the nucleotide sequence is determined, its amino acid sequence can be correctly understood. The producer strain of pediocin L50 grows and produces the bacteriocin at 8°C. This is quite different than the lower growth temperature for *Ped. acidilactici* that we have found (≥ 12°C; Ray, unpublished data). Species in genus *Pediococcus* have undergone several changes in the

Table 2. Pediocin producing *Pediococcus acidilactici* strains and their identity

Ped. acidilactici strain	Designation of pediocin	Source(s), comments and references
PAC 1.0	PA-1	Original designation NRRL-B 5627 (from NRRL collection, Peoria, IL). Later also designated as NRRL-B-18050 (Gonzales & Kunka, 1987; Gonzalez, 1989)
H, E, F	AcH	Fermented sausages (Bhunia *et al.*, 1987b; Ray *et al.*, 1989a)
LB42-923	AcH	Derivative of H (Ray *et al.*, 1989b)
M	AcH	Probably same as strain PC, obtained from a commercial source (Kim *et al.*, 1992)
JBL 1095	AcH	Originally strain H from Bhunia *et al.*, 1987b. (Degnan *et al.*, 1992; Luchansky *et al.*, 1992)
JBL 1096	PA-1	Original strain PAC 1.0 (Luchansky *et al.*, 1992)
JBL 1097	None	Original strain PO_2 (Luchansky, 1992)
PO_2, PC, HP	None	J. Bacus, FL (Hoover *et al.*, 1988,1989)
B 5627		From NRRL, Peoria, IL. Same as PAC 1.0 (Hoover *et al.*, 1988)
MCO_3	None	From fermented sausage. Designated as *Ped. pentosaceus* but most likely *Ped. acidilactici*[a] (Hoover *et al.*, 1988)
CCF-1 (also 6 other clinical isolates)	None	Clinical isolate. CCF-1 is found to be a non-producer[a]. (Hoover *et al.*, 1989)
FBB 63	None	R.N. Castilow, Michigan State University. Originally designated as *Ped. cerevisiae*
Chr Hansen commercial starter	None	Same as PAc 1.0 (Lozano *et al.*, 1992)
JD 1-23	JD	Commercial starter (Berry et *al.*, 1990; Christensen & Hutkins, 1992)
SJ-1	SJ-1	Fermented sausage (Schved et *al.*, 1993)
UL-5	None	Same as PAC 1.0 (Originally designated as *Leu. mesenteroides* UL-5 and mesenteroicin; Daba et *al.*, 1994)

Ped. acidilactici strain	Designation of pediocin	Source(s), comments and references
AB6, RS2	RS_2	Fermented sausage. Originally thought to have 1 to 2 amino acid different from AcH (Bhunia *et al.*, 1993). Later found to have same amino acid sequence as AcH[a] (Yang, 1994)
L50	L50	Has amino acid sequence different from PA-1 and AcH (Cintas *et al.*, 1995)

[a]Test in our laboratory (Ray, unpublished data). It will be important to examine if other clinical isolates are also pediocin non-producers.

last few years. It will be interesting to compare the 14S rRNA homology of this isolate with the standard strain of *Ped. acidophilus*. Similar suggestions can also be made for those *Ped. acidilactici* isolated from clinical cases and include pediocin producers, and non-producers (Hoover *et al.*, 1989).

The main objective of this article is to present the results of the large number of studies that have been conducted on pediocin PA-1 and pediocin AcH by many researchers. As indicated, for convenience sake a term pediocin PA-1/AcH will be used instead of using the two terms separately.

Pediocin PA-1/AcH: Most studied bacteriocin, besides nisin, of lactic acid bacteria (LAB)

The published and unpublished results for pediocin PA-1 and pediocin AcH generated by the two research groups and others who used the specific *Ped acidilactici* strains are quite extensive. Most of the information, except for animal feeding trials, necessary for regulatory approval are available (Table 3). Besides nisin, no other bacteriocin of LAB has been so thoroughly studied. In addition, there are reasons to believe that some of those designated differently are most likely identical to pediocin PA-1/AcH (Table 2). In the following section brief discussions in each area as listed in Table 3 for pediocin PA-1/AcH is presented. Effort is made to include most of the available publications for pediocin PA-1 and pediocin AcH in the citations. Several book chapters have recently been published on pediocins of *Pediococcus acidilactici* which also contain detailed reviews of pediocins with a large number of citations (Ray, 1992, 1994, 1995a; Ray & Hoover, 1993).

Antibacterial spectrum of pediocin PA-1/AcH

The antibacterial effectiveness of pediocin PA-1 or pediocin AcH against different gram-positive bacteria was studied initially either with neutralized supernatant of culture broths or with $(NH4)_2$ SO_4-precipitated crude pediocin from the culture supernatant. The cells of a test (indicator) bacterium were mixed with a melted and

Table 3. Areas studied for pediocin PA-1/AcH of *Pediococcus acidilactici*

a.	Antibacterial spectrum
b.	Mode of antibacterial action
c.	Bactericidal action against sublethally-injured bacteria
d.	Influence of several factors on bactericidal effect
e.	Safety and antigenic property
f.	Immunity and resistance
g.	Characteristic of the peptide
h.	Genetic control in production
i.	Optimizing production and purification
j.	Effectiveness in food systems as biopreservatives

tempered soft agar medium and overlayed on the surface of an agar medium plate. The test material (supernatant or dissolved crude pediocin) was then applied on the surface of the overlay directly ($\approx 5\ \mu l$) or on a disc placed on the surface ($\approx 25\ \mu l$) or in a well cut on the agar medium ($\geq 50\ \mu l$). The plates were then incubated generally up to 24 h at a temperature for the test organism to grow and examined for a zone of growth inhibition around the site where pediocin had been applied.

Results of these studies have shown that among the gram-positive spoilage and pathogenic bacteria many are sensitive to pediocin PA-1/AcH (Table 4). Many other gram-positive bacterial species from the genera listed in Table 4 also tested sensitive. They have been compiled and published separately (Ray, 1992, 1994). Initially, Bhunia *et al.* (1988) reported that some gram-negative bacteria as well as several *Staphylococcus aureus* strains showed sensitivity to crude preparations of pediocin AcH. Recent studies with highly purified pediocin AcH indicated that under normal conditions either the gram-negative bacteria or *Staphylococcus aureus* are not sensitive to pediocin PA-1/AcH. *Listeria monocytogenes* strains are very sensitive, although strains show differences in sensitivity. Non-proteolytic *Clostridium botulinum* E was found to be sensitive. Among the spoilage bacteria psychrotrophic *Clostridium laramie* associated with spoilage of meat and meat products was also very sensitive to pediocin AcH. *Lactobacillus bifermantas*, associated with spoilage of salad dressing was found to be sensitive to Pediocin PA-1. For specific information the published results can be consulted (Bhunia *et al.*, 1988; Bhunia, 1989; Ray *et al.*, 1989c; Kalchayanand, 1990; Holla, 1990; Motlagh *et al.*, 1991, 1992a; Motlagh, 1991).

Mode of antibacterial action of pediocin PA-1/AcH

Pediocin PA-1/Ach is bactericidal to sensitive cells of gram-positive bacteria. Cells are killed probably in less than a minute. The death occurs from the functional destabilization and structural integrity of the cytoplasmic membrane. As soon as pediocin PA-1/AcH molecules come in contact with the cell wall side

Table 4. Some Gram-positive pathogenic and spoilage bacteria sensitive to pediocin PA-1/AcH

Pathogenic bacteria[a]	Spoilage bacteria[a]
Listeria monocytogenes	*Lactobacillus sake*
Clostridium perfringens	*Lactobacillus curvatus*
Clostridium botulinum E	*Lactobacillus viridescens*
Bacillus cereus	*Lactobacillus bifermentas*
	Unidentified *Lactobacillus* spp[b]
	Leuconostoc carnosum
	Leuconostoc mesenteroides
	Unidentified *Leuconostoc* spp[b]
	Enterococcus faecalis
	Pediococcus damnosus
	Pediococcus cerevisiae
	Brochothrix thermosphacta
	Clostridium laramie

[a]One or more strains were found to be sensitive among the species listed. Some strains were resistant. Many other species/strains were also found to be sensitive but not included in the list as they are not regarded as pathogenic or spoilage bacteria (such as *Listeria innocua* or *Lactobacillus plantarum*).
[b]Many *Lactobacillus* and *Leuconostoc* spp. isolated from spoiled food products were sensitive to pediocin PA-1/AcH. However, the species of the isolates could not be determined by API-50 kit (Ray, unpublished data).

of the cytoplasmic membrane, the transmembrane electric potential is dissipated. This is produced in a voltage-independent manner and required specific protein (receptor) in the cytoplasmic membrane. Due to functional destabilization and structural integrity the membrane loses barrier functions allowing uncontrolled entrance and exist of small molecules to and from the cells (Bhunia *et al.*, 1991; Chikindas *et al.*, 1993; Bruno *et al.*, 1993). Pediocin from a separate *Pediococcus acidilactici* strain also was found to cause collapse of the proton motive force of the membrane affecting its function and causing cell death (Christensen & Hutkins, 1992).

Before the pediocin molecules come in contact with the cytoplasmic membrane of the sensitive cells of a gram-positive bacterium they have to pass through the cell wall from the environment. There are some disagreement on the mechanism(s) by which the pediocin molecules enter through the cell wall. One opinion is that the cell wall of gram-positive bacteria are permeable to molecules, such as pediocin (and other bacteriocins of gram-positive/bacteria); thus the molecules are transported freely through the wall and they come in contact with the cytoplasmic membrane (receptor) of the sensitive cells. A resistant cell does not have the specific receptor for pediocin in the membrane and thus is not affected and killed by the pediocin. The other opinion is that cationic pediocin molecules are adsorbed on the cell surface anionic molecules (such as lipoteichoic acid) in both resistant and sensitive cells of gram-positive bacteria in a nonspecific manner. However, in sensitive cells only, this adsorption causes specific conformational changes in the wall making it permeable to pediocin molecules (probably other bacteriocin molecules also), enabling the molecules to come in contact with the membrane (receptors) and producing bactericidal action (Fig. 2). In the resistant cells of gram-positive bacteria adsorption of pediocin molecules on the cell wall surface failed to bring about the specific conformational change. The resistance of gram-negative bacteria is due to the inability of pediocin molecules (and other bacteriocins of gram-positive bacteria) to

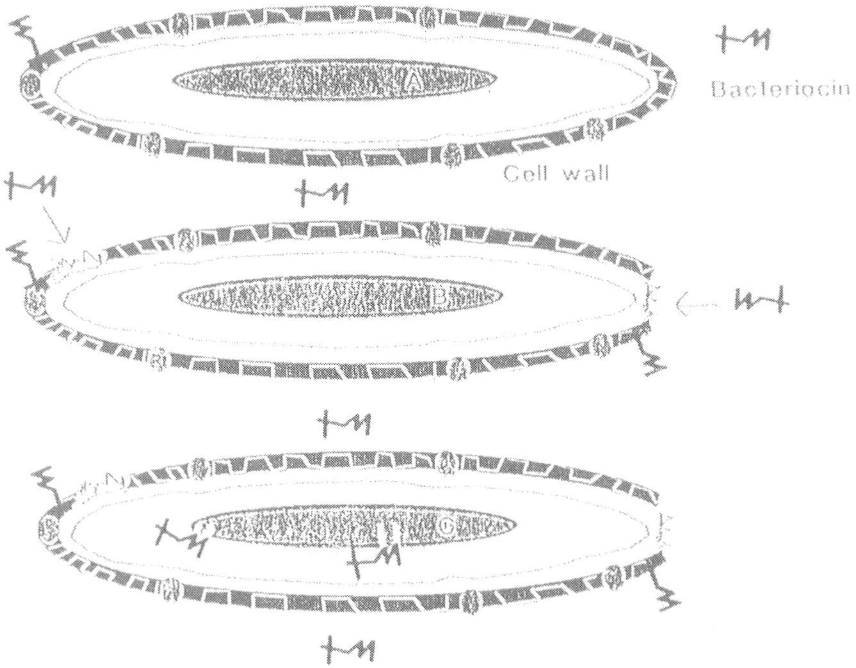

Fig. 2. Bactericidal mode of action of bacteriocins. The molecules initially adsorb on cell surface. In sensitive cells it results in alteration in cell wall properties which helps other molecules to pass through the cell wall, come in contact with the membrane, cause destabilization of its functions and cell death.

bring about specific conformational change following adsorption on the cell surface. In favor of this opinion these are the two observations: (a) the resistant cells of gram-positive and gram-negative bacteria become sensitive to pediocin (also nisin) following sublethal stresses. So their membranes are sensitive to the bacteriocin molecules, but could not come in contact with the molecules; and (b) even a strain very sensitive to pediocin PA-1/AcH has a few cells in the population that are resistant to pediocin PA-1/AcH. These cells retain resistance to pediocin as long as they are grown and maintained in the presence of pediocin. In the absence of pediocin, they revert back to the original population, i.e., most cells are sensitive and a few are resistant to pediocin PA-1/AcH. In these resistant cells due to some mutation, the cell wall is modified in such a way that following adsorption of pediocin molecules it does not undergo conformational change. However, as opposed to a resistant strain, the phenotype in these cells is not stable. This aspect will be discussed later (Bhunia *et al.,* 1991; Kalchayanand, 1992, 1994; Ray, 1992, 1993; Norles & Ray, 1994; Jack *et al.,* 1994).

Bacterial spores are not killed by pediocin PA-1/AcH. However, following germination the outgrowing cells of sensitive strains of *Bacillus* and *Clostridium* species are killed by pediocin PA-1/Ach (Ray *et al.,* 1989c; Ray, 1992, 1995b).

The viability loss of sensitive cells of gram-positive bacteria by pediocin PA-1/AcH results from the dissipation of proton motive force of the cytoplasmic membrane in a voltage-independent manner and loss of functions of the membrane. This is regarded to be the primary function of pediocin PA-1/AcH towards the sensitive cells. In addition, there are reports that some strains of sensitive cells of gram-positive bacteria also undergo lysis following treatment with this bacteriocin (Pucci *et al.,* 1988; Bhunia *et al.,* 1991; Motlagh *et al.,* 1991). This secondary action is produced through the induction of autolytic enzyme activity of cells by pediocin. Not all sensitive strains are lysed when exposed to pediocin and the strains differ in the level of lysis by the same amount of pediocin PA-1/AcH. Finally, maximum lysis of a sensitive strain occurs if pediocin is added during exponential growth phase as compared to stational growth phase (Jack *et al.,* 1995; Kalchayanand & Ray, 1995).

Bactericidal effect of pediocin PA-1/AcH against sublethally injured bacteria

Many species and strains of gram-positive bacteria and all gram-negative bacteria are normally resistant to bactericidal action of pediocin PA-1/AcH. Following exposure to many physical and chemical stresses bacterial cells develop sublethal-injury. The injury is manifested by alterations in different structural and functional components of cells that include both cell wall or outer membrane and cytoplasmic membrane or inner membrane. One of the manifestations of injury in the cell wall or outer membrane of bacterial cells is loss of their barrier functions allowing different types of molecules to get inside the cell (Fig. 3). The cells that are normally resistant to several antimicrobial chemicals become sensitive to them and lose viability (Ray, 1993).

Limited studies have shown that gram-positive and gram-negative bacterial cells, sublethally injured by freezing, low heat treatment, hydrostatic pressure, some organic acids, and EDTA, develop sensitivity to pediocin PA-1/AcH and lose viability (Ray, 1992; Kalchayanand *et al.,* 1992, 1994; Ray, 1995b). It is assumed that the cell wall or outermembrane of sublethally injured bacterial cells facilitated entrance of pediocin PA-1/AcH molecules from outside to enable them to come in contact with the membrane and destabilize it to produce cell death. The data presented in Table 5 showed that a greater viability loss of cells of gram-positive and gram-negative bacteria occurred when the cell suspensions containing pediocin PA-1/AcH are exposed to pediocin and hydrostatic pressure combination than with hydrostatic pressure treatment alone. Hydrostatic pressure, besides killing some cells, induces sublethal injury among the survivors. These injured cells are then killed by bacteriocins (Kalchayanand *et al.,* 1994). Treatment of bacterial spores to lower hydrostatic pressure range has been shown to induce germination and outgrowth. The outgrowing cells can then be killed by exposing them to pediocin (Ray, 1995b).

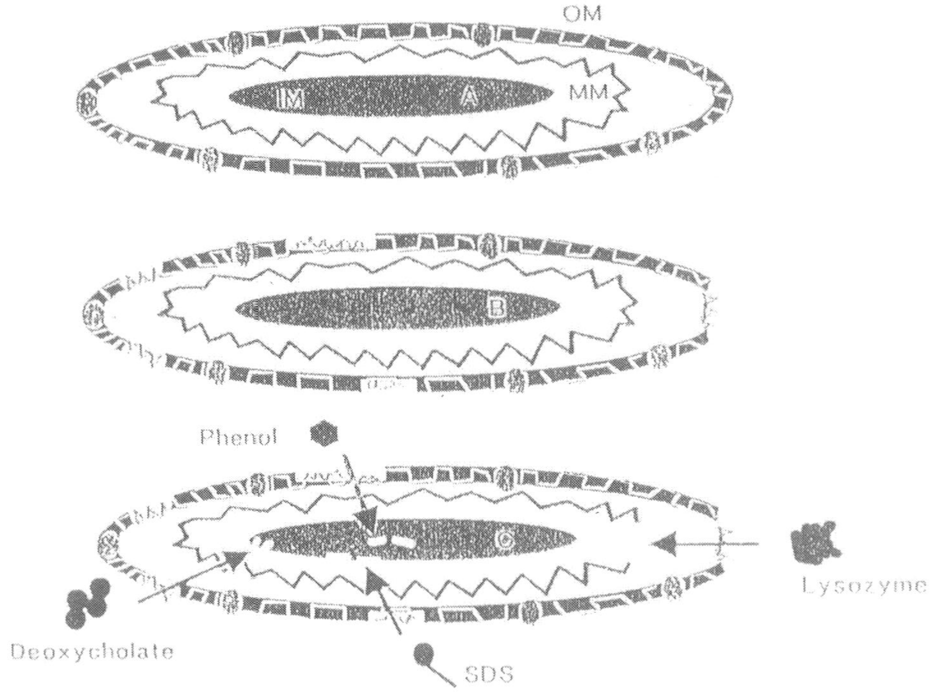

Fig. 3 Bacterial cells, initially resistant to different chemicals, become sensitive to them following stress due to sublethal injury of the barriers.

Table 5. Increased viability loss of gram-positive and gram-negative bacteria in combination of hydrostatic pressure and pediocin PA-1/AcH.

Bacterial species	Viability loss (log_{10}) by HP[b]	HP + pediocin[b]
Leuconostoc 03[a]	4.2	4.5
Lactobacillus 23[a]	8.7	10.1
Listeria monocytogenes Scott A	7.5	10.0
Listeria monocytogenes CA	3.8	5.0
Salmonella typhimurium M1	4.1	4.7
Escherichia coli 0157:H7 #932	4.2	5.5

[a]Isolates from spoiled processed meat products. Could not be identified to species level by recommended biochemical reaction profile methods.
[b]The cell suspension in 0.1% peptone were subjected to hydrostatic pressure (HP) of 40,000 to 50,000 psi for 1 min at 25°C without and with 5,000 AU/ml of pediocin PA-1/AcH (data adopted from Kalchayanand *et al.*, 1994).

Influence of several factors on bactericidal effect of pediocin PA-1/AcH

pH: The bactericidal property of pediocin PA-1/AcH is most effective at pH between 3 to 7 and partially lost at pH below 3 and above 7.0. At pH above 9.0 the activity is lost within 24 h at 25°C and in 15 min after heating at 93°C. The loss of activity at a higher pH could be due to degradation of the molecule (Gonzalez & Kunka, 1987; Bhunia *et al.*, 1988). The higher activity at pH below 6 to 7 is thought to be due to reduction in aggregation from higher positive charge in these cationic molecules making higher numbers of molecules available for bactericidal action (Jack *et al.*, 1995).

Temperature: The pediocin PA-1/AcH molecules are quite heat stable and retain fairly high bactericidal activity even after exposing to 121°C for 15 min. However, the activity level drops to about 60 to 70% level. The exact mechanism of this activity loss is not known. Heating at 70 to 80°C for up to 15 min at pH 5 to 7 did not cause any loss in activity (Gonzalez & Kunka, 1987; Kalchayanand, 1990; Bhunia *et al.*, 1991; Biswas *et al.*, 1991; Chen, 1995).

Enzymes: The bactericidal property of pediocin PA-1/AcH is completely lost following incubation with proteolytic enzymes, such as, α-chymotrypsin, trypsin, papain, ficin and proteases IV, XIV, XXIV. Treatment with RNase, DNase, lysozyme, lipase, phospholipase C, and catalase did not affect the activity (Gonzalez & Kunka, 1987; Bhunia et al., 1988; Ray et al., 1989a, Kim, 1990).

Organic and inorganic chemicals: Treatment of pediocin PA-1/AcH with the following compounds did not cause loss of its antibacterial activity: sodium chlorine, sodium phosphate, ammonium sulfate, sodium dodecyl sulfate, ethyl alcohol, isopropanol, acetone, chloroform and formaldehyde. However, both sodium chloride and sodium phosphate seemed to interfere with the bactericidal efficiency of pediocin in a concentration dependent manner. This is probably due to interactions of the Cl^- or PO_4^{3-}

with cationic pediocin molecules which in some unknown concentration dependent manner interferes with the bactericidal action of pediocin PA-1/AcH (Bhunia et al, 1991). In contrast, some organic acids, EDTA and lysozyme appeared to increase bactericidal action of pediocin PA-1/AcH both against gram-positive and gram-negative bacteria. This could be because organic acids, like acetic, propionic and lactic acids, and EDTA induce bacterial cell injury. Lysozyme also acts on the sublethally-injured cells increasing the total viability loss when used together with pediocin (Ray, 1992; Kalchayanand *et al.*, 1992; Ray, unpublished data).

Storage conditions: Pediocin PA-1/AcH either in heat treated broth cultures or in purified form and in liquid or powder forms retain full activity over 6 months at -20°C. At refrigerated (4°C) and room (25°C) temperatures the dried products also did not show loss of activity up to 6 months. The liquid preparations, however, showed about 50% activity loss in 1 month and about 60% activity loss in 6 months (Kalchayanand, 1990; Holla, 1990; Chen, 1995). The preparations retained higher activity when stored under vacuum than in air. The exact reason(s) on activity loss of pediocin in liquid during storage at 4° or 25°C is not known. However, aggregation of the molecules could apparently reduce activity level by the simple dilution method used to assay activity units or arbitrary units of pediocin AcH and pediocin PA-1 (Gonzalez & Kunka, 1987; Bhunia *et al.*, 1987, 1988; Biswas *et al.*, 1991; Ray, 1992; Yang *et al.*, 1992; Yang & Ray, 1994a).

Hydrostatic pressure: Purified pediocin PA-1/AcH in aqueous suspension was subjected to 50,000 psi at 25°C for 30 min. Following treatment and during subsequent storage at 25°C up to 5 months there was no loss in activity (Kalchayanand, unpublished data).

Combination with nisin: Limited studies have shown that the bactericidal efficiency and antibacterial spectrum of pediocin PA-1/AcH can be increased by combining it with nisin (Hanlin *et al.*, 1993). In all sensitive strains, there are some cells that are resistant to one but not to another bacteriocin. Thus, a combination of two or more bacteriocins can cause greater viability loss of sensitive bacteria.

Indicator bacteria: The bactericidal effectiveness of pediocin PA-1/AcH vary greatly to other gram-positive bacterial species and strains. With the same concentration of pediocin and under the same conditions of testing viability loss of a *Listeria monocytogenes* strains ranged between $\log_{10}$ 1.5 to 4.1. Such a variation has also been observed among sensitive species (Gonzalez & Kunka, 1987; Pucci *et al.*, 1988; Bhunia *et al.*, 1991, Motlagh *et al.*, 1991; Ray, 1992). Also, in a strain the exponentially growing cells are more sensitive to pediocin than the stationary phase cells of gram-positive bacteria. The proportion of resistant cells in a sensitive population seemed to increase when a culture is allowed to grow or kept longer without subculturing (Ray, 1992). The bactericidal efficiency of pediocin PA-1/AcH is directly proportional to its concentrations against a given population of sensitive cells of a gram-positive bacterial strain (Motlagh *et al.*, 1992a; Ray, 1992).

Food types: Pediocin PA-1/AcH molecules are proteinaceous, cationic, hydrophobic molecules. Due to proteinaceous nature the molecules are hydrolyzed by proteolytic enzymes. Thus food, especially raw and unheated food products with active proteolytic enzyme(s) can reduce or destroy the bactericidal effectiveness of pediocin molecules. Also as the pediocin molecules are positively charged, food molecules with negative charges can bind them and make pediocin molecules unavailable for bactericidal purposes. Finally, as the pediocin molecules are hydrophobic, they can be absorbed by the lipid components of a food. The antibacterial effectiveness can be reduced for this reason in high lipid containing foods (Ray, 1992).

Safety and antigenic property of pediocin PA-1/AcH

Pediocin producing *Ped. acidilactici* H was initially isolated from a fermented sausage while *Ped. acidilactici* PAC-1.0 is used as a starter culture for fermented food products. *Ped. acidilactici* H was used in a study to produce fermented mutton sausage and the products were used in taste panel studies for over a period of one year involving 12 to 14 people each time. No health problems were reported from these individuals (Wu *et al.*, 1991). Pediocin AcH was also administered to mice and rabbits by subcutaneous, intravenous and interperitoneal injections. No local or general adverse symptoms were observed in these animals (Bhunia *et al.*, 1991). Murine hybridoma cells in cell cultures were also treated with pediocin AcH. There was no loss of viability of cells during 48 h incubation (Bhunia, unpublished data). All these data suggest that pediocin PA-1/AcH is safe. In addition, the molecules are hydrolyzed and inactivated by the proteolytic enzymes trypsin and α-chymotrypsin, of the gastrointestinal (GI) tract. Thus, if consumed it is not going to upset the normal GI tract microflora.

Initial studies involving injecting partially purified pediocin AcH to mice and rabbits by several routes for 4 weeks showed that it did not stimulate production of antibody in the animals. Similar results were obtained by injecting pediocin conjugated with bovine serum albumin (Bhunia *et al.*, 1991). Later, Bhunia (1994) reported the development of monoclonal antibody by injecting mice for 12 weeks with pediocin conjugated with polyacrylamide gel. Studies showed the antibody specifically reacted with pediocin PA-1/AcH but not with any other bacteriocins of LAB with different amino acid sequences. Bhunia and Johnson (1992a) also reported that injection of the broken heat killed cells of *Ped. acidilactici* H in mice resulted in the production of a monoclonal antibody (Ped-2B2) that specifically reacted with a 116 KDa protein present only on the cell surface of pediocin PA-1/AcH producing strains of *Ped. acidilactici*. Both the monoclonal antibodies developed by them are useful in differentiating pediocin producing and non-producing strains of *Ped. acidilactici*.

Immunity and resistance against pediocin PA-1/AcH

Pediocin PA-1 or AcH producing strains are immune to their own bacteriocin. Like other bacteriocins of gram-positive bacteria, a pediocin producing strain carries a specific gene that encodes for an immunity protein. Initial studies of Gonzalez and Kunka (1977) suggested that the pediocin immunity gene is encoded in the chromosome, since curing of the plasmid that encodes for pediocin production did not make the pediocin nonproducer variants sensitive to pediocin. Ray *et al.* (1989a,b), however, showed that curing pediocin producing strains that encode for pediocin production made the pediocin non-producing variants sensitive to pediocin. They suggested that gene for pediocin immunity in *Ped. acidilactici* is present in the same plasmid that encodes for pediocin production. Subsequent genetic studies showed that the structural gene for pediocin PA-1/AcH and immunity gene against pediocin are located in the same plasmid. This aspect is discussed in a later section.

A recent study by Noerlis and Ray (1994) showed that while the immunity of the producer strains against pediocin PA-1/AcH is encoded by a plasmid linked gene the resistance of non-producing strains to pediocin can be due to some modification, probably in the cell wall. If a sensitive strain of gram-positive bacteria is grown in the presence of pediocin PA-1/AcH, a few resistant cells present in the population survive and grow and the population becomes resistant to pediocin as long as they are cultured in its presence. When these resistant populations are subcultured in the absence of pediocin most cells in a population become sensitive. Thus, resistance of these cells while growing in the presence of pediocin is the result of some changes in the physiological makeup of the cells rather than a change in the genetic materials. There are other gram-positive bacterial species and strains that are resistant to pediocin PA-1/AcH even when they are grown for a long time in the absence of pediocin. These strains may have genetic determinants that confer resistance to them and it is different from the immunity gene and immunity protein.

Characteristics of the peptide (pediocin PA-1/AcH)

Bhunia *et al.* (1987a,b) initially showed on SDS-PAGE gel that the pediocin produced by *Ped. acidilactici* H is a peptide of about 2700 Da. They developed a procedure by which the pediocin, obtained by $NH_4(SO_4)_2$-precipitation from culture broth supernatant, could be analyzed on SDS-PAGE gel and the band with antibacterial activity could be identified by a bioassay. Later, Bhunia and Johnson (1992a) modified this procedure to stain as well as detect antibacterial activity of pediocin band on the same gel. Gonzalez and Kunka (1987) using gel filtration chromatography with Sephacryl S-200, determined the molecular weight of pediocin of *Ped. acidilactici* PAC 1.0 to be 16,500 Da. This controversy was resolved in 1992 following the publication of amino acid sequence of pediocin from both strains of *Ped. acidilactici* by the two groups (Henderson *et al.,* 1992; Motlagh *et al.,* 1992). These results showed that the pediocins from the two strains have identical amino acid sequence. The molecule has 44 amino acids and a molecular mass of 4628 Da (Fig. 1). The active molecule has four cysteine units at positions 9, 14, 20 and 44 which form two disulfide bonds, one between cysteines at 9 and 14 and the other between cysteines at 20 and 44. Later studies showed that when pediocin was treated with reducing agent like dithiothreitol to break the disulfide bonds, its bactericidal property was lost. This suggests that the disulfide bonds are important for the antibacterial action of pediocin PA-1/AcH (Chikindas *et al.,* 1993; Jack *et al.,* 1995). Treatment with ß-mercaptoethanol also reduces bacteridical activity of pediocin. However, following removal of the ß-mercaptoethanol, the activity is restored probably due to reformation of disulfide bonds. Other studies showed that of the two disulfide bonds the bond between cysteines at 20 and 44, but not between cysteines at 9 and 14, is more important for bactericidal action of pediocin PA-1/AcH (Ray, unpublished data).

Among the many strains of *Ped. acidilactici* reported to produce pediocin(s), two others have been analyzed for amino acid sequence and found to have the same sequence as pediocin PA-1/AcH (Lozano *et al.,* 1992; Daba *et al.,* 1994). In a separate study, Bhunia *et al.* (1993) reported that analysis of amino acid sequence of pediocin from *Ped. acidilactici* RS_2 revealed that it has -His-His- at positions 13 and 14 as compared to -Ser-Cys- in pediocin PA-1/AcH. However, nucleotide sequence determination of the structural gene from *Ped. acidilactici* RS2 showed that pediocin from this strain is identical to that of strains producing pediocin PA-1/AcH (Yang, 1994).

Analysis of the nucleotide sequence of the structural genes associated with pediocin production in strain PAC 1.0 and H revealed that at the translation level the prepediocin molecule has 62 amino acids consisting of an N-terminal 28 amino acid leader sequence and a C-terminal 44 amino acid propediocin (Marugg *et al.*, 1992; Motlagh *et al.*, 1992). The prepediocin molecule following translation in the reduced environment of the cytoplasm probably do not have two disulfide bonds and may not have antibacterial property. A recent study has indicated that prepediocin might have some bactericidal activity (Venema *et al.*, 1995). However, this study was not conducted under strict reducing environment. Following translation the 18 amino acid leader sequence is removed from the prepediocin releasing the 44 amino acid pediocin from the cell interior in biologically active form which finally is excreted in the environment. The processing of prepediocin to pediocin may be influenced at a pH ≤5.0. It was proposed that the specific proteolytic enzyme involved in the removal of leader peptide from the prepeptide might be an acid protease and has greater catalytic action at lower pH (Ray *et al.*, 1992). Depending upon the pH of the environment, the pediocin molecules may remain bound to the cell surface of the producer strain (at higher pH) or remain free in the environment. A model of the translation, processing, transmembrane translocation and excretion in the environment of pediocin PA-1/AcH has recently been proposed (Jack *et al.*, 1995). These aspects are discussed more in the next section dealing with genetics of pediocin PA-1/AcH.

Genetic control in pediocin PA-1/AcH production

Gonzalez and Kunka (1987) first reported that pediocin PA-1 production phenotype in *Ped. acidilactici* PAC 1.0 is linked to a 6.2 MDa or 9.4 Kb plasmid, designated as pSRQ11. They also suggested that the immunity of producer strain to pediocin PA-1 is linked to chromosome. Studies by Ray *et al.* (1989a) showed that both pediocin AcH production and immunity against it in *Ped. acidilactici* H are linked to a 7.4 MDa plasmid, pSMB74. Linkage of pediocin production to apparently different sizes of plasmids in several other *Ped. acidilactici* strains were also reported that are listed in Table 2. Many of these plasmids could be essentially identical; differences in size could be due to difference in standards used. Restriction enzyme cleavage patterns of the plasmids from several pediocin producing *Ped. acidilactici* strains were found to be the same (Ray *et al.*, 1992). Also, a nucleotide probe made from structural gene encoding for pediocin production in *Ped. acidilactici* H was found to hybridize by southern hybridization test with the plasmids isolated from several pediocin producing *Ped. acidilactici* strains (Bhunia *et al.*, 1994).

One study reported that pSMB74, the plasmid encoding for pediocin production and producer strain immunity phenotypes in *Ped. acidilactici* H, was conjugally transferred to a plasmidless and pediocin sensitive strain of *Ped. acidilactici* (Ray *et al.*, 1989b). Subsequent studies to transfer the plasmid encoding for pediocin production, however, were not successful both by conjugation and electroporation. One reason for this failure, especially electroporation, could be due to inability of detecting the recipient cell from the background (Kim *et al.*, 1992).

Restriction enzyme mapping of pSMB74 and pSRQ11 showed that both plasmids might be very similar (Vandebergh *et al.*, 1990; Ray *et al.*, 1992). Both plasmids have unique sites for: *Sal* I, *Bgl* I, *Xba* I, *Ned* I, *EcoR* I, *Sac* I, and *Sst* I. The plasmid pSMB74 also has three *Hind* III sites that cut it into three fragments, one 3.5 Kb and two about 2.7 Kb, making the size of the plasmid 8.9 Kb (Ray *et al.*, 1992). In contrast, initial results showed pSRQ11 to be 9.4 Kb with two *Hind* III sites (Vandenbergh *et al.*, 1990). A later study although showed that pQRS11 has three *Hind* III sites (Marugg *et al.*, 1992). Also, complete nucleotide sequencing of pSMB74 revealed that it has a total of 8,877 bp (Motlagh, 1994).

In an effort to detect the location of structural gene and other gene(s) associated with pediocin production, Marugg *et al.* (1992) cloned a 5.6 Kb *Sal* I-*EcoR* I fragment from plasmid pSRQ11 in a *Escherichia coli* cloning vector and conducted nucleotide sequencing of the fragment. Their analysis showed that a segment in the *Sal* I-*EcoR* I fragment there are four open reading frames (ORF; designated as *ped* A, B, C and D) with a common promotor upstream of *ped* A. They encode for four proteins, Ped A, B, C and D, containing 62, 112, 174 and 724 amino acids, respectively. The ORF *ped* A is the structural gene encoding for 62 amino acid prepediocin containing an N-terminal 18 amino acid leader sequence and a C-terminal 44 amino acid propeptide as discussed earlier. They also showed by mutation analysis that Ped D protein is required for pediocin production. Due to similarities of Ped D protein to ATP-dependent transport proteins, they predicted that it is involved in the membrane translocation of pediocin (Marugg *et al.,* 1992). About the same time, Motlagh *et al.* (1992b) cloned the three *Hind* III fragments of pSMB74 in *E. coli* cloning vector and started sequencing them. In one fragment they identified the structural gene (designated as *pap* A), that encodes for 62 amino acid prepediocin and is identical to *pap* A. Later studies by them also showed that the plasmid pSMB74 contains a cluster of four ORF with a common promoter and a common terminator in a stretch of 3,500 bp and are identical in all respects to *ped* A, B, C and D (Motlagh *et al.,* 1994). The 66 amino acid prepeptide, encoded by ORF 1, following translation undergoes processing by a specific endoprotease that recognizes the Gly-Gly at -1 and -2 positions and separates the 18 amino acid leader peptide from the 44 amino acid propediocin. Before being bactericidal the two disulfide bonds form in propediocin between four cysteines at positions 9 and 14 and 20 and 44 (Fig. 4). Later, mutation analysis studies by Bukhtiyarova *et al.* (1994) showed that the 724 amino acid protein, encoded by ORF 4 is necessary for the processing of prepediocin to active pediocin and also involved in the translocation of the molecules through the membrane. They also indirectly showed that the 112 amino acid protein, encoded by ORF 2, is associated with host immunity against pediocin. They also suggested that the 174 amino acid protein encoded by ORF 3 is probably a helper protein necessary, along with the 724 amino acid protein, for efficient translocation of pediocin through the cell membrane (Bukhtiyarova *et al.,* 1994; Jack *et al.,* 1995). Recently, Venema *et al.* (1995)

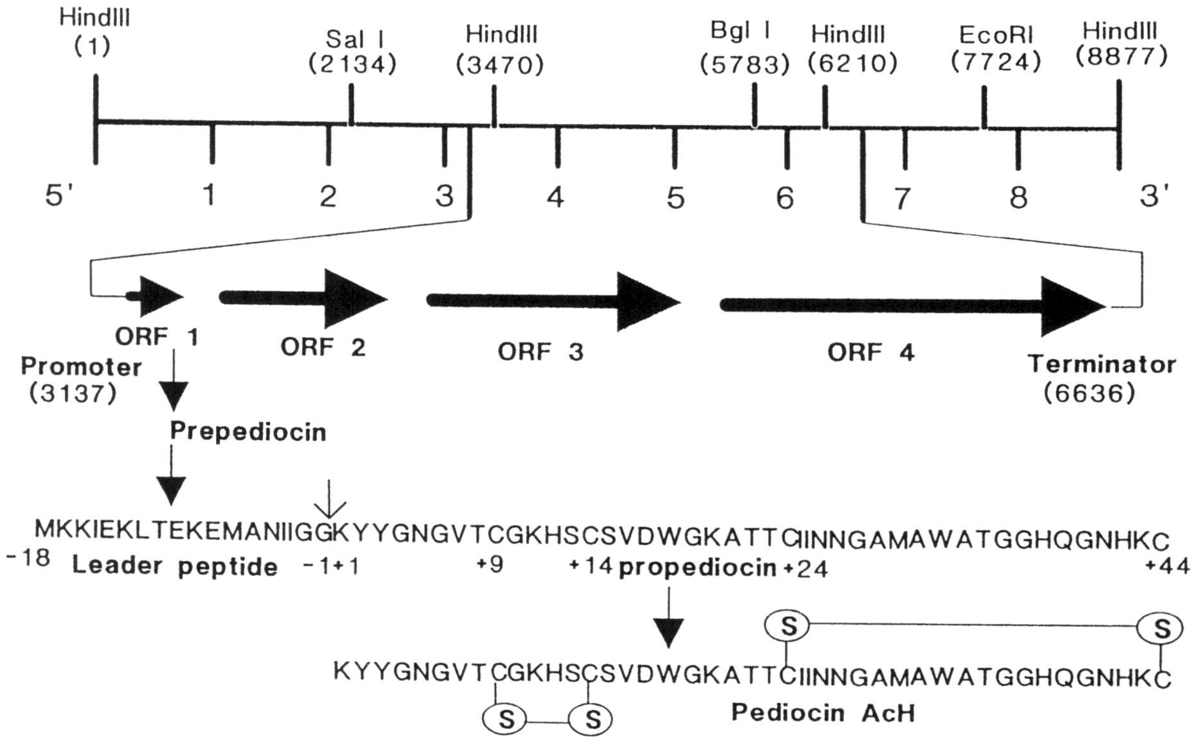

Fig. 4. Restriction map of the plasmid encoding pediocin, the four open reading frames (ORF) with individual ribosome binding sites (rbs) and common promoter and terminator, 62 amino acid prepediocin molecules with leader sequence, propediocin and site of endopeptidase cleavage (arrow) and the 44 amino acid active pediocin with two disulfide bonds.

also reported that the protein encoded by ORF 2 is the immunity protein, both proteins from ORF 3 and 4 are necessary for the transportation of active pediocin outside the producer cells and an N-terminal domain of the 724 amino acid protein, encoded by ORF 4, is associated with processing of prepediocin to pediocin. They also suggested that in mutants clones with only ORF 1 and 4, prepediocin can show antibacterial activity. However, this could be due to processing of prepediocin to pediocin by the ORF 4 protein (Bukhtiyarova *et al.,* 1994). Venema *et al.* (1995) also reported that transcription of pediocin (ped) operon result in the formation of two overlapping transcripts (mRNA), one for ORF 1, 2 and 3 and another for all four ORFs.

Previous reports showed that pediocin can be produced by *E. coli* carrying suitable plasmid vectors in which the promoter, all four ORFs and the terminator from the plasmid of pediocin producing *Ped. acidilactici* strains have been cloned (Marugg *et al.,* 1992). Later studies showed that both *E. coli* and *Ped. acidilactici* carrying a vector cloned with the four ORF, even without the specific promoter from the plasmid in *Ped. acidilactici* are capable of producing pediocin (Bukhtiyarova *et al.,* 1994).

Chikindas *et al.* (1995) recently reported the cloning of ped-gene cluster in a lactococcal plasmid vector with lactococcal promoter. When this cloned vector was introduced in a suitable *Lactococcus lactis* strain it produced pediocin PA-1/AcH, but at a reduced level.

Luchansky *et al.* (1992) analyzed the genomic DNA of three strains of pediocin producing *Ped. acidilactici* strains that included both PAC 1.0 and H (producers of pediocins PA-1 and AcH, respectively). They treated the total DNA, following extraction from the cells, with a low-frequency-cleavage endonuclease, *Asc* I and then analyzed the fragment profiles on an agarose gel following pulsed-field electrophoresis. The DNA fragment profiles of both strains (as well as another pediocin producer strain, PO_2) were identical indicating genomic similarities between the two strains (also among other pediocin producing strains).

Optimizing production and purification of pediocin PA-1/AcH

Studies have shown that pediocin PA-1/AcH can be produced by the producer strains by growing in a suitable liquid medium or a agar medium as well as in different food systems. In this section pediocin production in liquid media will be discussed. One way pediocin PA-1/AcH can be used as a food biopreservative is by growing the producer strains in a suitable liquid medium and then either adding the culture broth in the food in some form (dry or liquid) or purifying pediocin from the culture broth and adding the purified material to the food. In the selection of a liquid medium several factors need to be considered: (a) the ingredients should be food-grade and used in concentrations recommended by the regulatory agencies, (b) it should allow high production of pediocin quickly, and (c) it should be economical.

In most studies MRS-broth, MRS-broth supplemented with other nutrients, and other complex media were used for the production of pediocin (Gonzalez & Kunka, 1987; Bhunia *et al.*, 1987b, 1988; Vandenbergh, 1990). Biswas *et al.* (1991) compared the influence of MRS and other complex media as well as some simple media on the production of pediocin by *Ped. acidilactici* H. They also studied the influence of various parameters, such as various nutrients, growth temperature, incubation time, initial media pH, and final (terminal) media pH on the level of pediocin production. The results showed a higher level of pediocin is produced in a broth that has no or least amount of buffering salts. A high level of pediocin was produced in a medium, designated as TGE broth, that contains: tryptone or trypticase, glucose, yeast-extract - each 1%, Tween 80-0.2% and Mg^{2+} and Mn^{2+} - each 0.005% with pH 6.8. By increasing the tryptone/trypticase, glucose and yeast-extract to 2% the production level was increased by 15 to 20%. Studies in their laboratory also revealed that incubation at 30 to 37°C, for 16 to 18 h, with terminal pH 3.6 to 3.7 produced the highest amounts of pediocin. They also observed that *Ped. acidilactici* strains differ greatly in the level of pediocin PA-1/AcH production. As the plasmid encoding for pediocin is quite unstable, loss of the specific plasmid by some cells in the population of a producing strain can reduce the production of pediocin greatly. Finally, as the indicator strains differ greatly in sensitivity to pediocin, a strain used to assay the activity or arbitrary units (AU) of pediocin/ml of a culture broth can cause differences in the apparent production level of bacteriocin (Ray, 1992, 1993, 1994; Yang & Ray, 1994a). Recently, Yang and Ray (1995) showed that pediocin producing strains of *Ped. acidilactici* that are normally unable to hydrolyze lactose can grow in a dairy based medium and produce high levels of pediocin, only if the medium is supplemented with a commercial preparation of ß-galactosidase prior to incubation.

Methods of purification of pediocin PA-1/AcH has been studied to obtain pure pediocin for the analysis of amino acid sequence, determination of mode of bactericidal action and similar analysis. The amounts used in these studies were quite small and the cost of preparation was not a factor. However, if purified pediocin is used as a food preservative, the cost of purification should be low and the method(s) have to be efficient. Yang *et al.* (1992) developed a very simple and economical method of purification of pediocin PA-1/AcH based on a previous study in the same laboratory on the adsorption and desorption of pediocin to the cell surface of the producer strain and in other gram-positive bacteria (Bhunia *et al.*, 1991). Pediocin molecules are adsorbed on the cell surface of gram-positive bacteria at a higher pH in the environment with about 90 to 100% absorption at about pH 6.0. However, when the pH of the environment is lowered to about 1.5 to 2.0, most of the pediocin molecules are released from the cell surface. Yang *et al.* (1992) following growing the cells in a suitable broth for optimum pediocin AcH production and heating to kill the cells; adjusted the pH of the culture broth to pH 6.0 and centrifuged the broth to harvest the cells. The cells were then resuspended in a low pH suspending solution in one-tenth of the original volume, mixed well and kept overnight. The suspension was then centrifuged and the cells were discarded. The liquid was dialyzed and freeze-dried to get fairly pure pediocin with very high activity units (AU; over $1x10^6$ AU/g). When stored frozen this product retains activity without any loss over 6 months.

Effectiveness of pediocin PA-1/AcH in food systems as biopreservatives

The antibacterial effectiveness of both pediocin PA-1 and AcH has been studied in many food systems against gram-positive spoilage and pathogenic bacteria (Table 6). Pucci *et al.* (1988) grew *Ped. acidilactici* PAC 1.0 in MRS broth and used the supernatant of the culture broth as pediocin source. Cottage cheese, half and half cream and cheese sauce were mixed with pediocin and inoculated with *Listeria monocytogenes* cells. After 7 d at 4°C the population increased from initial $3x10^2$ cells/ml to $5.4x10^6$ cells/ml in the half and half cream without pediocin (control), but in the samples with pediocin it increased to $5.9x10^3$ cells/g. Similarly, in the control cheese sauce samples the population increased from an initial $7x10^2$ cells/ml to $1.7x10^7$ cells/ml in 7 d at 4°C but in the pediocin treated samples the population remained at about $1x10^2$ cells/ml. In the cottage cheese the initial population of $7.5x10^3$ cells/ml was reduced to $<10^2$/ml in both the control and pediocin treated samples, probably due to acid in the product. Gonzalez (1989) also reported that a pediocin preparation from strain PAC 1.0 when mixed in a salad dressing and challenged with *Lactobacillus bifermentans* the population level dropped from an initial $3.5x10^3$ cells/ml to <10/ml in 5 d at 25°C but the control samples had $1.5x10^3$ cells/ml. Nielsen *et al.* (1990) reported that application of pediocin from strain PAC 1.0 reduced the population of attached *Lis. monocytogenes* cells in fresh beef by 0.5 to 2.2 log cycles in 2 min. Foegeding *et al.* (1992) demonstrated that when strain PAC 1.0 cells were mixed with a sausage mix inoculated with *Lis. monocytogenes* cells and incubated, *Ped. acidilactici* produced both pediocin and organic acid and effectively controlled the pathogen in the finished product. A pediocin non-producing *Ped. acidilactici* strain, however, failed to do so at the same level.

Wu (1989) used *Ped. acidilactici* H as a starter to produce fermented lamb sausage. The levels of different bacterial groups

Table 6. Food systems in which antibacterial effectiveness of pediocin PA-1/AcH was studied

Food system	Bacteriocin tested	Form of pediocin	Reference
From strain PAC 1.0			
a. Cottage cheese, half and half cream, cheese sauce	*Listeria monocytogenes*	Powder from MRS culture broth supernatant + 10% nonfat dry milk	Pucci *et al.* (1988)
b. Salad dressing	*Lactobacillus bifermentans*	Ammonium sulfate precipitate from MRS culture broth supernatant	Gonzalez (1989)
c. Fresh beef	*Listeria monocytogenes*	MRS culture broth supernatant	Nielsen *et al.* (1990)
d. Fermented sausage	*Listeria monocytogenes*	In situ production in sausage mix during fermentation	Foegeding *et al.* (1992)
From strain H			
e. Fermented mutton sausage	Normal meat bacteria	In situ production in sausage mix during fermentation	Wu (1989)
f. Fresh beef	*Leuconostoc mesenteroides* & normal meat bacteria	Concentrated TGE culture broth	Kalchayanand (1990)
g. Fresh beef	Normal meat bacteria	Concentrated TGE culture broth	Rozbeh (1990)

Food system	Bacteriocin tested	Form of pediocin	Reference
h. Ground beef, sausage mix, cottage cheese, ice cream	*Listeria monocytogenes*, *Leuconostoc mesenteroides*	Concentrated TGE culture broth	Holla (1990)
i. Pork products (hot dogs, ham)	*Listeria monocytogenes*, *Leuconostoc mesenteroides*	Concentrated TGE culture broth	Zhang (1991)
j. Beef by-products (heart, kidney, liver, tongue)	*Listeria monocytogenes*	Concentrated TGE culture broth	Owens (1992)
k. Roast beef, chicken hot dogs, turkey rolls	*Listeria monocytogenes*, *Clostridium botulinum* B (nonproteolytic) *Lactobacillus sake* *Leuconostoc mesenteroides* *Clostridium laramie*	Concentrated TGE culture broth	Ray (unpublished data)
l. Wiener exudate	*Listeria monocytogenes*	In situ production by cells or a dried preparation of TGE culture broth supernatant	Yousef *et al.* (1991)
m. Wieners	*Listeria monocytogenes*	In situ production by cells during incubation	Degnan *et al.* (1992)
n. Turkey summer sausage	*Listeria monocytogenes*	In situ production by cells during fermentation	Luchansky *et al.* (1992)

Food system	Bacteriocin tested	Form of pediocin	Reference
o. Slurries of nonfat dry milk, butter fat, beef tissue, beef tallow	*Listeria monocytogenes*	Partially purified or encapsulated	Degnan *et al.* (1993)

in the finished products were relatively lower when pediocin producing strain was used. Rozbeh (1990) used a preparation of pediocin from *Ped. acidilactici* H to inoculate fresh beef at 1% level. The meat samples were vacuum-packaged, stored at 4°C and enumerated for different bacterial groups up to 8 wk. The results showed that the populations of psychrotrophs, lactic acid bacteria and *Brochothrix thermosphacta* were 1 to 3 log cycles lower in the treated samples as compared to the untreated controls at the end of 8 wk. In a similar study Kalchayanand (1990) inoculated fresh beef with a meat spoilage isolate of *Leuconostoc mesenteroides* at a level of $\log_{10}$ 3.4/g along with 1% of a pediocin preparation from strain H, vacuum-packaged the samples and stored at 4°C. After 8 wk the control samples had a *Leuconostoc* population of log 5.3 but the pediocin treated samples had about log 3.3 (Table 6). Similar storage studies were conducted using pediocin preparations from strain H in sterile ground beef slurry, milk, hot dogs, chopped ham and roast beef inoculated with either *Lis. monocytogenes* strains or *Leuconostoc mesenteroides*. Enumeration data of the inoculated bacterial species after the storage period of 4 to 9 wk at 4°C, revealed that pediocin treated samples, as compared to the untreated controls, had much lower population levels (Table 6; Holla, 1990; Kalchayanand, 1990; Zhang, 1991; Ray & Field, 1992; Ray, unpublished data). Limited studies with roast beef also have indicated that the pediocin preparation effectively controlled growth of *Clostridium Botulinum* B and E (nonproteolytic) and *Clostridium laramie* strains (Ray, unpublished data). Pediocin preparations in different forms were also used in wiener exudate, vacuum-packaged wieners, and model food systems inoculated with *Lis. monocytogenes* strains. At the end of the storage periods pediocin-containing products had much less *Listeria* cells as compared to the controls (Yousef *et al.*, 1991; Degnam *et al.*, 1992, 1993). When *Ped. acidilactici* H cells were used as starter to make turkey summer sausage, it controlled *Lis. monocytogenes* much more effectively than a pediocin non-producing strain of *Ped. acidilactici* (Luchansky *et al.*, 1992).

Berry *et al.* (1990) also reported that pediocin producing *Ped.*

Table 7. Effectiveness of pediocin from strain H in controlling growth of psychrotrophic spoilage (*Leuconostoc mesenteroides*) and pathogenic (*Listeria monocytogenes*) bacteria in several refrigerated foods during storage at 4°C.

Food	Bacteria	Pediocin (- or +)	Log_{10} cfu/g or ml (d or wk)	
Sterile ground beef[a]	*Listeria monocytogenes*	Ped-	2.0 (1 d)	6.0 (4 wk)
		Ped+	2.0 (1 d)	<1.0 (4 wk)
Sterile milk[a]	*Listeria monocytogenes*	Ped-	2.0 (1 d)	6.0 (4 wk)
		Ped+	2.0 (1 d)	<1.0 (4 wk)
Beef[b]	*Leuconostoc mesenteroides*	Ped-	3.4 (1 d)	5.3 (8 wk)
		Ped+	3.4 (1 d)	3.3 (8 wk)
Hot dog[c]	*Leuconostoc mesenteroides*	Ped-	3.8 (1 d)	8.6 (9 wk)
		Ped+	3.8 (1 d)	4.1 (9 wk)
Chopped ham[c]	*Listeria monocytogenes*	Ped-	3.2 (1 d)	5.5 (6 wk)
		Ped+	3.2 (1 d)	1.9 (6 wk)
Roast beef[d]	*Listeria monocytogenes*	Ped-	2.7 (1 d)	9.5 (8 wk)
		Ped+	2.7 (1 d)	4.3 (8 wk)

Results are from: [a]Holla (1990); [b]Kalchayanand (1990); [c]Zhang (1991); [d]Ray (unpublished data).

acidilactici strain JD1-23 effectively controlled growth of *Lis. monocytogenes* during fermentation of semidry sausage. They also reported that addition of cells of pediocin producing strain JD1-23 in the packaged frankfurters controlled growth of *Lis. monocytogenes* at 4°C (Berry *et al.,* 1991). El-Khateib *et al.* (1993) showed that treatment of a pediocin preparation from *Ped. acidilactici* PO2 reduced the population of *Lis. monocytogenes* cells attached to the surface of fresh beef. In a meat model system a pediocin product by *Ped. acidilactici* PO2 effectively controlled growth of *Lactobacillus curvatus* associated with spoilage of refrigerated meat products (Coventry *et al.,* 1995).

CONCLUDING REMARKS

The material presented here justify the statement that besides nisin, pediocin PA-1/AcH is the most well studied bacteriocin of lactic acid bacteria. Results from the product inoculation studies showed it effectively controls viability and growth of psychrotrophic spoilage and pathogenic bacteria. Most studies have been conducted with different pathogenic strains of *Listeria monocytogenes*. Only limited results showed that pediocin also inhibits growth of some pathogenic and spoilage *Clostridium* species/strains and psychrotrophic spoilage *Lactobacillus* and *Leuconostoc* species in vacuum-packaged refrigerated products. It is necessary to study the effectiveness of pediocin of *Ped. acidilactici* in controlling other pathogenic and spoilage bacteria in different food systems.

REFERENCES

Anonymous (1988) Safety considerations of new generation of refrigerated foods: refrigerated foods and microbiological criteria committee of the National Food Processors Association. Dairy Food Sanitation 8:5-10.

Berry ED, Liewen MB, Mandigo RW, Hutkins RW (1990) Inhibition of *Listeria monocytogenes* by bacteriocin-producing *Pediococcus* during the manufacture of fermented sausage. J Food Prot 53:194-197.

Berry ED, Hutkins RW, Mandigo RW (1991) The use of bacteriocin-producing *Pediococcus acidilactici* to control post-processing *Listeria monocytogenes* contamination of frankfurters. J Food Prot 54:681-686.

Bhunia AK, Kim WJ, Johnson MC, Ray B (1987a) Partial purification and characterization of antimicrobial substances from *Pediococcus acidilactici* and *Lactobacillus plantarum*. Abstract No. 143. Annual Meeting, Institute of Food Technologists, Las Vegas. June 16-19.

Bhunia AK, Johnson MC, Ray B (1987b) Direct detection of an antimicrobial peptide of *Pediococcus acidilactici* in sodium dodecyl sulfate-polyacrylamide electrophoresis. J Industrial Microbiol 2:312-319.

Bhunia AK, Johnson MC, Ray B (1988) Purification, characterization and antimicrobial spectrum of a bacteriocin produced by *Pediococcus acidilactici*. J Applied Bacteriol 65:261-268.

Bhunia A (1989) Characteristics and mode of action of an antimicrobial peptide produced by *Pediococcus acidilactici* strain H. Ph.D. Thesis, University of Wyoming, Laramie.

Bhunia AK, Johnson MC, Ray B, Belden EL (1990) Antigenic property of pediocin AcH produced by *Pediococcus acidilactici* H. J Applied Bacteriol 69:211-215.

Bhunia AK, Johnson MC, Ray B, Kalchayanand N (1991) Mode of action of pediocin AcH from *Pediococcus acidilactici* H on sensitive bacterial strains. J Applied Bacteriol 70:25-33.

Bhunia AK, Johnson MG (1992a) Monoclonal antibody-colony immunoblot method specific for isolation of *Pediococcus acidilactici* from foods and correlation with pediocin (bacteriocin) production. Appl Environ Micribiol 58:2315-2320.

Bhunia AK, Johnson MG (1992b) Modified method to directly detect in SDS-PAGE the bacteriocin of *Pediococcus acidilactici*. Letters Appl Microbiol 15:5-7.

Bhunia AK, Bly LA, Johnson MG (1993) Characterization of new pediocins of *Pediococcus acidilactici*: inhibitory spectrum, isoelectric point and N-terminal amino acid sequence. Abstract No. 018. 93rd Annual Meeting American Society Microbiology, Washington, DC.

Bhunia AK, Bhowmik TK, Johnson MG (1994) Determination of bacteriocin-encoding plasmids of *Pediococcus acidilactici* strains by Southern hybridization. Letters Appl Microbiol 18:168-170.

Bhunia AK (1994) Monoclonal antibody-based enzyme immunoassay for pediocins of *Pediococcus acidilactici*. Appl Environ Microbiol 60:2692-2696.

Biswas SR, Ray P, Johnson MC, Ray B (1991) Influence of growth conditions on the production of a bacteriocin, pediocin AcH, by *Pediococcus acidilactici* H. Appl Environ Microbiol 57:1265-1267.

Bukhtiyarova M, Yang R, Ray B (1994) Analysis of the pediocin AcH gene cluster from plasmid pSMB74 and its expression in a pediocin-negative *Pediococcus acidilactici* strain. Appl Environ Microbiol 60:3405-3408.

Chen Y (1995) Stability of bacteriocin-based biopreservatives and application in foods. M.S. Thesis, University of Wyoming, Laramie.

Chikindas ML, Garcia-Garcera JM, Driessen, AJM, Ledeboer AM, Nissen-Meyer J, Nes IF, Abee T, Konings WN, Venema G (1993) Pediocin PA-1, a bacteriocin from *Pediococcus acidilactici* PAC 1.0, forms hydrophilic pores in the cytoplasmic membrane of target cells. Appl Environ Microbiol 59:3577-3584.

Chikindas ML, Venema K, Ledeboer AM, Venema G, Kok J (1995) Expression of lactococcin A and pediocin PA-1 in heterologous hosts. Letters Appl Microbiol 21:183-189.

Christensen DP, Hutkins RW (1992) Collapse of the proton motive force in *Listeria monocytogenes* caused by a bacteriocin produced by *Pediococcus acidilactici*. Appl Environ Microbiol 58:3312-3315.

Cintas LM, Rodriguez JM, Fernandez MF, Sletten K, Nes IF, Hernandez PE, Holo H (1995) Isolation and characterization of pediocin L50, a new bacteriocin from *Pediococcus acidilactici* with a broad inhibitory spectrum. Appl Environ Microbiol 61:2643-2648.

Coventry MJ, Muirhead K, Hickey MW (1995) Partial characterization of pediocin PO_2 and comparison with nisin for biopreservation of meat products. Int J Food Microbiol 26:133-145.

Daba H, Lacroix C, Huang J, Simard RE, Lemieux L (1994) Simple method of purification and sequencing of a bacteriocin produced by *Pediococcus acidilactici* UL 5. J Appl Bacteriol 77:682-688.

Degnan AJ, Yousef AE, Luchansky JB (1992) Use of *Pediococcus acidilactici* to control *Listeria monocytogenes* in temperature-abused, vacuum-packaged wieners. J Food Prot 55:98-103.

Degnan AJ, Luchansky JB (1992) Influence of beef tallow and muscle on the antilisterial activity of pediocin AcH and liposome-encapsulated pediocin AcH. J Food Prot 55:552-554.

DeVuyst L, Vandamme E (eds) (1994) Bacteriocins of lactic acid bacteria: microbiology, genetics and application. Chapman and Hall, LTd., London.

El-Khateib T, Yousef AE, Ockerman HW (1993) Inactivation and attachment of *Listeria monocytogenes* on beef muscle treated with lactic acid and selected bacteriocins. J Food Prot 56:29-33.

Foegeding PM, Thomas AB, Pilkington DH, Klaenhammer TR (1992) Enhanced control of *Listeria monocytogenes* by in situ-produced pediocin during dry fermented sausage production. Appl Environ Microbiol 58:884-890.

Gonzalez CF, Kunka BS (1987) Plasmid-associated bacteriocin production and sucrose fermentation in *Pediococcus acidilactici*. Appl Environ Microbiol 53:2534-2538.

Gonzalez CF (1989) Methods for inhibiting bacterial spoilage and resulting compositions. U.S. Patent No. 4,883,673.

Graham DC, McKay LL (1985) Plasmid DNA in strains of *Pediococcus cerevisiae* and *Pediococcus pentosaceus*. Appl Environ Microbiol 50:532-534.

Hanlin MB, Kalchayanand N, Ray P, Ray B (1993) Bacteriocins of lactic acid bacteria in combination have greater antibacterial activity. J Food Protection 56:252-255.

Henderson JT, Chopko AL, Van Wassenaar PD (1992) Purification and primary structure of pediocin PA-1 produced by *Pediococcus acidilactici* PAC 1.0. Archives Biochem Biophys 295:5-12.

Holla S (1990) Efficiency of pediocin AcH on viability loss of pathogenic and spoilage bacteria in food. M.S. Thesis, University of Wyoming, Laramie.

Hoover DG, Walsh PM, Kolaetis KM, Daly MM (1988) A bacteriocin produced by *Pediococcus* species associated with 5.5 MDa plasmid. J Food Prot 51:29-31.

Hoover DG, Dishart KJ, Hermes MA (1989) Antagonistic effect of *Pediococcus* spp. against *Listeria monocytogenes*. Food Biotechnol 3:183-196.

Hoover DG, Steenson LR (eds) (1993) Bacteriocins of lactic acid bacteria. Academic Press Inc., San Diego, CA.

Jack RW, Tagg JR, Ray B (1995) Bacteriocins of gram-positive bacteria. Microbiol Rev 59:171-200.

Jung G, Sahl H-G (eds) (1991) Nisin and novel lantibiotics. Escom Publishers, Leiden, The Netherlands.

Kalchayanand N (1990) Extension of shelf-life of vacuum-packaged refrigerated fresh beef by bacteriocins of lactic acid bacteria. Ph.D. Thesis. University of Wyoming, Laramie.

Kalchayanand N, Hanlin MB, Ray B (1992) Sublethal injury makes gram-negative and resistant gram-positive bacteria sensitive to the bacteriocins, pediocin AcH and nisin. Letters Appl Microbiol 15:239-243.

Kalchayanand N, Sikes T, Dunne P, Ray B (1994) Hydrostatic pressure and electroporation have increased bactericidal efficiency in combination with bacteriocins. Appl Environ Microbiol 60:4174-4177.

Kalchayanand N, Ray B (1995) Bacteriocins activate lytic system(s) of sensitive gram-positive bacteria to induce cell lysis. Abstract No. P274. 9th World Congress of Food Science and Technology. July 30 to August 4. Hungarian Scientific Society for Food Industry, Budapest.

Kim W-J (1990) Development of plasmid transfer systems in lactic acid bacteria used in meat fermentation. Ph.D. Thesis, University of Wyoming, Laramie.

Kim, WJ, Ray B, Johnson MC (1992) Plasmid transfer by conjugation and electroporation in *Pediococcus acidilactici*. J Appl Bacteriol 72:201-207.

Lozano JCN, Nissen-Meyer J, Sletten K, Pelaz C, Nes IF (1992) Purification and amino acid sequence of bacteriocin produced by *Pediococcus acidilactici*. J Gen Microbiol 138:1985-1990.

Luchansky JB, Glass KA, Harsono KD, Degnan AJ, Faith NG, Cauvin B, Baccus-Taylor G, Arihara K, Bater B, Maurer AJ, Cassens RG (1992) Genomic analysis of *Pediococcus* starter cultures used to control *Listeria monocytogenes* in turkey summer sausage. Appl Environ Microbiol 58:3053-3059.

Marugg JD, Gonzalez CF, Kunka BS, Ledeboer AM, Pucci MJ, Toonen MY, Walker SA, Zoetmudler LCM, Vandenbergh PA (1992) Cloning, expression, and nucleotide sequence of genes involved in production of pediocin PA-1, a bacteriocin from *Pediococcus acidilactici* PAC 1.0. Appl Environ Microbiol 58:2360-2367.

Motlagh, AM, Johnson MC, Ray B (1991) Viability loss of foodborne pathogens by starter culture metabolites. J Food Protection 54:873-878.

Motlagh A (1991) Antimicrobial efficiency of pediocin AcH and genetic studies on the plasmid encoding for pediocin AcH. Ph.D. Thesis, University of Wyoming, Laramie.

Motlagh AM, Holla S, Johnson MC, Ray B, Field, RA (1992a) Inhibition of *Listeria* spp. in sterile food systems by pediocin AcH, a bacteriocin produced by *Pediococcus acidilactici* H. J Food Protection 55:337-343.

Motlagh AM, Bhunia AK, Szostek F, Hansen TR, Johnson MG, Ray B (1992b) Nucleotide and amino acid sequence of *pap*-gene (pediocin AcH production) in *Pediococcus acidilactici* H. Letters Appl Microbiol 15:45-48.

Motlagh AM, Bukhtiyarova M, Ray B (1994) Complete nucleotide sequence of pSMB74, a plasmid encoding the production of pediocin AcH in *Pediococcus acidilactici*. Letter Appl Microbiol 16:305-312.

Nielsen JW, Dickson JS, Crouse JD (1990) Use of bacteriocin produced by *Pediococcus acidilactici* to inhibit *Listeria monocytogenes* associated with fresh meat. Appl Environ Microbiol 56:2142-2145.

Noerlis Y, Ray B (1994) Factors influencing immunity and resistance of *Pediococcus acidilactici* to the bacteriocin, pediocin AcH. Letters Appl Microbiol 18:138-143.

Owens J (1994) Biopreservation of vacuum-packaged refrigerated beef by-products. M.S. Thesis, University of Wyoming, Laramie.

Pucci MJ, Vedamuthu ER, Kunka BS, Vandenbergh PA (1988) Inhibition of *Listeria monocytogenes* by using bacteriocin PA-1 produced by *Pediococcus acidilactici* PAC 1.0. Appl Environ Microbiol 54:2349-2353.

Ray B (1989) Characteristics of pediocin AcH, an antibacterial peptide of *Pediococcus acidilactici* H. Abstract No. P136. Prospects of Protein Engineering, Anniversary Congress of the University of Groningen. August 14 to 18. Haren, The Netherlands.

Ray B, Kalchayanand N, Field RA (1989c) Isolation of a *Clostridium* species from vacuum-packaged refrigerated beef and its susceptibility to bacteriocins from *Pediococcus acidilactici*. pp 285-290. Proceedings 35th International Congress of Meat Science and Technology. August 20-25, Copenhagen. Danish Meat Research Inst, DK 4000 Roskilde, Denmark.

Ray B, Kalchayanand N, Field RA (1992) Meat spoilage bacteria: are we prepared to control them? The National provisioner 206(2):22.

Ray B, Daeschel MA (eds) (1992) Biopreservatives of microbial origin. CRC Press Inc., Boca Raton, FL.

Ray B (1992) Pediocin(s) of *Pediococcus acidilactici* as food biopreservatives. In: "Food Biopreservatives of Microbial Origin", Ray B and Daeschel MA (eds), pp 265-322. CRC Press Inc., Boca Raton, FL.

Ray B, Motlagh A, Johnson MC (1992) Processing of prepediocin in *Pediococcus acidilactici*. FEMS Microbiol Rev 12:191.

Ray B, Field RA (1992) Antibacterial effectiveness of a pediocin AcH-based biopreservative against spoilage and pathogenic bacteria from vacuum-packaged refrigerated meat. pp 731-734. Proceedings 38th International Congress of Meat Science and Technology. August 23-28. Clermont Ferrand, INRA, 63122 Cey RT, France.

Ray B, Motlagh AM, Johnson MC, Bozoglu F (1992) Mapping of pSMB74, a plasmid encoding bacteriocin, pediocin AcH production (Pap^+), by *Pediococcus acidilactici* H. Letters Appl Microbiol 15:35-37.

Ray, B (1993) Sublethal injury, bacteriocins and food microbiology. ASM News 59:285-291.

Ray B, Hoover DG (1993) Pediocins. In: "Bacteriocins of Lactic Acid Bacteria". Hoover DG and Steenson LR (eds), pp 181-210. Academic Press Inc., San Diego, CA.

Ray B (1994) Pediocins of *Pediococcus* species. In: "Bacteriocins of Lactic Acid Bacteria". DeVuyst L and Vandamme EJ (eds), pp 465-495. Blackie Academic and Professional, NY.

Ray B (1995a) *Pediococcus* and fermented foods. In: "Food Biotechnology". Hui YH and Khachatourians GG (eds), pp 745-795. VCH Publishers, Inc, NY.

Ray B (1995b) Combined antibacterial efficiency of bacteriocins and ultrahigh hydrostatic pressure. Abstact No. L078. Proceedings of 9th World Congress of Food Science and Technology. July 30 to August 4. Budapest, Hungary.

Ray B, Kalchayanand N, Means W, Field R (1995) The spoiler: *Clostridium laramie*. Meat and Poultry 41(7):12-14.

Ray SK, Johnson MC, Ray, B (1989a) Bacteriocin plasmid of *Pediococcus acidilactici*. J Industrial Microbiol 4:163-171.

Ray SK, Johnson MC, Ray B (1989b) Conjugal transfer of a plasmid encoding bacteriocin production and immunity in *Pediococcus acidilactici* H. J Applied Bacteriol 66:393-399.

Rozbeh M (1990) Extension of shelf-life in vacuum-packaged fresh beef. M.S. Thesis, University of Wyoming, Laramie.

Rozbeh M, Kalchayanand N, Field RA, Johnson MC, Ray B (1993) The influence of biopreservatives on the bacterial level of refrigerated vacuum-packaged meat. J Food Safety 13:99-111.

Sahl H-G, Jack RW, Bierbaum G (1995) Biosynthesis and biological activities of lantibiotics with unique post-translation modifications. Eur J Biochem 230:827-853.

Schved F, Lalazar A, Henis Y, Juven BJ (1993) Purification, partial characterization and plasmid-linkage of pediocin SJ-1, a bacteriocin produced by *Pediococcus acidilactici*. J Appl Bacteriol 74:67-77.

Schved F, Lindner P, Juven BJ (1994) Interaction of the bacteriocin pediocin SJ-1 with the cytoplasmic membrane of sensitive bacterial cells as detected by ANS fluorescens. J Appl Bacteriol 76:30-35.

Szostek F (1993) Isolation, characterization, bacteriocin production and sensitivity of bacteriocin associated with spoilage of vacuum-packaged meat products. M.S. Thesis, University of Wyoming, Laramie.

Tagg JR, Dajani AS, Wannamaker LW (1976) Bacteriocins of gram-positive bacteria. Bacteriol Rev 40:722-756.

Vandenbergh PA, Pucci MJ, Kunka BS, Vedamuthu ER (1990) Methods for inhibiting *Listeria monocytogenes* using a bacteriocin. U.S. Patent No. 4,929,445.

Vandenbergh PA (1992) Lactic acid bacteria, their metabolic products and interference with microbial growth. FEMS Microbiol Rev 12:221-238.

Venema K, Kok J, Marugg JD, Toonen MY, Ledeboer AM, Venema G, Chikindas MA (1995) Functional analysis of the pediocin operon of *Pediococcus acidilactici* PAC 1.0: Ped B is the immunity protein and Ped D is the precursor processing enzyme. Mol Microbiol

Wu WH (1989) Evaluation of fermented mutton sausages made with the different lactic acid starter cultures. M.S. Thesis, University of Wyoming, Laramie.

Wu WH, Rule DC, Busboom JR, Field RA, Ray B (1991) Starter culture and time/temperature of storage influences on quality of fermented mutton sausage. J Food Sci 56:916-919.

Yang R, Johnson MC, Ray B (1992) Novel method to extract large amounts of bacteriocins from lactic acid bacteria. Appl Environ Microbiol 58:3355-3359.

Yang R (1994) The bacteriocins of lactic acid bacteria: Production, purification, genetic control and application. Ph.D. Thesis, University of Wyoming, Laramie.

Yang R, Ray B (1994a) Factors influencing production of bacteriocins by lactic acid bacteria. Food Microbiol 11:281-291.

Yang R, Ray B (1995) Influence of fermentation parameters on optimum bacteriocin production by lactic acid bacteria in a dairy based medium. Abstract No. 273. 9th World Congress of Food Science and Technology. July 30-August 4. Hungarian Food Science Society for Food Industry, Budapest.

Yousef AE, Luchansky JB, Degnan AJ, Doyel MP (1991) Behavior of *Listeria monocytogenes* in wiener exudates in the presence of *Pediococcus acidilactici* H or pediocin AcH during storage at 4 or 25°C. Appl Environ Microbiol 57:1461-1467.

Zhang H (1992) Control of spoilage and pathogenic bacteria in processed pork products by biopreservatives. M.S. Thesis, University of Wyoming, Laramie.

POLYSACCHARIDES, OLIGOSACCHARIDES, SPECIAL SUGARS AND ENZYMES VIA *LEUCONOSTOC MESENTEROIDES* SP. FERMENTATIONS

VANDAMME, E.J., RAEMAEKERS[a], M., VEKEMANS, N. & SOETAERT, W.
Laboratory of Industrial Microbiology and Biocatalysis
Department of Biochemical and Microbial Technology
University of Gent
Coupure Links 653
B-9000 Gent
Belgium

Abstract

Leuconostoc mesenteroides, a common lactic acid bacterium, displays a wide range of biocatalytic properties, which are potentially useful for industrial carbohydrate modifications. The use of *L. mesenteroides* for the production of dextran via whole cell fermentation, for leucrose synthesis with dextransucrase, for the synthesis of alternan and gluco-oligosaccharides with alternansucrase, for mannitol fermentation with viable *L. mesenteroides* cells, and for the synthesis of α-D-glucose-1-phosphate using sucrose phosphorylase, are discussed.

Introduction

Leuconostoc mesenteroides strains are particularly well adapted in nature to sugary niches and consequently possess a wide spectrum of biocatalytic properties useful in carbohydrate modifications. A lot of research has been directed towards the industrial exploitation of these useful properties. Several new processes have recently been optimised up to the pilot scale level and in a few cases, commercial processes have resulted from it (Vandamme and Soetaert, 1995).

L. mesenteroides strains and their enzymes can be used to produce carbohydrates and derivatives as diverse as dextran and alternan (biopolymers), fructose, mannitol (a polyol), leucrose (a non-cariogenic disaccharide), gluco-oligosaccharides, α-D-glucose-1-phosphate, and many others. The various (potential) industrial applications of *L. mesenteroides* will be discussed below with emphasis on key enzymes involved, metabolic and genetic deregulation, fermentation optimisation aspects and (oligo)sugar or polymer characterisation.

Dextran

Dextran is a well-known glucan, produced by *L. mesenteroides* strains. It is only produced when cultured in a sucrose-containing medium, since sucrose is the only-known inducer of the

---

[a] Research Assistant of the Belgian National Fund for Scientific Research

NATO ASI Series, Vol. H 98
Lactic Acid Bacteria:
Current Advances in Metabolism, Genetics and Applications
Edited by T. Faruk Bozoğlu and Bibek Ray

enzyme involved, dextransucrase (E.C. 2.4.1.5). Dextrans are high-molecular-weight homopolysaccharides composed of glucose units. They contain a substantial number of consecutive α-(1→6)-D-glucosidic linkages in the main chains (figure 1), together with various proportions of α-(1→2), α-(1→3) and α-(1→4) linkages, depending on the strain (Misaki et al., 1980; Robyt, 1992). Jeanes et al. (1954) reported that the α-(1→6)-glucosidic linkages from 96 different dextrans constituted from 50 to 97% of the total linkages. The non-(1→6) linkages usually occur as the origin of branch points, as in the case of dextran B-512. However, certain glucans, such as the less water-soluble glucan fractions from *L. mesenteroides* NRRL B-1299 and the soluble glucan fraction S of *L. mesenteroides* NRRL B-1355, contain α-(1→3) linkages in the main chain (Misaki et al., 1980). The latter glucan was referred to as alternan, because of the alternating α-(1→3) and α-(1→6) linkages. The varying molecular weight (15 - 20 million Da), the type of branch linkages and the branching frequency cause considerable differences in physicochemical properties between the dextrans, produced by different *L. mesenteroides* strains (Brooker, 1979). Dextrans are also formed by oral *Streptococci* and certain *Lactobacilli* via their respective dextransucrases (Kim and Robyt, 1994).

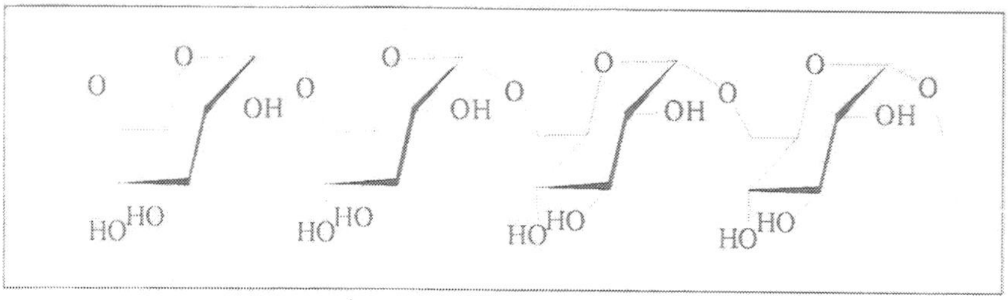

Figure 1. Backbone structure of dextran.

Commercial dextran, mainly produced using *L. mesenteroides* NRRL B-512(F), consists for 95% of an α-(1→6) glucan backbone, with 5% of α-(1→3) linkages. The sole enzyme involved in its synthesis, dextransucrase is a glucosyltransferase, which has been characterised in detail (Robyt & Taniguchi, 1976). The enzyme polymerises the glucose moiety of sucrose into dextran, thereby leaving the fructose moiety untouched. So essentially the enzyme converts sucrose into dextran and fructose. The reaction can be represented as follows:

$$\text{n Sucrose} + H_2O \longrightarrow \underset{\text{(Dextran)}}{\text{H-(Glucosyl)}_n\text{-OH}} + \text{n D-Fructose}$$

The reaction mechanism of dextransucrase action was elucidated by Robyt et al. (1974) via "pulse and chase" experiments with immobilised enzyme. The active site of dextransucrase possesses two equivalent reaction sites which hydrolyse the sucrose molecule, bind the glucosyl residue and build the chain via an insertion mechanism (figure 2). During the chain elongation reaction the glucan chain stays attached to the enzyme and moves from one reaction site to the other and vice versa, while glucose units are transferred to the non-reducing end of the dextran chain. Sucrose is the only substrate and no ATP or cofactors are involved since all energy is delivered by the glycosidic bond between glucose and fructose in sucrose.

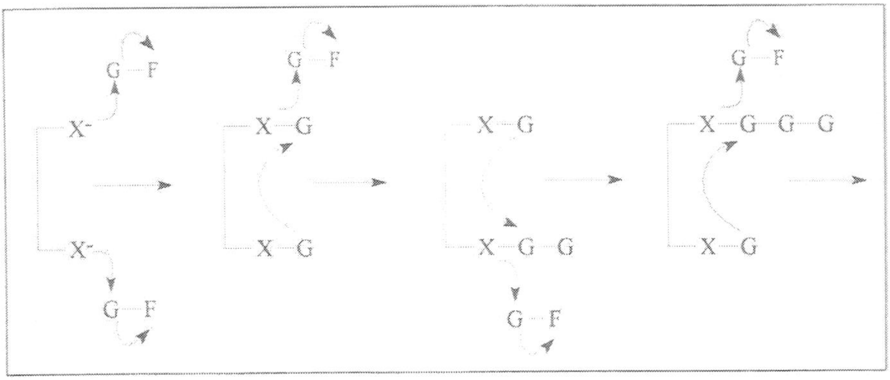

Figure 2. Two-site reaction mechanism for the biosynthesis of dextran chains by dextransucrase. G-F represents sucrose, G-G α-(1→6)-linked glucose residues (adapted from Robyt et al., 1974).

Dextransucrase is secreted extracellularly by the cells during their growth phase, so the dextran synthesis occurs completely extracellularly. The industrially applied production method uses a non-aerated whole cell fermentation. According to De Belder (1990), the major factors influencing dextransucrase secretion and dextran biosynthesis are pH, initial sucrose concentration, calcium concentration, temperature and process time.

Dextran fermentation has been performed on a commercial scale at Pfeifer & Langen in Dormagen (Germany) since 1950. In this anaerobic fermentation the cells are cultured on a medium containing an excess of sucrose (10-12 %) and limited quantities of nitrogen source and trace elements. During the fermentation, crude dextran, having a high molecular weight of several millions, accumulates besides fructose, low-molecular-weight dextran and small quantities of mannitol. In a following step the dextran is separated from the broth by precipitation in ethanol. This crude dextran is the starting product for a whole range of dextrans and derivatives. The native dextran is partially hydrolysed by acid hydrolysis, followed by fractionation to obtain a specific molecular weight fraction.

An important application is as clinical dextran, a bloodplasma substitute having a molecular weight of 40 or 70 kDa.

Dextran is also frequently chemically modified for several other applications :

- in veterinary medicine (Fe^{2+}-dextran as a source of Fe^{2+}),
- in human medicine as a cholesterol-lowering agent (DEAE-dextran),
- in separation technology, as molecular sieve (cross-linked dextran) and in aqueous two phase systems (dextran/polyethyleneglycol),
- as a microcarrier in tissue or cell culture.

Due to process technological problems caused by the broth viscosity during the fermentation, enzymatic (cell-free) processes have also been developed (Brown & McAvoy, 1990). In this approach the enzyme is first produced by fermentation. The enzyme is then added to a sucrose solution to synthesise the dextran.

Alternan

Alternan is a glucan with alternating α-(1→6)- and α-(1→3)-glucosidic linkages and is synthesised by alternansucrase (E.C. 2.4.1.140), an enzyme produced by *Leuconostoc mesenteroides* NRRL B-1355. Similar alternating α-(1→6)- and α-(1→3)-glucans are also produced by glucansucrases from *L. mesenteroides* NRRL B-1498 and B-1501 (Seymour and Knapp, 1980; Côté and Robyt, 1982a). The structure of alternan was elucidated via various methods: periodate oxidation (Jeanes et al., 1954; Côté and Robyt, 1982a), acid hydrolysis, acetolysis, Smith-degradation (Goldstein and Whelan, 1962), enzymatic hydrolysis (Sawai et al., 1978; Misaki et al., 1980; Côté and Robyt, 1982a; Côté, 1992; Côté and Bielly, 1994), methylation-fragmentation analysis (Seymour et al., 1977; Misaki et al., 1980) and ^{13}C-N.M.R. (Seymour and Knapp, 1980).

Misaki et al. (1980) found by methylation-fragmentation analysis that alternan contains 10% of α-(1→6) end-linkages, 10% of α-(1→3) branch points, 45% of α-(1→6) linkages and 35% of α-(1→3) linkages in linear positions. Methylation-fragmentation results obtained by Seymour et al. (1977) were similar. From ^{13}C-N.M.R. data, Seymour and Knapp (1980) concluded that alternan would have the following structure (figure 3):

-{-[-Glc*p*-(1→3)-Glc*p*-(1→6)-]$_n$-Glc*p*-(1→3)-Glc*p*-(1→6)-}$_x$-

Glc*p*-(1→6)-[-Glc*p*-(1→6)]$_q$-Glc*p*-(1→6)

with n = 5.1 and q = 0.7

Figure 3. Alternan structure as proposed by Seymour and Knapp (1980), deduced from ^{13}C-N.M.R. analysis.

From oligosaccharide fragments obtained by acid hydrolysis, acetolysis and enzymatic hydrolysis of alternan, combined with the methylation-fragmentation results, Misaki et al. (1980) concluded that the alternan structure is as shown in figure 4.
Because of the alternating character of the α-(1→6)- and α-(1→3)-glucosidic linkages in alternan, few enzymes are able to hydrolyse alternan. Unlike dextran, alternan is resistant to the action of *Penicillium* sp. endodextranase (Côté and Robyt, 1982a): only 3% was hydrolysed. However, the exo-(1→6)-α-D-isomaltodextranase (E.C. 3.2.1.94) from *Arthrobacter globiformis* T6 was able to (slowly) degrade alternan up to 85% (Sawai et al., 1978). Another alternan-hydrolysing enzyme is the isomaltodextranase from *Actinomadura* sp.

Both enzymes release isomaltose from the non-reducing ends of the alternan molecule (Côté, 1992). Recently, a third enzyme able to hydrolyse alternan was isolated from a *Bacillus* sp. by Côté and Bielly (1994). The enzyme is an endo-α-1,3-α-1,6-D-glucanase, referred to as alternanase, and releases a series of gluco-oligosaccharides. The oligosaccharide produced at the highest extent is a non-reducing, cyclic tetrasaccharide, linked in an alternating α-1,3-α-1,6 fashion.

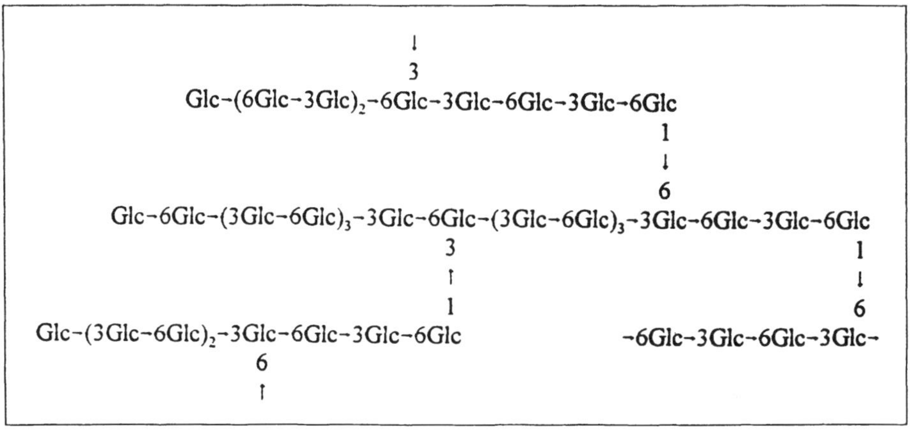

Figure 4. Alternan structure as proposed by Misaki et al. (1980).

Because of its resistance against most carbohydrases, CÔTÉ (1992) suggested that alternan and its low-molecular-weight derivatives may be of use as low- or non-digestible (low- or non-caloric) additives or ingredients, e.g. in artificially sweetened food preparations. They could eventually replace polydextrose, maltodextrins, and arabic gum, used as soluble fillers, binders, bulking agents, and extenders for food products, and as components of inks, adhesives, cosmetic creams, and ointments (CÔTÉ, 1991). Alternan might also be chemically modified to yield new gums, gels, or films that could have applications in food preparations, pharmaceuticals, and drug or pesticide delivery systems (KIM and ROBYT, 1994).

The molecular weight of alternan was determined by light-scattering analysis. Seymour et al. (1979b) found a value of 40 million Da, while Côté measured 10 ± 1 million Da. Jeanes et al. (1954) reported that the intrinsic viscosity of alternan is very low (0.19) in comparison with native B-512(F) dextran (0.95).

Similar to dextran, alternan is synthesised from sucrose by alternansucrase, which transfers the glucose units from sucrose to alternan, while fructose is released. Alternansucrase is an extracellular enzyme only synthesised by L. mesenteroides NRRL B-1355. Since it is an inducible enzyme, sucrose must be present in the growth medium to serve as inducer. Optimal batch-fermentation conditions resulted in an alternansucrase activity in the culture broth of 1.5 U/ml at a controlled pH 6.5 and dissolved oxygen concentration of 75% at 27°C, after 12 h of fermentation (Raemaekers, 1995). One unit of alternansucrase was defined to convert one µmole of glucose from sucrose into alternan, per minute, at pH 5.5, 30°C and 100 g/l sucrose.

However, *L. mesenteroides* NRRL B-1355 secretes both alternansucrase and (at least two types of) dextransucrase in the fermentation broth. However, the majority of the alternan- and dextransucrase activity remains attached to the cells. The cell-associated alternansucrase (and dextransucrase) can be isolated via centrifugation of the culture broth. To isolate the soluble fractions of alternansucrase and dextransucrase, an aqueous two-phase extraction of the culture supernatant with polyethylene glycol can be used. A selective heat treatment at 48°C allows to inactivate the dextransucrase in the enzyme preparation or the alternansucrase-containing cell suspension (Côté and Robyt, 1982a; Raemaekers, 1995). The time necessary to inactivate the dextransucrase can be evaluated via determination of:

- the extent of endodextranase resistance of the polysaccharides produced by the various heat-treated enzyme solutions or cell suspensions, against the action of *Penicillium sp.* endodextranase. Since alternan is almost completely resistant to the endodextranase action, a constant level of hydrolysis indicated the absence of dextransucrase.
- the amount of oligodextrans, i.e. oligosaccharides containing consecutively α-(1→6)-linked glucose residues, produced from maltose and sucrose. Alternansucrase produces oligoalternans, i.e. oligosaccharides with alternating α-(1→6)-linked and α-(1→3)-linked glucose residues, together with a little amount of oligodextrans. Since the dextransucrase only synthesises oligodextrans, the level of oligodextrans can be used to determine the absence of dextransucrase.

Under the applied heat treatment conditions, an incubation of 10 minutes at 48°C is sufficient to remove the dextransucrase activity from an enzyme solution, leaving 40% of the initial sucrase activity, while 15 minutes at 48°C inactivated the cell-bound dextransucrase in a cell suspension, leaving 60% of the initial sucrase activity.

The synthesis of alternan from sucrose with heat-treated cell suspensions was studied by Raemaekers (1995). It was found that increasing amounts of leucrose were formed with increasing initial sucrose concentrations. Leucrose is an isomer of sucrose consisting of a glucose and a fructose moiety, both in the pyranose configuration (α-D-glucopyranosyl-(1→5)-D-fructopyranose). At 400 g/l initial sucrose, a maximal alternan yield of 130 g/l was obtained. Leucrose synthesis was highest at 500 g/l sucrose: 130 g/l.

Leucrose

The presence of small quantities of leucrose in the dextran fermentation broth was already reported by Stodola et al. in 1952. In 1986, the German company Pfeifer & Langen was granted a patent for a biotechnical process that permits leucrose to be produced in large quantities from sucrose (Schwengers & Benecke, 1986). The process is based on systematic studies towards the reaction mechanisms of dextransucrase. Leucrose is formed from sucrose by the glucosyl acceptor reaction with dextransucrase (figure 5). Instead of being incorporated into dextran, glucose is used for the synthesis of leucrose. Since fructose is a low-effective glucosyl acceptor, high concentrations of fructose with respect to sucrose have to be used to compete with the normal dextran formation. In the industrial process, almost no net formation of fructose (and dextran) is observed, so that in essence sucrose is converted to leucrose.

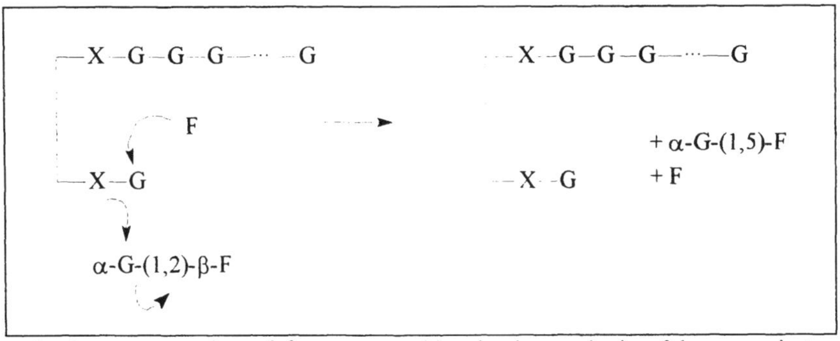

Figure 5. Acceptor reaction of fructose, resulting in the synthesis of leucrose, instead of dextran.

The enzyme dextransucrase is obtained by fermentation with a *L. mesenteroides* strain. After fermentation, the extracellularly secreted enzyme is separated from the cells by centrifugation and the enzyme is further concentrated by ultrafiltration. The enzymatic reaction is performed at 25°C in a concentrated solution of 65% (w/v) consisting of 1/3 sucrose and 2/3 fructose at 25°C. The conversion efficiency is about 90%. After the conversion is finished the leucrose is separated from fructose by chromatography. Leucrose is then obtained in pure form by crystallisation.

Leucrose displays desirable characteristics as a bulk sweetener. It is essentially non-cariogenic and is resorbed easily without any incompatibility problems, as it is broken down to its glucose and fructose units as for sucrose. It has a sweetness of about 50% compared to sucrose. It is a reducing sugar with a high stability towards acid of the glucose-fructose bond, contrary to the acid-labile sucrose.

Gluco-oligosaccharides

In the presence of a glucosyl acceptor (like maltose), dextransucrase and alternansucrase transfer glucosyl residues from sucrose to the acceptor and eventually to its acceptor products, giving rise to a series of gluco-oligosaccharides (oligomers with 2 to 10 glycose units). These glucosyl acceptor reactions compete with each other and with the normal chain elongation reaction.

The glucosyl acceptor act as a nucleophile and releases the bound glucose unit from the enzyme by attack at the reducing end (figure 6). In the case of dextransucrase, most glucosyl transfers occur on the C-6 hydroxyl group of the acceptor, resulting in an α-(1→6) glycosidic linkage. Exceptions to this rule are D-mannose, D-galactose and D-cellobiose (Robyt and Eklund, 1982). Various saccharides and saccharide derivatives are reported to serve as glucosyl acceptors: e.g. maltose, isomaltose, nigerose, methyl-α-D-glucoside and low-molecular-weight dextran (10 kDa). Homologous series of gluco-oligosaccharides may be produced by sequential transfer of glucose units to the initial acceptor and its acceptor product(s). This is the case for maltose and isomaltose with dextran- and alternansucrase. The quantitative effects of various glucosyl acceptors with dextransucrase were studied by Robyt and Eklund (1983).

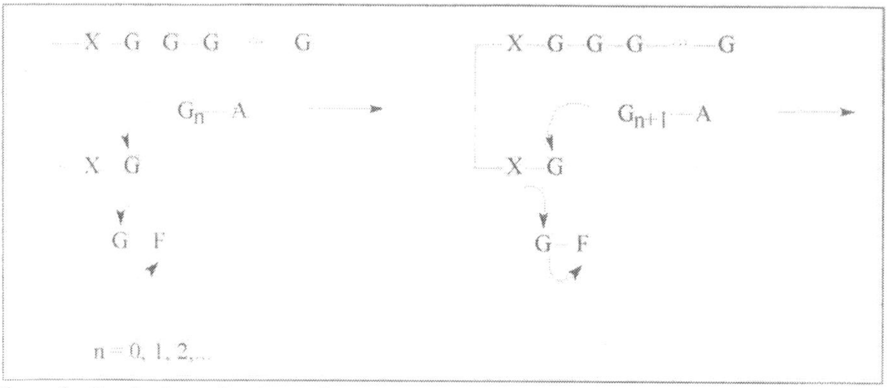

Figure 6. Synthesis of gluco-oligosaccharides by glucosyl acceptor reactions with dextransucrase and alternansucrase. A = Acceptor (glucose, maltose, isomaltose, nigerose,...).

In the presence of sucrose, high-effective acceptors increase the overall reaction rate (i.e. fructose release) and decrease the polysaccharide yield in favour of oligosaccharide synthesis. The most effective glucosyl acceptors for alternansucrase were maltose and nigerose, while maltose and isomaltose were the best acceptor disaccharides with dextransucrase (Côté and Robyt, 1982b). Low-effective glucosyl acceptors, like fructose decrease the overall reaction rate, while the polysaccharide yield is less affected. Glucose, being the acceptor product of water, is also formed in small amounts, both with dextran- and alternansucrase (i.e. the so-called hydrolytic activity).

Côté and Robyt (1982b) studied the acceptor reactions of alternansucrase. Similar as most dextransucrase acceptor products, the alternansucrase acceptor products contained the original acceptor molecule at the reducing end of the molecule. They observed that α-(1→3) linkages were formed only in the case when the acceptor had the non-reducing glucosyl residue linked by an α-(1→6) linkage: glucose gave isomaltose, and not nigerose, maltose gave only panose and not α-D-Glc-(1→3)-α-D-Glc-(1→4)-D-Glc.
In addition, they found that once an α-(1→6) linkage was present at the non-reducing end, two acceptor products were formed, resulting from the transfer of D-glucose to either the C-3 or

the C-6 hydroxyl group of the non-reducing D-glucosyl residue of the acceptor: panose was converted into 6^2-α-D-isomaltosyl-maltose (α-D-Glc-(1→6)-α-D-Glc-(1→6)-α-D-Glc-(1→4)-D-Glc) and 6^2-α-D-nigerosyl-maltose (α-D-Glc-(1→3)-α-D-Glc-(1→6)-α-D-Glc-(1→4)-D-Glc). Isomaltose acceptor products were isomaltotriose and α-D-Glc-(1→3)-α-D-Glc-(1→6)-D-Glc.

The acceptor products of alternansucrase which contained α(1→3) glucosidic linkages were referred to as oligoalternans by Pelenc et al. (1991). The smallest oligoalternan produced from maltose is therefore 6^2-α-D-nigerosyl-maltose. Acceptor products that contained consecutive α(1→6) linkages were called oligodextrans. It was observed that after sucrose depletion, the distribution of gluco-oligosaccharides did not remain constant, due to disproportionation reactions (Lopez-Munguia et al., 1993): especially 6^2-α-D-nigerosyl-maltose is very sensitive to this type of reaction.

The effect of sucrose and maltose concentration on the final concentrations of gluco-oligosaccharides was studied by Raemaekers (1995), using heat-treated suspensions of *L. mesenteroides* NRRL B-1355 cells. At 25 to 150 g/l maltose, initial reaction rates increased 2.2- to 4.5-fold in comparison with maltose-free reactions (between 100 and 500 g/l sucrose). Increasing sucrose:maltose ratios gave rise to an increasing average degree of polymerisation (d.p.) and a decreasing oligosaccharide yield. Additionally, at a constant sucrose:maltose ratio, higher oligosaccharide yields and average d.p.'s were observed at increasing substrate concentrations. Oligosaccharide yields of up to 90% with respect to the initial available glucose, present in maltose and sucrose, were attained.

The mechanism of acceptor reactions is used by Pfeifer & Langen in the synthesis of PL1 dextran with dextransucrase, starting from sucrose and glucose. PL1 dextran (1000 Da) is used in medicine to reduce immunity responses that may be formed by clinical dextrans.

Mannitol

Mannitol is a common polyol derived from mannose and is extensively used. It is now produced by catalytic hydrogenation of fructose using a nickel catalyst and hydrogen gas. This hydrogenation yields mannitol, as well as its isomer sorbitol in about equal amounts, due to the poor selectivity of the nickel catalyst used. This leads to a less efficient production process as sorbitol can be produced cheaper by hydrogenation of glucose. Numerous process improvements to increase the ratio of mannitol/sorbitol formation have been suggested and patented (Makkee et al., 1985).

Mannitol is a common storage product of many fungi and yeasts and its production by fermentation has often been tempted, but the yields and productivities were too low to compete with the chemical hydrogenation process.
Recently, a new fermentation process capable of converting fructose quantitatively to mannitol

has been developed (Soetaert, 1991). The process makes use of the capacity of *L. mesenteroides* to use fructose as an alternative electron acceptor, thereby reducing it to mannitol, with the enzyme mannitol dehydrogenase (E.C. 1.1.1.138). In the process the reducing equivalents are generated by the conversion of glucose into D(-)-lactic acid and acetic acid. Based on the hydrogen balance the following (theoretical) fermentation equation can be derived:

$$2\ \text{Fructose} + \text{Glucose} \longrightarrow 2\ \text{Mannitol} + \text{D(-)-Lactic Acid} + \text{Acetic Acid} + CO_2$$

In this process there is no formation of sorbitol, only of limited quantities of D(-)-lactic acid. D(-)-lactic acid is an interesting by-product that finds application as a chiral synthon for organic synthesis, more particular in the industrial synthesis of chiral phenoxyherbicides. The influence of various factors on the fermentation have been studied in detail (Soetaert et al., 1995). A key factor thereby was the conversion efficiency, defined as the ratio of the produced amount of mannitol versus the consumed fructose. The conversion efficiency could be markedly increased to near quantitative conversion by choosing appropriate fermentation conditions. An optimised batch fermentation resulted in a conversion efficiency of 92%. Fundamental studies towards the fermentation mechanism enabled to devise an optimal fed batch fermentation procedure with automatic feeding strategy (fig. 7).

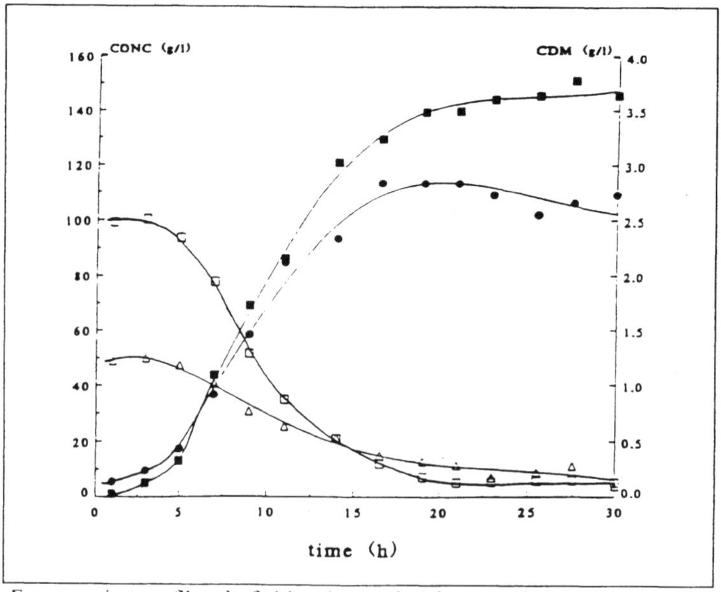

Figure 7. Fermentation profile of a fed-batch mannitol fermentation: (●) Cell Dry Mass (g/l); (△) Glucose (g/l); (□) Fructose g/l; (■) Mannitol (g/l).

A very fast and complete conversion is reached in less than 24 hours. The conversion

efficiency is 94% and the maximal conversion rate is 11 g mannitol/(L.h). The final mannitol concentration is 150 g/l, close to its solubility limit of 180 g/l. Thus a very high mannitol concentration can be produced in high yield using a fed-batch strategy in less than 24 hours of fermentation.
By selection of a genetically modified strain, better process characteristics could be obtained, resulting in quantatitative conversion and a further concentration increase up to 185 g/l mannitol by increasing the fermentation temperature to 35 °C.
The downstream-processing of the fermentation broth has also been optimised. The use of electrodialysis followed by crystallisation results in cost-effective recovery of highly pure crystalline D-mannitol and D(-)-lactic acid. The new process thus offers an attractive alternative to the presently used industrial synthesis routes of mannitol.

Sucrose phosphorylase

Sucrose phosphorylase (E.C. 2.4.1.7) is an intracellular enzyme obtained from *L. mesenteroides* (Vandamme et al., 1987). It is a glucosyltransferase transferring glucose from sucrose to a number of acceptor molecules, phosphate being the most effective acceptor, forming α-D-glucose-1-phosphate.

$$\text{Sucrose} + P_i \rightleftharpoons \alpha\text{-D-Glucose-1-phosphate} + \text{D-Fructose}$$

At this moment, α-D-glucose-1-phosphate is an expensive fine chemical. Technology has been developed - based on sucrose phosphorylase - to turn α-D-glucose-1-phosphate into a cheap commodity chemical. α-D-Glucose-1-phosphate is essentially a C_1-protected glucose molecule. It can be easily converted enzymatically into glucose-6-phosphate, a C_6-protected glucose molecule. Both can be used in glucose derivatisation reactions. For instance α-D-glucose-1-phosphate can be oxidised at C_6 position to glucuronic acid-1-phosphate. This can then be easily hydrolysed to glucuronic acid, an interesting intermediate for further synthesis reactions. α-D-Glucose-1-phosphate is used in infusion solutions as a supplier of glucose (energy) and phosphate. It can be used in immobilised cell reactors with calcium alginate cell carriers to supply the cells with phosphate, that would otherwise (in a free form) complex the calcium, thus destabilising the beads.
Sucrose phosphorylase can also be used for the synthesis of new disaccharides. Fructose and other similar carbohydrates such as L-sorbose and xylitol behave as good acceptors of the glucose moiety (Kitao & Sekine, 1992). Sucrose phosphorylase for instance also catalyses the following reaction:

$$\alpha\text{-D-glucose-1-phosphate} + \text{xylitol} \longrightarrow \text{4-O-}\alpha\text{-D-glucopyranosyl-xylitol} + P_i$$

In this way such complex disaccharides can be conveniently synthesised with the use of a single enzyme.
The process to produce α-D-glucose-1-phosphate starts with a fermentation using *L.*

mesenteroides. After this, the cells containing the enzyme are collected by centrifugation. The cells are then permeabilised and immobilised in gelatine beads. An immobilised sucrose phosphorylase is thus obtained that has a good mechanical stability and a half life time of 40 days. These beads are used in a continuous mode in a column. The substrate consisting of sucrose and phosphate is converted to α-D-glucose-1-phosphate and fructose. The phosphate conversion efficiency is about 80%. The substrate is then separated by chromatography, either in batch (semi-continuous) or continuous mode. This results in two product streams, a first stream consisting mainly of fructose, which can be marketed as fructose syrup. The second is the α-D-glucose-1-phosphate stream from which crystalline α-D-glucose-1-phosphate is readily obtained by crystallisation in high yield. Unconverted sucrose and phosphate are recycled, resulting in a very efficient process.

Conclusion

Although *L. mesenteroides* has well-recognised importance as starter culture in the food industry, these bacteria have rarely been used as production strains in industrial bioconversions, dextran synthesis being up till recently the only well-documented industrial use of the micro-organism. The here presented processes stress the underestimated potential of this lactic acid bacterium as a potent tool in tailor made carbohydrate bio-modification. Highly selective reactions can be performed with the micro-organism and its enzymes, resulting in very efficient, high yielding and very fast bioconversions of high concentrations of common sugars into specialty carbohydrates.

References

Brooker BE (1979) Electron microscopy of the dextrans produced by lactic acid bacteria. In: Berkely RCW, Gooday GW, Ellwood DC (eds). Microbial polysaccharides and polysaccharidases. London, Academic Press: 85-115

Brown DE, Mc Avoy A (1990) A pH-controlled fed-batch process for dextransucrase production. J Chem Technol Biotechnol 48 : 405-414

Côté GL (1991) Bacterial enzyme changes beet sugar into a versatile dextran. Industrial Bioprocessing, October, p 4

Côté GL (1992) Low-viscosity α-D-glucan fractions derived from sucrose which are resistant to enzymatic digestion. Carbohydr Polym 19 : 249-252

Côté GL, Bielly P (1994) Enzymically produced cyclic α-1,3-linked and α-1,6-linked oligosaccharides of D-glucose. Eur J Biochem 226 : 641-648

Côté GL, Robyt JF (1982a) Isolation and partial characterization of an extra-cellular glucansucrase from *Leuconostoc mesenteroides* NRRL B-1355 that synthesizes an alternating (16), (13)-α-D-glucan. Carbohydr Res 101 : 57-74

Côté GL, Robyt JF (1982b) Acceptor reactions of alternansucrase from *Leuconostoc mesenteroides* NRRL B-1355. Carbohydr Res 111 : 127-142

De Belder AN (1990) Dextran. Pharmacia LEO Therapeutics, Uppsala, Sweden

Goldstein IJ, Whelan WJ (1962) Structural studies of dextrans Part 1: A dextran containing α-1,3-glucosidic linkages. J Chem Soc : 170-175

Jeanes A, Haynes WC, Wilham CA, Rankin JC, Melvin EH, Austin MJ, Cluskey JE, Fisher BE, Tsuchiya HM, Rist CE (1954) Characterization and classification of dextrans from ninety-six strains of bacteria. J Am Chem Soc 76 : 5041-5052

Kim D, Robyt JF (1994) Production and selection of mutants of *Leuconostoc mesenteroides* constitutive for glucansucrases. Enzyme Microb Technol 16 : 659-664

Kitao S, Sekine H (1992) Transglucosylation catalysed by sucrose phosphorylase from *Leuconostoc mesenteroides* and production of glucosyl-xylitol. Biosci Biotech Biochem 56 : 2011-2014

Makkee M, Kieboom APG, Van Bekkum H 1985 Production methods of D-mannitol. Starch/Stärke 37 : 136-141

Misaki A, Torii M, Savai T, Goldstein IJ (1980) Structure of the dextran of *Leuconostoc mesenteroides* B-1355. Carbohydr Res 84 : 273-285

Pelenc V, Lopez-Munguia A, Remaud M, Biton J, Michel JM, Paul F, Monsan P (1991) Enzymatic synthesis of oligoalternans. Science des Aliments 11 : 465-476

Raemaekers M (1995) *In vitro* synthesis of alternan and oligosaccharides via *Leuconostoc mesenteroides* NRRL B-1355 alternansucrase. PhD thesis, University of Ghent, Belgium

Robyt JF (1992) Structure, biosynthesis, and uses of nonstarch polysaccharides: dextran, alternan, pullulan, and alginate. In: Alexander RJ and Zobel HF (eds) Developments in carbohydrate chemistry. St Paul, Minnesota, The American Association of Cereal Chemists: 261-292

Robyt JF, Eklund SH (1982) Stereochemistry involved in the mechanism of action of dextransucrase in the synthesis of dextrans and the formation of acceptor products. Bioorganic Chemistry 11 : 115-132

Robyt JF, Eklund SH (1983) Relative, quantitative effects of acceptors in the reaction of *Leuconostoc mesenteroides* B-512F dextransucrase. Carbohydr Res 121 : 279-286

Robyt JF, Taniguchi H (1976) The mechanism of dextransucrase action. Arch Biochem Biophys 174 : 129-135

Robyt JF, Kimble BK & Walseth TF (1974) The mechanism of dextransucrase action: direction of dextran biosynthesis. Arch Biochem Biophys 165 : 634-640

Sawai T, Tohyama T, Natsume T (1978) Hydrolysis of fourteen native dextrans by *Arthrobacter* isomaltodextranase and correlation with dextran structure. Carbohydr Res 66 : 195-205

Schwengers D, Benecke H (1985) Leucrose : sweetener and its use. European patent 185 302. Chem Abstr 105 : 77815 (1986)

Seymour F, Knapp R (1980) Structural analysis of α-D-glucans by ^{13}C-nuclear magnetic resonance, spin-lattice relaxation studies. Carbohydr Res 81 : 67-103

Seymour FR, Slodki ME, Plattner RD, Jeanes A (1977) Six unusual dextrans: methylation structural analysis by combined GLC-MS of per-O-acetyl-aldononitriles. Carbohydr Res 53 : 153-166

Soetaert W (1991) Synthesis of D-mannitol and L-sorbose by microbial hydrogenation and dehydrogenation of monosaccharides. PhD-thesis, University of Ghent, Belgium

Soetaert W, Buchholz K, Vandamme EJ (1995) Production of D-mannitol and D(-)-lactic acid by fermentation with *Leuconostoc mesenteroides*. Agro-Food-Industry Hi-Tech 6 (1) : 41-44

Stodola FH, Koepsell HI, Sharpe ES (1952) A new disaccharide produced by *Leuconostoc mesenteroides*. J Am Chem Soc 74 : 3202-3203

Vandamme EJ, Soetaert W (1995) Biotechnical modification of carbohydrates. FEMS Microbiol Rev 16 : 163-186

Vandamme EJ, Van Loo J, Machtelinckx L, De Laporte A (1987) Microbial sucrose phosphorylase : fermentation process, properties, and biotechnical applications. Adv Appl Microbiol 32 : 163-201

Role of Lactic Acid Fermentation in Bioconversion of Wastes

Antonio M. Martin
Department of Biochemistry
Memorial University of Newfoundland
St. John's, Newfoundland, Canada A1B 3X9

1. Introduction

Among the biological methods for transforming organic wastes, two types of fermentation processes are important, as they allow the recovery of nutrients for use in the production of feed, fertilizers, and food. The term "fermentation" will be used in this regard in the broad sense used by Prescott & Dunn (1959): "a process in which chemical changes are brought about in an organic substrate through the action of enzymes elaborated by microorganisms." One of those processes includes the hydrolysis (if required) of biological polymers to simple molecules, and the subsequent growth of microbial populations utilizing organic waste products as their source of energy and nutrients. As a consequence of this kind of process, products such as "single cell protein", "microbial biomass protein", "single cell oil", or such metabolites as alcohol and others, associated or not with the microbial growth, can be produced. Tewari *et al.* (1988) evaluated acids and cellulase enzymes for the hydrolysis of lignocellulosic residues. Processes for the growth of microbial biomass are generally aerobic, and there is an extensive literature about them (Moo-Young & Gregory, 1986).

NATO ASI Series, Vol. H 98
Lactic Acid Bacteria:
Current Advances in Metabolism, Genetics and Applications
Edited by T. Faruk Bozoğlu and Bibek Ray

The other main method consists of processes involving lactic acid fermentation. Traditionally, the role of lactic acid bacteria in the preservation of organic products has been acknowledged. The most classical examples of the use of lactic acid bacteria and lactic acid fermentation processes are their role in the preservation of foods, such as fermented milk, sauerkraut, cucumbers, and sausages, among others. In those processes, a stable product is produced due to the rapid lactic anaerobic fermentation, which reduces the pH to levels inhibitory to the growth of spoilage microorganisms (Brookes & Buckle, 1992). In addition, this fermentation enhances some other organoleptic characteristics of food products, such as flavour and texture. Mookherjee (1976) indicated that the microorganisms responsible for this fermentation are bacteria which predominantly convert sugars to lactic acid with minimal changes in other food components. One of the effects of the acid produced is to suppress the growth of microorganisms that could be deleterious to the food material (Pederson, 1978). Ingram *et al.* (1956) indicated that the following three effects should be taken into consideration to understand the preservative action of acidic substances: the pH effect, the effect of the dissociation of the acid, and specific effects which depend on the molecule of the preservative acid.

This operation is also valid for the preparation of animal feed by the preservation of by-products and wastes of animal and plant origin (McCullough, 1978; Woolford, 1984; Nash, 1985). These processes are generally identified as ensiling, or silage fermentation. In some instances the products from this kind of precess can also be employed as fertilizer. Also, other processes such as proteolysis could contribute to the characteristics of the final product.

1.1 ENSILING PROCESS

Ensiling is the process whereby materials, mostly of biological origin, which due to their composition and moisture content could be prone to decomposition by aerobic microorganisms, are stored anaerobically. The resulting product is called silage, and the structure in which the process occurs is termed a silo (Woolford, 1984). Silage is generally employed as animal feed.

During the ensiling process an acid fermentation occurs in which bacteria produce acids anaerobically. The outcome is a reduction of the pH of the medium, which inhibits the growth of spoilage microorganisms.

The term silage also applies to the product when microbial activity in a material such as grain or fodder has been prevented by the addition of acids or sterilizing agents at the ensiling. This treatment tends to decrease the loss of nutrients in the material. Animal by-products can be ensiled also, and the term silage is applied as well to the liquid resulting from the digestion of fish biomass under acid conditions (Woolford, 1984).

2. Lactic acid bacteria

The first attempt at a classification of the lactic acid bacteria was made in 1919 by Orla-Jensen, who defined them as micro-aerophilic, gram-positive, non-spore-forming organisms which ferment sugars to lactic acid. This author classified them as a) homofermentative lactic acid bacteria (homolactics), which ferment hexoses predominantly to lactic acid, and b) heterofermentative lactic acid bacteria (heterolactics), which ferment hexoses to lactic acid and other products such as ethanol and acetic acid (Orla-Jensen, 1919; Beck, 1978).

Commonly, the genera *Lactobacillus*, *Leuconostoc*, *Pediococcus* and *Streptococcus* have been considered representative of the lactic acid bacteria group. Axelsson (1993), in his study of the classification of the physiology of the lactic acid bacteria group, indicated that its boundaries have been the object of controversy. Levin (1994) presented the characteristics which can be considered common to the most important groups of lactic acid-producing bacteria: *Lactobacillus*, *Pediococcus* and *Streptococcus*.

It has been recognized that lactic acid bacteria create an unfavourable environment for food-spoiling or pathogenic microorganisms (Mossel, 1971). The effect of the lactic acid bacteria is primarily related to undissociated organic acids produced in the fermentation processes, although other mechanisms may be involved. Among the latter, some catalase-deficient lactic acid bacteria systems can produce toxic levels of hydrogen peroxide (Dahiya & Speck, 1968; Price & Lee, 1970).

It has been well documented that lactic acid bacteria require several nutrients for their metabolism, including amino acids, vitamins and metal ions. However, studies of their mechanisms of nutrient intake have been restricted to carbohydrates and, to a minor degree, to proteins. The first enzymes from lactic acid bacteria to make contact with milk proteins are extracellular proteases and, possibly, peptidases. Once inside the cell, peptides are hydrolysed by cytoplasmic peptidases (Marshall & Law, 1984).

Lactobacillus have specific characteristics that make them suitable for industrial processes, namely that they are aerotolerant, non-pathogenic, do not require aeration for growth, do not produce a toxic product or toxin, resist low pH, grow relatively rapidly, and their cultures are fairly stable.

Though the specific growth requirements vary among different organisms, *Lactobacillus* have complex nutritive requirements for amino acids, fatty acids, fermentable carbohydrates, nucleic acid derivatives, peptides, salts, and vitamins (Anonymous, 1974). Generally, however, the amino acids and other growth factors required by the lactic acid bacteria are available in adequate amounts in the substrates to which they are exposed, such as proteinaceous materials. This genus is probably the most commonly used in lactic acid fermentation processes.

Ray (1992) presented a thorough study of lactic acid bacteria as food biopreservatives, including an informative table indicating the current and previous names for the most important species of lactic acid bacteria employed in controlled fermentation of food materials.

2.1 CHARACTERISTICS OF MICROORGANISMS FOR SILAGE PROCESSES

The characteristics that a given microorganism should possess in order to be used in silage have been described by many authors. Among the lactic acid bacteria, *Lactobacillus plantarum* has been considered one of the most suitable microorganisms for silage inoculation. Because production of lactic acid by *L. plantarum* is only accelerated after the pH falls below 5, a mixed culture with another lactic acid-producing bacteria active in the pH 5 to 6.5 range has been suggested, such as *Streptococcus faecalis* (McDonald, 1981). Table 1 presents some significant species of microorganisms involved in the silage process, and some of their characteristics. The acid production by lactic acid bacteria contributes to the reduction of the growth of competing microbial populations and to the destruction of harmful microorganisms. Table 2 lists some of these.

Table 1: Microbial species important in the silage process*

Species	Distinguishing characteristics
Lactobacillus acidophilus *Lactobacillus delbrueckii*	homofermentative lactobacilli, produce lactic acid via the Embden-Meyerhof-Parnas pathway
Lactobacillus casei *Lactobacillus coryniformis* *Lactobacillus curvatus* *Lactobacillus farciminis* *Lactobacillus plantarum*	homofermentative / facultatively heterofermentative lactobacilli, based on previous characterization of streptobacteria, ferment pentoses via phosphoketolase induction (Brookes & Buckle, 1992)
Lactobacillus bifermentans *Lactobacillus brevis* *Lactobacillus buchneri* *Lactobacillus cellobiosus* *Lactobacillus fermentum* *Lactobacillus viridescens*	obligate heterofermentative lactobacilli, all metabolic pathways involving phosphoketolase
Leuconostoc cremoris *Leuconostoc dextranicum* *Leuconostoc lactis* *Leuconostoc mesenteroides* *Leuconostoc paramesenteroides*	obligate heterofermentative leuconostocs (cocci), metabolise glucose via a combination of hexose-monophosphate and phosphoketolase pathways
Pediococcus acidilactici *Pediococcus cerevisiae* *Pediococcus dextrinicus* *Pediococcus pentosaceus*	homofermentative pediococci. *P. acidilactici* and *P. pentosaceus* grow rapidly, with final pH ≤4. *P. acidilactici* tolerates 50°C (Garvie, 1986).
Streptococcus cremoris *Streptococcus (Enterococcus) faecalis* *Streptococcus (Enterococcus) faecium* *Streptococcus lactis* *Streptococcus raffinolactis*	homofermentative streptococci. Though the two *Enterococcus* spp. hydrolyse arginine, only *S. faecalis* uses arginine as an energy source. Metabolism of arginine by *S. faecalis* seems to be similar to that of *S. lactis* (Deibel, 1960). Fermentation of lactose and sucrose by *S. lactis* is plasmid mediated (Levin, 1994).

* From Brookes & Buckle (1992); McDonald (1981); Levin (1994).

Table 2: Competing microbial populations and harmful microorganisms destroyed during ensiling

Competing genera (representative species), in order of succession*	Displaced genera (representative species)**
Streptococcus (*lactis*)	*Aeromonas* (*salmonicida*)
Leuconostoc (*mesenteroides*)	*Bacillus* (*cereus*)
Pediococcus (*acidilactici*)	*Clostridium* (*botulinum, perfringens, sporogenes*)
Lactobacillus (*curvatus, plantarum*)	*Escherichia* (*coli*)
	Listeria (*monocytogenes*)
	Mycobacterium
	Salmonella
	Staphylococcus (*aureus*)
	Vibrio (*anguillarum*)

* From Woolford (1984)

** From Lindgren (1992).

Lindgren & Clevström (1978*a,b*) reported antibacterial activities of microorganisms isolated from lactic acid silages of fish and forage.

Processes employing lactic acid bacteria commonly utilize several of the lactic acid-producing species. Also, many of the applications of lactic acid bacteria and lactic acid fermentation processes involve their interrelation with other types of microorganisms. Therefore, the study of lactic acid bacteria in the context of mixed cultures is important. Vandevoorde *et al.* (1992) discussed the factors controlling the competitive behaviour of lactic acid bacteria in mixed cultures.

The ensiling efficiency is generally given by the ratio of lactic acid to butyric acid produced. A "high quality" silage is produced if lactic acid is predominant, and this is characterized by a resulting low pH. The inverse ("low quality" silage and a higher pH) happens when butyric acid is predominant (Woolford, 1984). This is a consequence of insufficient acidity being produced. Bacteria of the genus *Clostridium* will ferment sugars and lactic acid to butyric acid, which is weaker than lactic and acetic acids. Woolford (1994) indicates that the liberation of CO_2 will contribute to an increase of the pH, which will favour proteolytic clostridia. The latter will produce ammonia, amines and amides from proteins, further increasing the pH.

2.2 INOCULANTS FOR SILAGE

Raw materials low in lactic acid bacteria content generally benefit from inoculants with a good degree of viability. Their selection (species and concentration) is therefore an important consideration when designing an ensiling process (Brookes & Buckle, 1992). Nevertheless, there are reports that, if the raw material already has a high concentration of lactic acid bacteria, inoculants will not improve the process (Bolsen *et al.*, 1987). Also, it has been reported that inoculants will not be effective if adverse conditions for the ensiling process are present (Done, 1986). Table 3 summarizes some of the characteristics of inoculants.

3. Raw materials for silage processes

There is a great emphasis on the development of modern methods for the bioconversion of waste materials to industrial products (Martin, 1991). Litchfield (1987) reviewed biological methods for recovering and utilizing agricultural and food processing wastes.

A variety of plant materials can be employed in making silage. The main prerequisites for this are: a) the crop should contain an adequate concentration of water soluble (fermentable) carbohydrates (WSC); b) it should have a relatively low buffering capacity; and c) the dry matter content in the fresh crop should be above 200 g kg^{-1} (McDonald, 1981).

Because newly harvested crops are not abundant in microbial populations, the biochemical changes that occur during the first stages of ensiling are mostly due to activities of the plant enzymes, which are responsible for respiration and proteolysis. Table 4 summarizes some of the plant waste materials preserved by ensiling for use as animal feed.

Besides the use of nutrient supplements and inoculants, the use of other silage additives as means of enhancing the lactic acid fermentation is important. Table 5 summarizes the characteristics of some of them.

The silage fermentation of wastes and by-products is generally accomplished by a rapid reduction in pH to below 4.5. The acidity created by the fermentation should also offer a safety margin by creating the

Table 3: Some desirable characteristics of inoculants*

Categories	Characteristics
species criteria	homofermentative
	rapid growth
	rapid lactic acid production
	competitive with native species
	tolerant to pH ≈4
	viable at temperatures up to 50°C
	grow well on substrate provided
	ferment a variety of sugars
operating criteria	provide adequate cell concentration
	provide complementary species
	available in powdered or granular form
	stable when stored
	no further degradation of organic acids

* From Brookes & Buckle (1992).

minimum inhibitory concentration (MIC) of lactic acid for the preservation of the product against further contamination (Lindgren, 1992). The important conditions for a proper fermentation were discussed by Owens & Mendoza (1985). Keller & Gerhardt (1975) developed a mathematical model to simulate the continuous fermentation of whey for the production

Table 4: Some plant waste materials preserved by ensiling

Material	References
fibrous residues from sugar beet and sugar cane	McDonald *et al.* (1991)
sugar beet pulp	Watson & Nash (1960)
wastes from vegetable processing	Le Dividish *et al.* (1976); Moon (1981*a,b*); Ashbell & Lisker (1987)

of a protein-rich feed supplement. The authors found that lactic acid was produced mainly as a function of cell maintenance instead of by the metabolic reactions associated with growth. Samuel *et al.* (1980) developed unstructured kinetic models for cell growth, substrate consumption and lactic acid production in a batch system for the fermentation of crude sorghum extract. The bioconversion of potato processing wastes into a protein-enriched feed supplement has been reported by Forney & Reddy (1977). By neutralizing the lactic acid produced during the fermentation with ammonia, ammonium lactate was produced, which proved to be a good nitrogen source for ruminants.

Table 5: Additives used in silages

Additive(s)	Action	Reference(s)
amylases / amylolytic enzymes	break down starches into fermentable carbohydrates	Lassén *et al.* (1990); Brookes & Buckle (1992)
benzoic, propionic & sorbic acids	inhibit yeast & mould growth	Lindgren (1992)
clostridiaphages	kill strains of *Clostridium* spp.	Brookes & Buckle (1992)
ethoxyquin	prevent oxidation	Levin (1994); Martin & Bemister (1994)
formic acid & formaldehyde mixture	lower pH, inhibit spoilage organisms	Brookes & Buckle (1992)
formic, sulphuric, acetic, propionic acids	lower pH, inhibit spoilage organisms	Brookes & Buckle (1992)
malt enzymes	convert cereals to fermentable sugars	Nilsson & Rydin (1965)
plant fibre degrading enzymes (cellulases &/or hemicellulases)	rupture cell walls of plant materials to release fermentable carbohydrates	Setälä (1988-1989); Brookes & Buckle (1992)
potassium sorbate or propionic acid	inhibit moulds	Levin (1994)
sorbic & propionic acids	inhibit yeasts	Lindgren & Pleje (1983)
trypsin inhibitors	lessen proteolysis	Lindgren & Pleje (1983)

4. Fermentation of proteinaceous food wastes of animal origin

Two important factors are to be considered for the success of this process (Lindgren, 1992): a) the substrate composition; and b) the lactic acid bacteria present.

4.1 SUBSTRATE COMPOSITION

Proteinaceous food wastes of animal origin are generally characterized by low levels of carbohydrates content. This tends to thwart the lactic fermentation. Some methods have been developed to overcome this limitation, based on the addition of nutrient supplements. This could as well be conducted in any silage fermentation with the objective of strengthening the ensiling process. Table 6 presents some of the work reported on the addition of carbohydrate supplements. Most of the examples presented concern processes based on materials of animal origin, to highlight them.

The high concentration of protein and the occurrence of proteolytic reactions, which are desired in the production of silage of proteinaceous food wastes of animal origin, will tend to raise the pH of the medium (due to the eventual liberation of ammonia). This creates the need for producing additional lactic acid. For example, fish biomass has a high buffering capacity, due to the presence of protein and minerals. Therefore, large quantities of acid are required to produce a pH below 4, and additional amounts of carbohydrates must be added to this end (Raa *et al.*, 1983). This was also observed by Lindgren (1992), who indicated that fish offal is a high-buffering substrate, in contrast with grass, which is low-buffering.

Table 6: Supplements used to overcome low carbohydrate contents in silages

Method / supplement	Reference
malt and oatmeal	Nilsson & Rydin (1963)
cereal flour enriched with enzyme-rich brewers malt	Nilsson & Rydin (1965)
molasses	Roa (1965), Kompiang *et al.* (1980), Anonymous (1988)
tapioca starch and a rice and amylolytic mould mix (ragi)	Stanton & Yeoh (1977)
cereal flour	Lindgren & Pleje (1983)
cassava and glucose	Twiddy *et al.* (1987)
blackstrap molasses, corn syrup	Levin (1994)
peat extracts	Martin & Bemister (1994)

4.2 THE LACTIC ACID BACTERIA PRESENT

The population concentration and activity of the lactic acid-producing bacteria initially present in the substrate is of importance for a rapid and effective lactic acid production. Generally, wild-type microbial populations have been commonly present in these processes. Hassan & Heath (1986) reported that fish waste was fermented with levels of 10^3 cells/g, which is low compared to the level of 10^5 to 10^6 cells/g suggested by Lindgren (1992). The effects of prefermentation to increase the concentration of lactic acid bacteria before the addition of the fish waste have been reported (Cooke *et al.*, 1987). Lindgren & Pleje (1983) propagated *Pediococcus acidilactici* and *L. plantarum* in a cereal mixture before the addition of the fish waste. The prefermented cereals provided a concentration of 2×10^8 cells/g, producing a rapid fermentation process which resulted in a pH below 4.5 within 30 hours at 24°C. The authors also indicated that a lower level of lactic acid bacteria retarded the decrease of the pH value. Twiddy *et al.* (1987) evaluated the use of rice or cassava as a carbohydrate source with and without a 1-day prefermentation prior to mixing with the minced fish. Also, different fish/carbohydrate proportions were studied in the range of 20 - 100 % w/w of minced fish. The authors found that the use of prefermented cassava (20 % w/w) resulted in a rapid fermentation; the pH declined to less than 4.5 and the ratio of lactic acid bacteria to spoilage bacteria exceeded four log cycles of growth within 48 h. The addition of sugar (2 % w/w glucose) was required in order to prevent an undesirable pH increase after the first 2 days of fermentation.

Deibel & Niven (1960) studied a group of bacteria present in cured meat and meat curing brines, suggesting that they be identified as

Pediococcus homari. This species exhibited homofermentative characteristics, producing *dextro*-rotatory lactic acid from the fermentation of sugar. Adams *et al.* (1987) studied the lactic acid fermentation parameters in a minced fish-salt-glucose system.

5. Fermentation process characteristics of proteinaceous silage

A new process, derived from the traditional ensiling process, was applied for the wet conservation of slaughterhouse by-products with high protein and fat contents. A "meat pulp" was produced by the addition of lactic acid bacteria and fermentable carbohydrate to previously minced and autoclaved by-products. Approximately 3 to 5 % of lactic acid was produced in 3 to 4 days (Szakács *et al.*, 1988). The authors reported that two plants were producing this product, which was mixed with plant materials and fed to pigs. Sander *et al.* (1995) evaluated the fermentation of poultry carcasses with *Lactobacillus*, employing dried whey and cornmeal as nutrient additives for the process.

6. Bioconversion of fisheries wastes

The use of fisheries wastes and of non-utilized species has been generally conducted using processes with a low level of technological sophistication. Fisheries processing operations have lagged behind the level of development which has characterized other food industries. Increased attention has been recently focused on the application of biotechnological methods to the seafood industry with the objective of increasing its productivity. Pollution legislation and a better knowledge of the value of the marine biomass should encourage its recovery and the development of new products from it. The existing and potential technologies for the bioconversion of fisheries by-products have been presented by Martin & Patel (1991). Martin (1994) published a comprehensive treatise on the application of biotechnology to the seafood industry.

Biological processes imply the use of enzymes or microorganisms. The direct application of the latter in fisheries processing has been conducted to a lesser extent than the application of enzymes. Enzymes of varied origin have found application in such fish processing operations as liquefaction of fish protein and removal of skins. Microbial enzymes are probably the group of enzymes with more future potential applications in the food industry, including the seafood industry. Microbial enzymes are present, among enzymes of other origin, in the fermentation process for the production of fish sauces, although their role is still not well understood. It has been indicated that bacterial action is important in the development of fish sauce flavour.

Proteolytic microorganisms could be employed in the production of fish protein hydrolysates, solubilizing the fish biomass protein. Microorganisms of the *Bacillus* spp. were found to be efficient hydrolysing agents, resulting in the solubilization of a large portion of the fish protein.

6.1 FISH SILAGE

Fish silage is the hydrolysed product that results from fish wastes being ground and mixed with acid. The proteolytic enzymes present in the fish biomass break down the protein, liquifying it. The main role of the acid is to prevent microbial spoilage. This product, a liquified fish biomass rich in protein, is called fish silage, although it has been noted that it is in some way different from what customarily the term silage has come to mean (Woolford, 1984). Another option is to produce lactic acid by the addition of a carbohydrate source and lactic acid bacteria. This biological method more resembles the traditional ensiling process. The final product, from both the chemical and biological procedures, is a high quality protein and mineral source for animal feeding. It can be used to supplement the diets of several animal species (Haard *et al.*, 1985). Table 7 presents some of the positive characteristics of fish silage.

In general, two main processes are employed in the production of fish silage: a) preservation by acid addition, and b) fermentation by lactic acid bacteria. Due to its objectives, this paper will concentrate on the study of the lactic acid fermented silage. Figure 1 presents a diagram of the production of lactic acid fish silage.

6.1.1 *Comparison between fish meal and fish silage.* Fish meal is a product obtained by heat treatment, drying and grinding of fish biomass, including fish waste. Fish meal is used as a protein source in animal feeding in the aquaculture and poultry industries, among others. Because it is generally capital- and energy-intensive, the economy of this process is not always good. Fish meal is probably one of the most expensive ingredients in the formulation of feed diets.

Hardy & Masumoto (1990) indicated that the demand for high-quality products of marine origin for fish cultivation operations is expected to increase, therefore novel alternatives to fish meal, such as fish silage, will be needed to satisfy the increasing demands of the aquaculture industry. It has been acknowledged that by producing microbiologically stable products, the fermentation of fishery by-products and of biomass from

Table 7: Benefits of the production of fish silage

Advantage	Reference
does not putrefy, has a fresh, acidic smell	Raa & Gildberg (1982)
is almost sterile	Raa & Gildberg (1982)
production can be adjusted to the fish waste supply, and can be economical where fish waste availability is not constant	Raa & Gildberg (1982)
can be sun-dried without being infested by flies	Raa & Gildberg (1982)
simple, low capital technology	Dong *et al.* (1993)
good storage properties	Dong *et al.* (1993)
process can use biological methods, which could save the expense of mineral acids, and avoid the hazards and difficulties of using them	Dong *et al.* (1993)

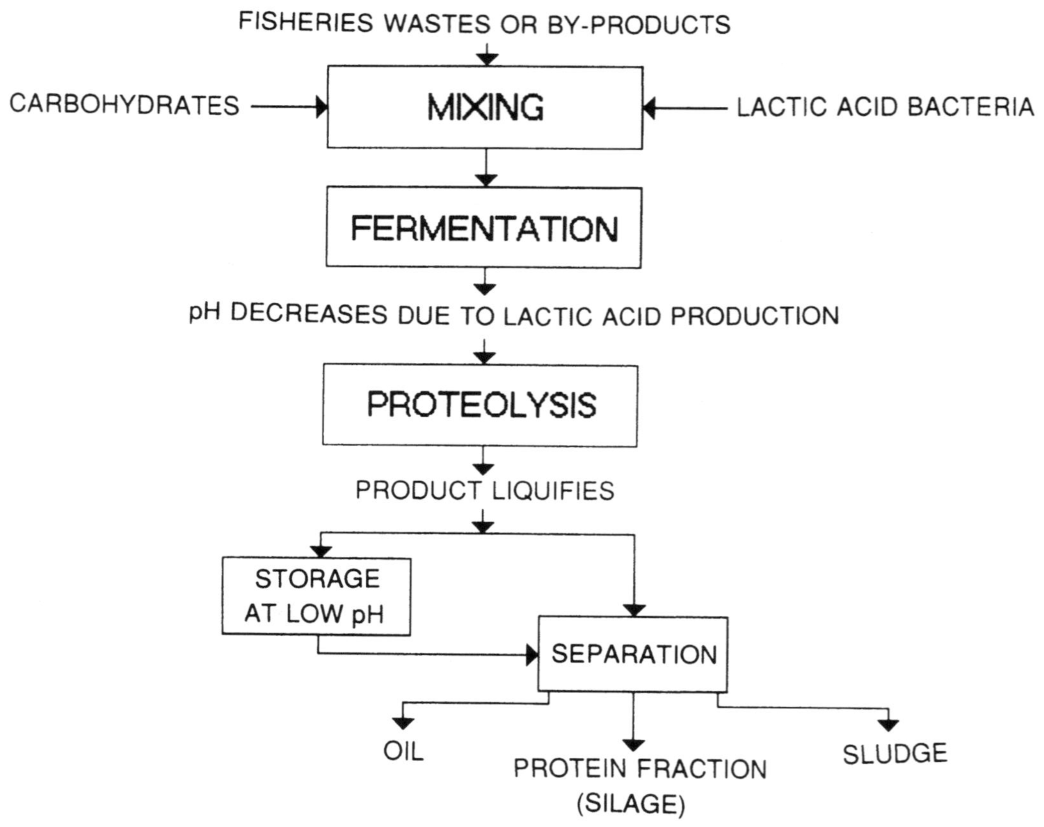

Figure 1: Diagram of the production of lactic acid fermented fish silage.

underutilized species could represent an alternative protein source to fish meal (Dong *et al.*, 1993). The same authors prepared salmon viscera silage using cultures of *L. plantarum* and *S. faecium*, and also by adding acid. They reported advantages in the use of bacteria instead of acid, such as: economic savings by avoiding the need to purchase acid; ease of maintenance and reproduction of bacterial cultures; avoidance of the need to handle acids; and that a lower moisture level results from the lactic acid silage process, facilitating further drying.

6.2 LACTIC ACID FERMENTED FISH SILAGE

It has been reported that the use of fermentation in fish silage production started in Scandinavia (Disney *et al.*, 1977). Table 8 presents some of the initial works reported on the production of fermented fish silage.

In this process, the fermentation is initiated by the addition of lactic acid bacteria and fermentable carbohydrate to fish biomass that has been previously minced. As indicated above, the bacteria will convert the carbohydrate to acid, lowering the pH and allowing the final liquid product to resist microbial spoilage.

During liquefaction, connective protein tissues are broken down by the enzymes present in the fish muscle or viscera. These enzymes are proteases that break down proteins into peptides and individual amino acids. Lipases that break down fats and oils into free fatty acids are also present (Jangaard, 1987). The majority of proteins are converted to short peptides, a portion of which may be further hydrolysed to free amino acids. This autolysis process results in the fraction of nitrogen present as free amino acids increasing while the fraction present as polypeptides decreases. In spite of this, there is always a quantity of protein that is not

Table 8: Early work on the production of fermented fish silage

Material added to fish	References
sulphuric acid and molasses	Peterson (1953)
molasses and lactic acid bacteria	Peterson (1953)
cereal meals as the starch source and malt meal as the amylase source	Nilsson & Rydin (1965)
tapioca starch and ragi (mixed culture of amylolytic moulds grown on rice)	Stanton & Yeoh (1977)

hydrolysed by the enzymes. Jangaard (1987) indicated that probably the rate of reaction decreases when the most favourable binding or splinting sites have been occupied.

Although lactic acid bacteria are natural inhabitants of fish, they are present in low numbers, for example 10^1 - 10^4/g (Knøchel, 1981). Fish also contain only small amounts of free sugar, which is the essential substrate for growth of such bacteria (Raa *et al.*, 1983; Mackie, 1971). Therefore, to preserve fish or any other animal product by fermentation, it is essential to add a fermentable carbohydrate source, and a starter culture of proper lactic acid bacteria. Both spoilage bacteria and lactic acid bacteria will contribute to the initial acid production because the conditions are anaerobic and sugars are available, but growth of the lactic acid bacteria will be favoured as the silage becomes more acidic. If the pH falls below 4, *Lactobacilli* will become the predominant organism present. On the other hand, harmful bacteria, such as coliform and enterococci, and also spores, such as those of *Clostridium botulinum*, will be destroyed in such a silage (Raa *et al.*, 1983).

It has been reported that at temperatures in the range of 23 - 30°C, approximately 80 % of the protein in fish usually becomes solubilized after one week in a silage process (Tatterson & Windsor, 1974; Backhoff, 1976). Fagbenro & Jauncey (1993), working with tilapia silage, reported that the silage process was temperature-dependent.

Martin & Bemister (1994) studied peat as a carbohydrate source for lactic acid bacteria in the production fish offal silage. Although acid peat extracts were insufficient as an agent for the reduction of the pH, they were found to be an appropriate carbohydrate source for the fermentation of fish offal. The authors also reported that no significant effect on the preservation process resulted from heating or freezing the raw material prior to ensiling.

Table 9 presents some of the progress reported in studies on the production of fermented fish silage.

7. Nutritional Characteristics of Fermented Fish Silage

Fish silage has been incorporated into animal feeds with different degrees of success depending upon the animal being fed and the method used to

Table 9: Progress reported in the production of fermented fish silage

Advances	References
use of molasses as carbohydrate additive	Kompiang *et al.* (1980)
use of antifungal additives	Lindgren & Pleje (1983)
prefermentation to increase the concentration of lactic acid bacteria	Lindgren & Pleje (1983); Cooke *et al.* (1987); Twiddy *et al.* (1987)
whey powder a good carbohydrate substrate	Van Wyk & Heydenrych (1985)
the silage process was temperature dependent	Fagbenro & Jauncey (1993)
peat as a carbohydrate source	Martin & Bemister (1994)

make the silage. It appears that the main drawback affecting its nutritional value is the consequence of a high level of protein hydrolysis (Hardy *et al.*, 1983; Stone & Hardy, 1986).

Espe *et al.* (1992*a*), studying the growth of young rats on diets to which silage, with different degrees of hydrolysis, had been added, established that the absorption of amino acids was faster when a hydrolysed protein was ingested.

Extensive research has been conducted using fish silage as feed, and many of these experiments have confirmed that fish silage is a useful protein source for many animals, such as pigs (Hyller *et al.*, 1976; Green *et al.*, 1988), poultry (Krogdahl, 1985; Ologhobo *et al.*, 1988), cattle (Nicholson & Johnson, 1991), and fish (Rungrangsak & Utne, 1981; Hardy *et al.*, 1983; Jackson *et al.*, 1984; Wood *et al.*, 1985; Espe *et al.*, 1992*b*; Fagbenro & Jauncey, 1993, 1994; Heras *et al.*, 1994).

Table 10 presents reported observations from research conducted on some nutritional characteristics of fish silage.

It has been reported that feeding animals silage produced using inorganic acids could cause calcium deficiency, among other injurious effects. Also, this kind of product can be stored only 5 to 10 days before deteriorating. On the other hand, organic acids are not only more expensive than inorganic acids, but can also create problems in animals, for example the development of ulcers and the damaging of mucous membranes by formic acid (Szakács *et al.*, 1988). However, lactic acid can be readily metabolized by such animals as fish (Dong *et al.*, 1993).

8. Current developments

The process of ensiling offers a practical and economical alternative for the recovery and use of agricultural and food industry wastes by employing the beneficial effects of lactic acid fermentation. From well-established traditional processes such as grass ensiling and the preservation of carbohydrate-rich wastes and by-products, technologies for the preser-

Table 10 : Observations from research conducted on some nutritional characteristics of fish silage

Research conducted	References
the length of storage of fish silage influenced its nutritional value	Hardy *et al.* (1983)
silage in which the liquefaction process has been discontinued by heating can be used in diets without affecting nutrient utilization	Gildberg *et al.* (1984); Fagbenro & Jauncey (1993)
the proximate composition of the silage did not change with the storage period	Wee *et al.* (1986)
L. plantarum and *S. faecium* were used in ensiling salmon viscera into a co-dried product for salmonid diets	Dong *et al.* (1993)
fermented tilapia silage showed good storage properties	Fagbenro & Jauncey (1993)
fermented tilapia silage and soy bean meal blend was used in dry diets for juvenile male Nile tilapia, *Oreochromis niloticus* and the African clariid catfish *Clarias gariepinus*	Fagbenro & Jauncey (1994)

vation of proteinaceous materials are evolving. It appears that ensiling could be a good solution for small scale operations for the bioconversion of specific wastes to industrial products, which can then find suitable markets. For example, fish silage could be an alternative to dwindling fish meal resources. Basic and applied studies in lactic acid fermentation should result in improved technologies for silage operations.

Increasing attention is being given to the biochemistry and genetics of the lactic acid bacteria. The manipulation of their metabolic reactions and developments in genetic engineering will make possible the tailoring of inoculants for specific ensiling processes. New studies in the area of additives will also exemplify the developments in the production of lactic acid silages (Brookes & Buckle, 1992). This technology offers great possibilities for meeting the increasing need for bioconversion of agricultural and food processing wastes and by-products to feeds and foods.

9. References

Adams MR, Cooke RD, Twiddy DR (1987) Fermentation parameters involved in the production of lactic acid preserved fish-glucose substrates. Int J Food Sci Technol 22: 105-114.

Anonymous (1974) Part 16: Gram-positive, asporogenous, rod-shaped bacteria. In: Buchanan RE, Gibbons NE (eds) Bergey's Manual of Determinative Bacteriology, 8th ed, Williams & Wilkins, Baltimore, MD, pp 576-598.

Anonymous (1988) Silage additives - The ADAS guide to choice value and application. Farmers Weekly, Supplement, 17 February. Quoted in Brookes RM, Buckle AE (1992) Lactic acid bacteria in plant silage. In: Wood BJB (ed) The Lactic Acid Bacteria, vol 1, Elsevier Applied Science, London, UK, pp 363-386.

Ashbell G, Lisker N 1987 Chemical and microbiological changes occurring in orange peels and in the seepage during ensiling. Biological Wastes 21: 213-220.

Axelsson LT (1993) Lactic acid bacteria; classification and physiology. In: Salminen S, von Wright A (eds) Lactic Acid Bacteria. Marcell Dekker, New York, NY, pp 1-63.

Backhoff PH (1976) Some chemical changes in fish silage. J Food Technol 11: 353-363.

Beck T (1978) The Microbiology of Silage Fermentation. In: McCullough ME (ed) Fermentation of Silage - A Review. National Feed Ingredients Assoc, West Des Moines, IA, pp 61-116.

Bolsen K, Fung D, Ilg H, Laytimi A, Hart R, Chain V, Nuzback L (1987) Effect of commercial inoculants on the fermentation of alfalfa, corn, forage sorghum, and triticale silages. In: Kansas State University Report, Manhattan, KS, pp 107-120.

Brookes RM, Buckle AE (1992) Lactic acid bacteria in plant silage. In: Wood BJB (ed) The Lactic Acid Bacteria, vol 1, Elsevier Applied Science, London, UK, pp 363-386.

Cooke RD, Twiddy DR, Reilly PJA (1987) Lactic acid fermentation as a low-cost means of food preservation in tropical counties. FEMS Microbiological Review 46: 369-379.

Dahiya RS, Speck ML (1968) Hydrogen peroxide formation by lactobacilli and its effect on *Staphylococcus aureus*. J Dairy Sci 51: 1568-1572.

Deibel RH (1960) Artinine as an energy source for the growth of *Streptococcus faecalis*. In: Bacteriol Proc, Society of American Bacteriologists, pp 163-164.

Deibel RH, Niven CF (1960) Comparative study of *Gaffkya homari*, *Aerococcus viridans*, tetra-forming cocci from meat curing brines and the genus *Pediococcus*. J Bacteriol 79: 175-180.

Disney JG, Tatterson IN, Olley J (1977) Recent development in fish silage. In: Proceedings of the International Conference on the Handling, Processing and Marketing of Tropical Fish, London, Tropical Products Institute, London, UK, pp 223-240.

Done DL (1986) Silage inoculants. A review of experimental work. Res Devel Agric 3(2): 83-87.

Dong FM, Fairgrieve WT, Skonberg DI, Rasco BA (1993) Preparation and nutrient analyses of lactic acid bacterial ensiled salmon viscera. Aquaculture, 109: 351-366.

Espe M, Haaland H, Njaa LR, Raa J (1992*a*) Growth of young rats on diets based on fish silage with different degrees of hydrolysis. Food Chem 44: 195-200.

Espe M, Haaland H, Njaa LR (1992*b*) Autolysed fish silage as a feed ingredient for atlantic salmon (*Salmo salar*). Comp Biochem Physiol 103(A) V2: 369-372.

Fagbenro O, Jauncey K (1993) Chemical and nutritional quality of stored fermented fish (tilapia) silage. Food Chem 48: 331-335

Fagbenro O, Jauncey K (1994) Growth and protein utilization by juvenile catfish, *Clarias gariepinus* fed most diets containing autolysed protein from stored lactic acid fermented fish silage. Bioresource Technol 48(1): 43-48.

Forney LJ, Reddy CA (1977) Fermentative conversion of potato-processing wastes into a crude protein feed supplement by lactobacilli. Develop Ind Microbiol 18: 135-143.

Garvie EI (1986) Genus *Pediococcus* Claussen 1903, 68. In: Sneath PHA, Nair NS, Sharpe ME, Holt JG (eds) Bergey's Manual of Systematic Bacteriology, vol 2, Williams & Wilkins, Baltimore, MD, pp 1075-1079. Quoted in Levin RE (1994) Lactic acid and propionic acid fermentations of fish hydrolysates. In: Martin AM (ed) Fisheries Processing. Biotechnological Applications, Chapman & Hall, London, UK, pp 273-310.

Gildberg A, Espejo-Hermes J, Magno-Orejano F (1984) Acceleration of autolysis during fish sauce fermentation by adding acid and reducing the salt content. J Sci Food Agric 35: 1363-1369.

Green S, Wiseman J, Cole DJA (1988) Examination of stability, and its effect on the nutritive value, of fish silage in diets for growing pigs. Animal Feed Science and Technol 21: 43-56.

Haard NF, Kariel N, Herzberg G, Feltham LAW, Winter K (1985) Stabilisation of protein and oil in fish silage for use as a ruminant feed supplement. J Sci Food Agric 36: 229-241.

Hardy WR Shearer DK, Stone EF, Wieg HD (1983) Fish silage in aquaculture diets. J World Maricult Soc 14: 695-703.

Hardy RW, Masumoto T (1990) Specifications for marine by-products in aquaculture feeds. In: Proceedings Alaska Fish By-Products Conference, Alaska Sea Grant College Program, University of Alaska, Fairbanks, AL, Report No.90-07, pp 109-120.

Hassan TE, Heath JL (1986) Biological fermentation of fish waste for potential use in animal and poultry feeds. Agricultural Wastes 15: 1-15.

Heras H, McLeod CA, Ackman RG (1994) Atlantic dogfish silage vs herring silage in diets for atlantic salmon (*Salmo salar*): growth and sensory evaluation of fillets. Aquaculture 125: 93-106.

Hyller GM, Peers DG, Morrison R, Parry DA, Woods WP (1976) Evaluation of a farm use of de-oiled herring silage as a protein feed for growing pigs. In: Proc Torry Res Station, Part IV, Ministry of Agriculture, Fisheries and Food, Aberdeen, U.K.

Ingram M, Ottoway FJH, Coppock JBM (1956) The preservative action of acid substances in food. Chemistry and Industry 42: 1154-1163.

Jackson AJ, Kerr AK, Cowey CB (1984) Fish silages dietary ingredient for salmon. I Nutritional and storage characteristics. Aquaculture 38: 211-220.

Jangaard MP (1987) Fish silage: A review and some recent developments. In: Fish Silage Workshop, Church Point, Nova Scotia, DFO Canada, Halifax, NS, pp 8-33.

Keller AK, Gerhardt P (1975) Continuous lactic acid fermentation of whey to produce a ruminant feed supplement high in crude protein. Biotechnol Bioeng 17: 997-1018.

Knøchel S (1981) Mikrobiell fermentering af fisk ved hjelp av naturligt forekommende laktobaciller. Hovedoppgave, Fiskeriministeriets Forsøgslaboratorium, Køvenhavn Univ. Quoted in Raa J, Gildberg A, Strøm T (1983) Silage production: Theory and practice. In: Ledward DA, Taylor AJ, Lawrie RA (eds) Upgrading Waste for Feeds and Food, Butterworths, London, UK, pp 117-132.

Kompiang IP, Yushadi, Creswell DC (1980) Microbial fish silage: Chemical composition, fermentation characteristics and nutritional value. In: Disney JG, James D (eds) Fish Silage Production and its Use, Proc IPFC Workshop on Fish Silage, FAO Fish Rep 230, Food and Agriculture Organization of the United Nations, Rome, pp 38-43.

Krogdahl A (1985) Fish viscera silage as a protein source for poultry. I Experiments with layer-type chickens and hens. Acta Agric Scand 35: 3-23.

Lassén TM, Hillemann G, Fors F (1990) Fisk och slaktavfall konserverade mad mjölksyrabakterier och enzympreparaten Pelzyme och Marilzil i foder till pälsdjur, NJF Seminarium, 185 (in Swedish). Quoted in Lindgren S (1992) Storage of waste products for animal feed. In: Wood BJB (ed) The Lactic Acid Bacteria, vol 1, Elsevier Applied Science, London, UK, pp 387-407.

Le Dividish J, Seve B, Geoffroy F (1976) Banana silage in animal feeding. Annales Zootechnology 25: 313-323. Quoted in Lindgren S (1992) Storage of waste products for animal feed. In: Wood BJB (ed) The Lactic Acid Bacteria, vol 1, Elsevier Applied Science, London, UK, pp 387-407.

Levin RE (1994) Lactic acid and propionic acid fermentations of fish hydrolysates. In: Martin AM (ed) Fisheries Processing. Biotechnological Applications, Chapman & Hall, London, UK, pp 273-310.

Lindgren S (1992) Storage of waste products for animal feed. In: Wood BJB (ed) The Lactic Acid Bacteria, vol 1, Elsevier Applied Science, London, UK, pp 387-407.

Lindgren SE, Clevström G (1978*a*) Antibacterial activity of lactic acid bacteria. 1. Activity of fish silage, a cereal starter and isolated organisms. Swedish J Agri Res 8: 61-66.

Lindgren SE, Clevström G (1978*b*) Antibacterial activity of fish silage. Swedish J Agri Res 8: 62-73.

Lindgren SE, Pleje M (1983) Silage fermentation of fish or fish waste products with lactic acid bacteria. J Food Agric 34: 1057-1067.

Litchfield JH (1987) Microbiological and enzymatic treatments for utilizing agricultural and food processing wastes. Food Biotechnol 1: 29-57.

Mackie IM (1971) Fermented fish products. FAO Fisheries Report No 100. Food and Agriculture Organization of the United Nations, Rome, 54 pp.

Marshall VME, Law BA (1984) The physiology and growth of dairy lactic-acid bacteria. In: Davies FL, Law BA (eds) Advances in the Microbiology and Biochemistry of Cheese and Fermented Milk, Elsevier Appl Sci Publ, London, UK, pp 67-98.

Martin AM (ed) (1991) Bioconversion of Waste Materials to Industrial Products. Elsevier Science Publishers, London, UK.

Martin AM (ed) (1994) Fisheries Processing, Biotechnological Applications. Chapman & Hall, London, UK.

Martin AM, Bemister PL (1994) Use of peat extract in the ensiling of fisheries wastes. Waste Manage Res 12: 467-479.

Martin AM, Patel TR (1991) Bioconversion of wastes from marine organisms. In: Martin AM (ed) Bioconversion of Waste Materials to Industrial Products, Elsevier Applied Science, London, UK, pp 417-440.

McCullough ME (ed) (1978) Fermentation of Silage. A Review. National Feed Ingredients Assoc, West Des Moines, Iowa.

McDonald P (1981) The Biochemistry of Silage, John Wiley & Sons, Chichester, UK.

McDonald P, Henderson AR, Heron SJE (1991) The Biochemistry of Silage. Chalcombe Publications, Marlow, UK, pp 19-47.

Mookherjee BD (1976) Lactic acid fermentation. In: Peterson MS, Johnson AH (eds) Encyclopedia of Food Science, AVI, Westport, CT, pp 443-447.

Moon NJ (1981*a*) Development of microbial populations in fermented wastes from frozen vegetable processing. J Food Protection 44: 288-193.

Moon NJ (1981*b*) Effect of inoculation of vegetable processing wastes with *Lactobacillus plantarum* on silage fermentation. J Sci Food Agric 32: 675-683.

Moo-Young M, Gregory KF (eds) (1986) Microbial Biomass Protein. Elsevier Applied Science Publishers, London, UK.

Mossel DAA (1971) Physiological and metabolic attributes of microbial groups associated with foods. J Applied Bacteriol 34: 95-118.

Nash MJ (1985) Crop Conservation and Storage. Pergamon Press, Oxford, UK.

Nicholson JWG, Johnson DA (1991) Herring silage as a protein supplement for young cattle. Can J Anim Sci 71: 1187-1196.

Nilsson R, Rydin C (1963) Acta Chem Scand 17: 174. Quoted in Raa J, Gildberg A, Strøm T (1983) Silage production: Theory and practice. In: Ledward DA, Taylor AJ, Lawrie RA (eds) Upgrading Waste for Feeds and Food, Butterworths, London, UK, pp 117-132.

Nilsson R, Rydin C (1965) A new method of ensiling foodstuff and feedstuff of vegetable and animal origin. Enzymologia 29: 126-142.

Ologhobo AD, Balogum AM, Bolarinwa BB (1988) The replacement value of fish silage for fish meal in practical broiler rations. Biol Wastes 25: 117-125.

Orla-Jensen SH (1919) The lactic acid bacteria. Mem Acad Royal Soc Denmark Ser 8(5): 81-197.

Owens JD, Mendoza LS (1985) Enzymatically hydrolysed and bacterially fermented fishery products. J Food Technol 20: 273-293.

Pederson CS (1978) Food Fermentations. In: Peterson MS, Johnson AH (eds) Encyclopedia of Food Science, AVI, Westport, CT, pp 311-314.

Petersen H (1953) Acid preservation of fish and fish offal. FAO Fish Bull 6(1-2): 18-26.

Prescott SM, Dunn CG (1959) Industrial Microbiology (3rd ed), McGraw-Hill, New York, NY.

Price RJ, Lee JS (1970) Inhibition of *Pseudomonas* species by hydrogen peroxide producing Lactobacilli. J Milk Food Technol 33: 13-18.

Raa J, Gildberg A (1982) Fish silage: A review. CRC Crit Rev Food Sci Nutr 16: 383-419.

Raa J, Gildberg A, Strøm T (1983) Silage production: Theory and practice. In: Ledward DA, Taylor AJ, Lawrie RA (eds) Upgrading Waste for Feeds and Food, Butterworths, London, UK, pp 117-132.

Ray B (1992) Cells of lactic acid bacteria as food biopreservatives. In: Ray B, Daeschel M (eds) Food Biopreservatives of Microbial Origin. CRC Press, Boca Raton, FL, pp 81-101.

Roa PD (1965) Ensilage of fish by microbial fermentation. Fishing News Int 4: 283-286.

Rungruangsak K, Utne F (1981) Effect of different acidified wet feeds on protease activities in the digestive tract and on growth rate of rainbow trout (*Salmo gairdneri* Richardson). Aquaculture 22: 67-79.

Samuel WA, Lee YY, Anthony WB (1980) Lactic acid fermentation of crude sorghum extract. Biotechnol Bioeng 22: 757-777.

Sander JE, Cai T, Barnhart HM Jr (1995) Evaluation of amino acids, fatty acids, protein, fat, and ash in poultry carcasses fermented with *Lactobacillus* bacteria. J Food Agric Chem 43: 791-794.

Setälä J (1988-1989) Enzymes in grass silage production. Food Biotechnology 2: 211-225.

Stanton WR, Yeoh QL (1977) Low fermentation method for conserving trash fish waste under SE Asian conditions. In: Proceedings of the Conference on the Handling, Processing and Marketing of Tropical Fish, London, Tropical Products Institute, London, UK, pp 227-282.

Stone EF, Hardy RW (1986) Nutritional value of acid established silage and liquified fish protein. J Sci Food Agric 37: 797-803.

Szakács G, Radvánszky B, Gyenes J (1988) Large-scale production of animal feed from meat industry by-products by lactic acid fermentation. In: Biotechnology and Food Industry, Proc Int Symposium, Budapest, pp 609-615.

Tatterson NI, Windsor LM (1974) Fish Silage. J Sci Food Agric 25: 369-379.

Tewari HK, Marwaha SS, Kennedy JF, Singh L (1988) Evaluation of acids and cellulase enzyme for the effective hydrolysis of agricultural lignocellulosic residues. J Chem Technol Biotechnol 41: 261-275.

Twiddy DR, Cross SJ, Cooke RD (1987) Parameters involved in the production of lactic acid preserved fish - starchy substrate combinations. Intern J Food Sci Technol 22: 115-121.

Vandevoorde L, Vande Woestyne M, Bruyneel B, Christiaens H, Verstraete W (1992) Critical factors governing the competitive behaviour of lactic acid bacteria in mixed cultures. In: Wood BJB (ed) The Lactic Acid Bacteria, vol 1, Elsevier Applied Science, London, UK, pp 447-475.

Van Wyk HJ, Heydenrych MSC (1985) The production of naturally fermented fish silage using various *Lactobacilli* and different carbohydrate sources. J Sci Food Agric 36: 1093-1103.

Watrous WL (1978) Lactic acid fermentation. In: Peterson MS, Johnson AH (eds) Encyclopedia of Food Science, AVI, Westport, CT, pp 443-445.

Watson SJ, Nash MJ (1960) The Conservation of Grass and Forage Crops, Oliver & Boyd, Edinburgh, UK, pp 299-300.

Wee KL, Kerdchuen N, Edwards P (1986) Use of waste grown tilapia silage as feed for *Clarias batrachus* L. J Aqua Trop 1: 127-137.

Whittenbury R (1965) Microbiology of grass silage. Proc Biochem 3: 27-31.

Wood JF, Carper BS, Nicolaides L (1985) Preparation and evaluation of diets containing fish silage, cooked fish preserved formic acid and low temperature dried fish meal as protein source for mirror carp *Cyprinus carpio*. Aquaculture 44: 27-40.

Woolford MK (1984) The Silage Fermentation. Marcel Dekker, New York, NY, USA.

Indigenous lactic acid bacteria of various food commodities and factors affecting their growth and survival

Friedrich-Karl Lücke
Mikrobiol. Lab., FB Haushalt & Ernährung,
Fachhochschule Fulda
P. O. Box 1269
D-36012 Fulda
Germany

(1) Introduction

All lactic acid bacteria share the following properties:

- They are GRAM-positive bacteria that do not form spores
- They do not contain heme-dependent catalase (although some strains are capable of synthesizing the protein moiety of this enzyme and incorporating exogenous porphyrines into it)
- They tolerate the ambient partial pressure of molecular oxygen (although some strains fail to form colonies on solid media incubated under aerobic conditions)
- They ferment sugars with lactic acid as a major end product: under anaerobic conditions with glucose in excess, at least one mole of lactic acid is formed from one mole of hexose.

Their importance to foods is due to several reasons:

- Most foods support growth of lactic acid bacteria.
- Most food-borne lactic acid bacteria are capable of lowering the pH to an extent no longer permitting growth of pathogens and spoilage bacteria.
- Of the lactic acid bacteria found in foods, only few (*Streptococcus*) species are pathogenic.

(2) Genera of lactic acid bacteria

Until the 1980's, lactic acid bacteria were divided into genera on the basis of the shape of their cells and their ability to form visible gas (carbon dioxide) during hexose fermentation. While all

NATO ASI Series, Vol. H 98
Lactic Acid Bacteria:
Current Advances in Metabolism, Genetics and Applications
Edited by T. Faruk Bozoğlu and Bibek Ray

rods were assigned to the genus *Lactobacillus*, „homofermentative" cocci (i.e. forming more than 1.7 moles lactic acid from hexoses under anaerobic conditions with excess glucose) were included into the genus *Streptococcus* while CO_2 producing („heterofermentative") cocci were included in the genus *Leuconostoc*. Homofermentative cocci forming tetrads during cell division were separated and assigned to the genus *Pediococcus*. Since the 1980's, insight into DNA-DNA homology, sequences of ribosomal RNA and pathways of sugar metabolism, lead to the subdivision of the streptococci and to the definition of new genera (see Devriese *et al.*, 1992). The recognized genera of lactic acid bacteria (as of mid-1995) are listed in Table 1.

Table 1: Recognized genera of lactic acid bacteria (status 1995)

Genus	Relevant to		
	food fermentations	food spoilage	food-borne diseases[1]
Carnobacterium	-	+	-
Enterococcus	-	+	unlikely
Lactobacillus	++	++	-
Lactococcus	++	-	-
Leuconostoc	+	++	-
Oenococcus	+	(+)	-
Pediococcus	+	+	-
Streptococcus	+	-	+
Tetragenococcus	+	(+)	-
Weissella	-	+	-
Aerococcus	-	-	-
Atopobium	-	-	-
Dolosigranum	-	-	-
Gemella	-	-	-
Vagococcus	-	-	-

[1]Some strains may cause problems by forming biogenic amines from amino acids

Genera of little if any relevance to foods will not be considered further in this paper. Table 2 suggests a key for the identification of food-relevant genera of the lactic acid bacteria. It also indicates that it is difficult to distinguish heterofermentative lactobacilli, *Weissella* and *Leuconostoc* species by simple morphological or physiological tests.

Table 2: Key for the identification of food-relevant genera of lactic acid bacteria (based on Holzapfel and Schillinger, 1992; Collins *et al.*, 1993)

(1)	Rods, final pH in MRS above 4.6	>*Carnobacterium*
	Rods, final pH below 4.6; coccobacilli or cocci	>(2)
(2)	CO_2 from glucose: yes	>(3)
	no	>(6)
(3)	Alanin in interpeptide bridge of murein: yes	>(4)
	no	>*Lactobacillus* (heterofermentative)
(4)	Ammonia from arginine: yes	>*Weissella*
	no	>*Leuconostoc* or *Weissella paramesenteroides* or *Weissella hellenica*
(6)	Cocci or egg-shaped cells	>(7)
	Rods	>*Lactobacillus* (homofermentative)
(7)	Cocci in tetrads: yes	>(8)
	no	>(9)
(8)	Growth in the presence of 15 % NaCl: yes	>*Tetragenococcus*
	no	>*Pediococcus*
(9)	Growth at 10 °C: yes	>(10)
	no	>*Streptococcus*
(10)	Growth at pH 9.6: yes	>*Enterococcus*
	no	>*Lactococcus*

The genus *Lactobacillus* is subdivided into the heterofermentative and the homofermentative species. Kandler and Weiss (1986) suggested to further subdivide the latter group into the „facultative heterofermentative“ group (fermenting pentoses and/or gluconate to about equimolar amounts of lactic acid and C_2 compounds) and the „obligately homofermentative“ group. These groups are similar - but not identical - to Orla-Jensen's „Streptobacteria“ and „Thermobacteria“, respectively. Food scientists stick to these „classical“ schemes of classification because it corresponds to some degree to different roles of the groups in food spoilage and food fermentations. However, Collins *et al.* (1991) showed that neither the Kandler and Weiss nor the original Orla-Jensen scheme of classification reflect phylogenetic relationships.

In studies related to food fermentations and food spoilage, it is often necessary to classify a large number of strains of lactic acid bacteria. Even for specialized laboratories, it is impractical to do so using DNA-DNA hybridization assays or ribosomal RNA sequencing. Hybridization (dot blot or reverse dot blot) using species-specific probes may provide a means of rapid identification of species (Hertel *et al.*, 1991; Ehrmann *et al.*, 1994). However, this still requires specialized reagents and, unless all species are known and specific probes have been developed for them, many strains will remain „untypable". Hence, „classical" biochemical tests such as determination of sugar fermentation patterns will remain important. Methods of identification using such properties employ numerical taxonomy (Shaw and Harding, 1984; Borch and Molin, 1988) or probability matrices (e.g. Döring *et al.*, 1988). Hierarchic (dichotomic) identification keys have also been published (Schillinger and Lücke, 1987; Hammes *et al.*, 1992). However, results of identification by physiological tests must be interpreted with caution because such traits are often unstable, and there is considerable variation within one species.

(3) Factors affecting survival and growth of lactic acid bacteria in foods

Various factors affect survival and growth of lactic acid bacteria and thereby also the composition of the lactic flora of foods:

- Lactic acid bacteria usually require at least some amino acids and vitamins and are thus unable to compete with other bacteria in nutrient-poor habitats.
- All lactic acid bacteria require a fermentable carbohydrate for growth and cannot compete well with other micro-organisms in foods containing very little carbohydrate. The type of carbohydrate present obviously selects for lactic acid bacteria capable of fermenting it. Accordingly, lactose fermenting strains dominate in milk and milk products while maltose fermenters are at advantage in cereal products such as doughs.
- Some foods - in particular, milk - contain very low concentrations of free amino acids. This select for strains capable of hydrolyzing proteins and peptides (see the contributions on proteolytic systems in this Volume).
- Some strains require specific growth factors that are present only in certain foods or produced by other micro-organisms adapted to a particular food. For example, *Lactobacillus homohiochii* from saké mash requires mevalonic acid, *Lactobacillus sanfrancisco* from sourdoughs requires growth factors present in bran and/or baker's yeast (Spicher, 1984) and some *Lactobacillus* species from kefir grains only grow well in the presence of culture filtrates of kefir yeasts (Takizawa *et al.*, 1994).
- At pH values between 4 and 5, lactic acid bacteria are, generally speaking, more competitive than most other food-borne bacteria, and their use in food fermentations is largely based on that property. However, different groups and species within groups differ in their tolerance of acids. For example, the final pH in broth cultures is always above 4.6 for

Carnobacterium and *Tetragenococcus* and below 4.5 for *Lactobacillus* and *Pediococcus* species. There are also differences in tolerance of undissociated acids such as acetic acid, lactobacilli being, as a rule of thumb, more resistant than leuconostocs, streptococci and lactococci. All these differences at least partially explain the microbial successions frequently observed in foods with ample carbohydrate content.

- Foods may contain „natural“ or „man-made“ antimicrobials that affect growth of lactic acid bacteria. Polyphenols from olives have been shown to inhibit *Lactobacillus plantarum* and must therefore be destroyed before olive fermentation (Ruiz-Barba *et al.*, 1991). In contrast, lactic acid bacteria are much more resistant than other bacteria towards sorbate and nitrite. Use of these preservative thus tends to select for a lactic flora, and there are even formulations of culture media made selective for lactic acid bacteria by including these compounds (Reuter, 1985).
- Growth of at least some species of lactic acid bacteria is possible at water activity values above 0.91 (only *Tetragenococcus halophilus* grows at even lower water activity). Some reduction of water activity (e.g. by adding salt or by pre-wilting crops before ensilage) may therefore contribute to the selection of a lactic flora.The minimum a_w value for growth varies considerably within one group; generally speaking, it is lower for *Lactobacillus, Pediococcus* and *Enterococcus* species than for *Lactococcus, Leuconostoc* and *Streptococcus* species.
- Lactic acid bacteria may react to molecular oxygen in different ways (Kandler, 1983; Condon, 1987). While growth of *Lactobacillus delbrueckii* and some other obligately homofermentative strains is inhibited at ambient partial pressure of oxygen, other strains use oxygen to regenerate NAD^+ and to gain additional ATP from converting acetyl phosphate into acetate. This gives them an advantage over other lactic acid bacteria in aerobic habitats such as doughs. However, in most cases, the presence of oxygen suppresses the development of a lactic flora because micro-organisms with a functional respiratory chain can make much more efficient use of oxygen for energy conservation.
- Storage temperature has a major effect on the composition of the lactic flora of foods. Psychrotrophic strains (i.e. capable of growth at 5 °C or below) include *Carnobacterium* and *Leuconostoc* species but also many strains of facultatively heterofermentative lactobacilli (in particular, *Lactobacillus sake* and *Lb. curvatus*). In contrast, the classical „yoghurt bacteria“ (*Streptococcus salivarius ssp. thermophilus* and *Lactobacillus delbrueckii ssp. bulgaricus*) as well as most obligately homofermentative lactobacilli grow at 45 but not at 15 °C.

- Interactions between strains may also affect the composition of the lactic flora of foods. Strains may inhibit other strains by formation of acid and antimicrobials (e.g. bacteriocins) but also by competing for fermentable carbohydrate.
- Heat treatment as usually applied when cooking or pasteurizing food is usually sufficient to inactivate all strains of lactic acid bacteria. However, if heat processing is marginal, e.g. in milk pasteurization or cooking of hams, and large numbers of cells are present in the raw product, some may survive and subsequently spoil the product. Heat resistant („thermoduric") strains are found, in particular, among *Enterococcus spp.* and *Streptococcus salivarius ssp. thermophilus* strains. A strain of *Enterococcus faecalis* is used as a reference strain for designing pasteurization processes (Reichert, 1985); its decimal reduction time was determined to be 3 minutes at 70 °C.

Of most relevance to foods is the capability of many lactic acid bacteria to grow in the presence of various inhibitory factors such as low pH, high concentrations of lipophilic undissociated acids (acetic, propionic, sorbic, benzoic acids) and nitrous acid. Organic acids interfere with the function of respiratory or other membrane-bound enzymes and co-factors, and nitrous acid blocks metabolic activities dependent on non-heme iron. It appears that these targets are either absent in lactic acid bacteria or their function is less essential for their growth. ATP-dependent efflux of inhibitory compounds may also be involved (Poolman, 1993). Table 3 provides some examples of lactic acid bacteria adapted to specific „ecological niches" in foods.

Table 3: Species of lactic acid bacteria adapted to special ecological niches

	Competitiveness at				
Species	pH < 4.0	a_w < 0.93	acetic acid >1%	ethanol > 10 %	temperature < 7 °C
Lactobacillus sake	-	+	-	-	++
Lactobacillus plantarum	++	+	+	(+)	-
Tetragenococcus halophilus	-	++	-	-	(+)
Oenococcus oenos	++	-		++	(+)
Leuconostoc mesenteroides	-	(+)	-	(+)	++

Table 4: Metabolic activities of lactic acid bacteria that affect the sensory properties of foods

Metabolic activity	desired in the manu-facture of	undesired in
lactic acid formation	fermented foods	most non-fermented foods
acetic acid formation	fermented vegetables; sourdough	most non-fermented foods; dairy and meat fermentations
CO_2 formation	sauerkraut and sourdough; kefir and some cheeses	non-fermented foods; meat fermentation; most dairy fermentations
slime formation	yoghurt and certain other sourmilks	non-fermented and most fermented foods
diacetyl formation	butter and certain cheeses	non-fermented and most fermented foods, e.g. beer
acetaldehyde formation	yoghurt	all other foods
hydrogen peroxide formation	fresh raw milk?	meats (discolouration, rancidity)
formation of peptides from proteins	-	cheeses (formation of bitter taste)
formation of hydrogen sulfide	-	all foods
formation of amino acids from proteins and/or peptides	cheeses; some fermented foods (formation of aroma precursors)	
degradation of malic and tartaric acids	certain wines and fruit juices	certain wines and fruit juices

(4) Changes in foods due to metabolic activities of lactic acid bacteria

Growth of lactic acid bacteria in all foods (except some wines) obviously leads to souring. Especially in proteinaceous foods, the subsequent coagulation of proteins causes changes in the texture of the product. Lactic acid has little effect on the aroma of the food; however, lactic acid bacteria may also form more volatile compounds with lower sensory thresholds. Such metabolic products include acetic acid, acetoin, diacetyl and acetaldehyde. Table 4 indicates metabolic products include acetic acid, acetoin, diacetyl and acetaldehyde. Table 4 indicates that acid formation as well as the other metabolic activities mentioned can be beneficial or detrimental to the sensory properties of a food, depending on their rate and extent and the nature of the food. Other undesired metabolic activities of lactic acid bacteria in foods include the decarboxylation of amino acids (formation of biogenic amines), and the formation of benzoic acid from hippuric acid.

Table 5: Food commodities that are changed by lactic acid bacteria

1	Dairy products	
	1.1 Raw milk	souring, coagulation, aroma formation (diacetyl, acetaldehyde...)
	1.2 Pasteurized milk	souring, coagulation, aroma formation (diacetyl, acetaldehyde...)
	1.3 Cheese	proteolysis; formation of biogenic amines and bitter peptides
2	Meat and meat products	
	2.1 Vacuum-packed meat	souring, cheesy odours
	2.2 Pasteurized meats	souring, greening, slime formation, pore formation, cheesy odours
	2.3 Raw sausages	souring, coagulation of soluble meat proteins; greening, formation of pores and/or biogenic amines
3	Fish, vacuum-packed, smoked	souring; formation of gas and/or biogenic amines
4	Products of plant origin	
	4.1 Syrups, sugar solutions	souring, slime formation
	4.2 Beer	souring, turbidity, diacetyl formation
	4.3 Brined cucumbers and olives	souring, gas and/or slime formation
	4.4 Wine and cider	degradation of malic and tartaric acid; formation of off-aromas (diacetyl)
	4.5 Fruit juices and mashes	gas and/or slime formation; formation of off-aromas
	4.6 Doughs	souring; proteolysis (-->aroma precursors)
	4.7 Grain mashes	souring
5	Vinegar-containing foods	turbidity, gas, off-odours

Table 5 lists the foods which are modified - regularly or occasionally, in a desired or undesired fashion - by lactic acid bacteria. These foods have in common:

- availability of fermentable carbohydrate and nutrients in sufficient concentrations
- absence of bacterial competitors (due to previous heat treatment) and/or (partial or total) inhibition of competitors by salt, acid and preservatives in various combinations
- pH above 3.0, water activity above 0.90.

Many of these foods are to be stored under refrigeration. The relevant lactic flora is therefore psychrotrophic.

In the following paragraph, emphasis is put on the spoilage of foods by lactic acid bacteria; food fermentations are dealt with in a separate contribution (Lücke, this volume).

(5) Lactic acid bacteria and food spoilage

(5.1) Milk and dairy products

Milk contains lactose as the only fermentable sugar, very little free amino acids, and molecular oxygen. To compete in milk, lactic acid bacteria should be able to ferment lactose, to liberate amino acids and to grow in the presence of oxygen. Since raw and pasteurized milk is usually kept cool, relevant lactic acid bacteria should also be able to grow at 10 °C. *Lactococcus lactis* and *Enterococcus spp.* fulfil these requirements and may spoil raw milk by souring and coagulation. Some strains of these species are, if present in large numbers, capable of surviving pasteurization.

Cheeses are preserved by different combinations of low pH and low water activity. Microbial spoilage is mostly caused by moulds. However, non-starter lactic acid bacteria may grow on cheeses, and some strains have been shown to occasionally cause problems because of their ability to produce bitter-tasting peptides and to decarboxylate amino acids. Strains of *Lactobacillus buchneri* have been shown to form histamine in cheese (ten Brink *et al.*, 1990).

(5.2) Meat and meat products

In contrast to milk, meat contains large amounts of free amino acids but only little fermentable carbohydrate. After *post-mortem* glycolysis, the concentration of the latter hardly exceeds 0.3 % and may be even lower in tissues with low initial glycogen content or rich in fat and collagen. This is why lactic acid bacteria compete poorly on fresh raw meat as long as molecular oxygen is available for aerobic degradation of carbohydrates and amino acids. On fresh meats, lactic acid bacteria usually dominate the spoilage flora only after it has been packaged under vacuum or in modified atmospheres (MA) containing high levels of carbon dioxide.

The fermentable carbohydrates of meat mainly consist of glucose, glucose-6-phosphate and small amounts of ribose. To compete successfully on meat under practical conditions, lactic acid bacteria should be capable to rapidly ferment these carbohydrates under chill conditions. It is not necessary to them to hydrolyze proteins or tolerate high concentrations of acids. Hence, the spoilage flora of vacuum or MA packed meat consist of the following psychrotrophic species (Shaw and Harding, 1984, 1989):

- *Lactobacillus sake, Lactobacillus curvatus*
- *Leuconostoc mesenteroides, Leuconostoc carnosum, Leuconostoc gelidum*
- *Carnobacterium divergens, Carnobacterium piscicola.*

On chilled fresh meat, the leuconostocs appear to grow most rapidly (Borch and Agerhem, 1992). The species of lactic acid bacteria that dominates the spoilage flora has little influence on the type of spoilage (Schillinger and Lücke, 1986). A notable exception to this statement are strains (mainly belonging to *Lactobacillus sake)* producing hydrogen sulfide from cysteine: they shorten the shelf life of the meat considerably by producing off-odours and green discolourations (Shay *et al.*, 1981).

In the manufacture of meat products, salt, curing agents, and, frequenty, sugar is added. The water activity of pasteurized meats is still high enough (range 0.96 - 0.98) to permit growth of most strains belonging to the species listed above (Borch and Molin, 1988; Dykes *et al.*, 1994). The composition of the spoilage flora of these products mainly depends on the degree of their (almost inevitable) post-process recontamination and the composition of the recontaminant flora. More heavy salting, such as in the manufacture of raw meat products, tends to suppress leuconostocs and carnobacteria; hence, the fermentation flora of raw sausage is usually dominated by *Lactobacillus sake* and *Lactobacillus curvatus*. These species, along with heterofermentative lactobacilli and *Weissella* species, may also be involved in the spoilage of raw sausages by excess souring as well as formation of pores, acetic acid and discolourations.

(5.3) Fish products

Basically, the information given for meats also applies for fish. Lactic acid bacteria predominate in the spoilage flora of lightly salted fish products (water activity range 0.95 - 0.98) that are either not heated or recontaminated after heat processing. Compared to meat, the pH value of these products is often higher, and preservatives such as sorbic and benzoic acid are used instead of nitrite. Nevertheless, the species of lactic acid bacteria dominating on these products appear to be similar to those on meats (Jeppesen and Huss, 1993). *Tetragenococcus („Pediococcus") halophilus* has been reported in heavily salted products such as anchovies (Villar *et al.*, 1985) and may cause problems because of its ability to decarboxylate free amino acids present in large amounts in many fish. Marinated (i.e. vinegar-preserved) fish is dealt with in part 5.5.

(5.4) Products of plant origin

On plant material, lactic acid bacteria can only compete if the following requirements are met:

- Liberation of fermentable sugars from the plant cells
- Availability of niches with reduced oxygen partial pressure
- pH and water activity still in the „permissible" range (i.e. pH > 3.0, a_w > 0.90).

On shredded leafy vegetables, *Leuconostoc spp.* predominate the lactic flora (Mundt, 1970), but their growth and spoilage potential is low and increases only somewhat under modified atmospere (low in oxygen, high in carbon dioxide). Only after the material is both shredded and firmly packed (such as for the sauerkraut and silage fermentation) or put in brines (as for the fermentation of olive and pickles such as cucumbers), lactic acid bacteria take over. In olive and pickle fermentation, the formation of too much gas is undesired („bloater" formation). Heterofermentative and malate-decarboxylating lactic acid bacteria may thus contribute to this defect and cause spoilage (see Daeschel *et al.*, 1987, for review). Strains of *Pediococcus* may also contribute to the formation of biogenic amines in sauerkraut (Kuensch *et al.*, 1990).

Fruit mashes such as prepared to manufacture fruit brandies may harbour a specialized lactic flora (*Lactobacillus suebicus;* some strains of *Lactobacillus plantarum*) that tolerates unusual high concentrations of acids and ethanol (Hammes *et al.*, 1992). Their growth may reduce the ethanol yield and lead to off-flavours in the final product.

In grain mashes, some growth and acid formation of lactic acid bacteria (species similar to those in sourdoughs) is desirable for special beers such as „lambic". However, excess activity of lactic acid bacteria may, as in fruit mashes, reduce the ethanol yield and lead to off-flavours in the final product.

Sugar solutions may be spoiled by dextrane-forming *Leuconostoc* strains if their water activity still permits growth of these organisms. Other lactic acid bacteria tolerating high concentrations of sugar and acid (e.g. strains of *Lactobacillus plantarum, Weissella confusa*) have been found to spoil cane juice by acidification and/or slime formation (see Hammes *et al.*, 1992).

Beer may be spoiled by lactic acid bacteria that cause turbidity, ropiness and off-flavours, mainly due to diacetyl. Most strains responsible have adapted to breweries as their habitat: they preferably ferment maltose (sometimes hardly any other sugar) and tolerate acid, CO_2 and humulones coming in from hop. The most important spoilage organisms are *Pediococcus damnosus* and specialized strains of the *Lactobacillus brevis*-group (Back, 1981).

In order to grow and compete in wine and cider, lactic acid bacteria must tolerate up to 12 % ethanol, pH values between 3 and 4, sulfites and certain antimicrobial ingredients of the plant cells (e.g. phenolic compounds). Many of the strains found in wine are capable of decarboxylating malic acid (malolactic fermentation). This reaction generates a proton-motive force across the cell membrane and thus adds to the competitiveness of strains carrying out this reaction (review: Poolman, 1993). It is desired in high-acid wines but tends to spoil low-acid wines. Other metabolic activities leading to undesired sensory changes include (Dittrich, 1993)

- slime formation by *Leuconostoc* strains and *Pediococcus damnosus*
- formation of acetic acid and mannitol by various heterofermentative lactic acid bacteria
- degradation of tartaric acid by certain strains of *Lactobacillus plantarum* and *Lactobacillus brevis*
- diacetyl formation.

Most lactic acid bacteria adapted to survival and growth in wines are heterofermentative, such as *Oenococcus oenos, Lactobacillus hilgardii,* and *Lactobacillus fructivorans.*

(5.5) Vinegar-preserved foods

Vinegar is the main preservative in mayonnaise, salad dressings and marinated products. If these are not pasteurized and subsequently handled aseptically, they are spoiled by vegetative acid-tolerant micro-organisms. Particularly in the absence of oxygen, lactic acid bacteria may compete well with yeasts and cause slime and off-odours. In marinated fish, formation of biogenic amines also can be a problem. Relevant strains must obviously be able to grow in the presence of high concentration (> 1 %) of undissociated acetic acid. In addition, they must tolerate preservatives such as benzoic and sorbic acids often added to the formulation, and in marinated fish, salt represents an additional „hurdle". The lactic spoilage flora found in vinegar-preserved foods comprises mainly strains of *Lactobacillus plantarum* (homofermentative) *Lactobacillus brevis* and *Lactobacillus fructivorans* (heterofermentative) (Baumgart *et al.*, 1983).

(6) Concluding remarks

Today, much is known about the composition of the lactic flora of various food commodities. The mechanisms that select for a specific group of lactic acid bacteria are also understood fairly well. However, it is still difficult or even impossible to predict, from the properties of the food and the processing methods applied, the patterns of species prevailing on a food and the rate of changes brought about by these bacteria. This will remain an exciting field of research.

(7) References

Back W (1981) Bierschädliche Bakterien. Taxonomie der bierschädlichen Bakterien - Gram-positive Arten. Monatsschr Brauerei 34:267-276

Baumgart J, Weber B, Hanekamp B (1983) Mikrobiologische Stabilität von Feinkosterzeugnissen. Fleischwirtschaft 63:93-94

Borch E, Agerhem H (1992) Chemical, microbial and sensory changes during the anaerobic cold storage of beef inoculated with a homofermentative *Lactobacillus* sp. or a *Leuconostoc* sp. Internat J Food Microbiol 15:99-108

Borch E, Molin G (1988) Phenetic taxonomy of lactic acid bacteria from meat and meat products. Antonie van Leeuwenhoek 54:301-323

Brink B ten, Damink C, Joosten HMLJ, Huis in't Veld JHJ (1990) Occurrence and formation of biologically active amines in foods. Internat J Food Microbiol 11:73-84

Collins MD, Rodrigues U, Ash C, Aguirre M, Farrow JAE, Martinez-Murcia A, Phillips BA, Williams AM, Wallbanks S (1991) Phylogenetic analysis of the genus *Leuconostoc* and related lactic acid bacteria as determined by reverse transcriptase sequencing of 16S rRNA. FEMS Microbiol Letters 77:5-12

Collins MD, Samelis J, Metaxopoulos J, Wallbanks S (1993) Taxonomic studies on some leuconostoc-like organisms from fermented sausages: description of a new genus *Weissella* for the *Leuconostoc paramesenteroides* group of species. J Appl Bacteriol 75:595-603

Condon S (1987) Responses of lactic acid bacteria to oxygen. FEMS Microbiol Rev 46:269-280

Daeschel MA, Andersson RE, Fleming HP (1987) Microbial ecology of fermenting plant materials. FEMS Microbiol Rev 46:357-367

Devriese L, Collins MD, Wirth R (1992) The genus *Enterococcus*. In: The Prokaryotes, 2nd ed. (Balows A, Trüper HG, Dworkin M, Harder W, Schleifer KH, eds) pp. 1465-1482. Springer, Berlin Heidelberg New York

Dittrich HH (1993) Mikrobiologie des Weines und des Schaumweines. In: Mikrobiologie der Lebensmittel - Getränke (Dittrich HH, ed) pp 183-259. Behr's Verlag Hamburg

Döring B, Ehrhardt S, Lücke FK, Schillinger U (1988) Computer-assisted identification of lactic acid bacteria from meats. System Appl Microbiol 11:67-74

Dykes GA, Britz TJ, von Holy A (1994) Numerical taxonomy and identification of lactic acid bacteria from spoiled vacuum-packaged Vienna sausages. J Appl Bacteriol 76:246-252

Ehrmann M, Ludwig W, Schleifer KH (1994) Reverse dot blot hybridization: a useful tool for the identification of lactic acid bacteria in fermented food. FEMS Microbiol Letters 117:143-150

Hammes WP, Weiss N, Holzapfel WH (1992) The genera *Lactobacillus* and *Carnobacterium*. In: The Prokaryotes, 2nd ed. (Balows A, Trüper HG, Dworkin M, Harder W, Schleifer KH, eds) pp. 1535-1594. Springer, Berlin Heidelberg New York

Hertel C, Ludwig W, Obst M, Vogel RF, Hammes WP, Schleifer KH (1991) 23S rRNA targeted oligonucleotide probes for the rapid identification of meat lactobacilli. System Appl Microbiol 14:173-177

Holzapfel WH, Schillinger U (1992) The genus *Leuconostoc*. In: The Prokaryotes, 2nd ed. (Balows A, Trüper HG, Dworkin M, Harder W, Schleifer KH, eds) pp. 1508-1534. Springer, Berlin Heidelberg New York

Jeppesen VF, Huss HH (1993) Characteristics and antagonistic activitiy of lactic acid bacteria isolated from chilled fish products. Internat J Food Microbiol 18:305-320

Kandler O (1983) Carbohydrate metabolism in lactic acid bacteria. Antonie van Leeuwenhoek 49:209-224.

Kandler O, Weiss N (1986) Regular, nonsporing Gram-positive rods. In: Bergey's Manual of Systematic Bacteriology, 9th ed, Vol 2 (Sneath PHA, Mair NS, Sharpe ME, Holt JG, eds) pp 1208-1234. Williams & Wilkins, Baltimore

Kuensch U, Schaerer H, Temperli A (1990) Study on the formation of biogenic amines during sauerkraut fermentation. In: Processing and quality of foods (Zeuthen P, Cheftel JC, Eriksson C, Gormley TR, Linko P, Paulus K, eds) pp2.240-2.243. Elsevier Applied Science London

Mundt JO (1970) Lactic acid bacteria associated with raw plant food material. J Milk Food Technol 33:550-553

Poolman B (1993) Energy transduction in lactic acid bacteria. FEMS Microbiol Rev 12:125-147

Reichert J (1985) Die Wärmebehandlung von Fleischwaren - Grundlagen der Berechnung und Anwendung. Holzmann Verlag, Bad Wörishofen

Reuter G (1985) Elective and selective media for lactic acid bacteria. Internat J Food Microbiol 2:55-68

Ruiz-Barba JL, Garrido Fernández A, Jimenez-Diaz R (1991) Bactericidal action of oleuropein extracted from green olives against *Lactobacillus plantarum*. System Appl Microbiol 13:199-205

Schillinger U, Lücke FK (1986) Milchsäurebakterien-Flora auf vakuumverpacktem Fleisch und ihr Einfluß auf die Haltbarkeit. Fleischwirtschaft 66:1515-1520

Schillinger U, Lücke FK (1987) Identification of lactic acid bacteria from meat and meat products. Food Microbiol 4:199-208

Shaw BG, Harding CD (1984) A numerical taxonomic study of lactic acid bacteria from vacuum-packed beef, pork, lamb and bacon. J Appl Bacteriol 56:25-40

Shaw BG, Harding CD (1989) *Leuconostoc gelidum* sp. nov. and *Leuconostoc carnosum* sp. nov. from chill-stored meats. Internat J System Bacteriol 39:217-223

Shay BJ, Egan AF (1981) Hydrogen sulphide production and spoilage of vacuum-packaged beef by a *Lactobacillus*. In: Psychrotrophic microorganisms in spoilage and pathogenicity (Roberts TA, Hobbs G, Christian JHB, Skovgaard N, eds) pp 241-251. Academic Press London

Spicher G (1984) Beiträge zur Vereinheitlichung der Ermittlung des Keimgehaltes von Getreide und Getreideprodukten. V. Mitteilung: Einleitende Untersuchungen über die Eignung verschiedener Kultursubstrate zum Nachweis von Milchsäurebakterien des Sauerteiges. Getreide Mehl Brot 38:261-264

Takizawa S, Kojima S, Tamura S, Fujinaga S, Benno Y, Nakase T (1994) *Lactobacillus kefirgranum* and *Lactobacillus parakefir* sp. nov., two new species from kefir grain. Internat J System Bacteriol 44:435-439

Villar M, de Ruiz Holgado AP, Sanchez JJ, Trucco RE, Oliver G (1985) Isolation and characterization of *Pediococcus halophilus* from salted anchovies (*Engraulis anchoisa*). Appl Environ Microbiol 49:664-666

Fermentation Processes for the Production of Lactic Acid

Antonio M. Martin
Department of Biochemistry
Memorial University of Newfoundland
St. John's, Newfoundland, Canada A1B 3X9

1. Introduction

As with many other biotechnological processes, lactic acid fermentation has evolved from an ancient "art," basically associated with the processing and preservation of foodstuffs, to a presently sophisticated technology.

Lactic acid is a natural organic hydroxy acid with many uses in food, chemical, and pharmaceutical industrial processes. Approximately half of the lactic acid produced is employed in the food industry as acidulant, and approximately half of the lactic acid produced worldwide is manufactured by fermentation processes (Ward, 1989).

Lactic acid is produced by animals, plants, and microorganisms. It exists in two optically active forms, the D(-), levorotatory, and the L(+), dextrorotatory. Lockwood *et al.* (1965) indicated that the esters and salts of L(+)-lactic acid are levorotatory. Both isomers exist in biological systems, and the lactic acid formed in fermentation processes is generally a racemic mixture (DL forms). Soccol *et al.* (1994) reported that the L(+) isomer is the most important for the food industry, as humans can assimilate it only, by producing L-lactate dehydrogenase.

Prescott & Dunn (1959), Holten *et al.* (1971) and Vick Roy (1985)

NATO ASI Series, Vol. H 98
Lactic Acid Bacteria:
Current Advances in Metabolism, Genetics and Applications
Edited by T. Faruk Bozoğlu and Bibek Ray

presented good reviews of the technology and biotechnology related to the production of lactic acid. Buchta (1981) and Erickson *et al.* (1991) also presented studies on lactic acid. Commercial production of lactic acid by fermentation has been developing since it started in the United States in 1881 (Holten *et al.*, 1971). Table 1 outlines the development of lactic acid production, while Table 2 profiles some of the chemical and physical properties of lactic acid.

2. Uses of Lactic Acid

Over the years, lactic acid has been applied in many processes and products. Because of its versatility, lactic acid has applications for which other products can be employed as well, and therefore new technologies could appear and cause the use of lactic acid to be discontinued in a given process, and vice versa. Vick Roy (1985) warns that care must be taken in distinguishing between potential applications, present uses, and past uses.

Commercial synthetic manufacture of lactic acid started *ca.* 1963 in the United States (Anon., 1963), and Thorne (1969) reported that this kind of production was also taking place in Japan. Van Ness (1981) reported on the synthetic production of lactic acid. Lactic acid can be converted to important chemicals such as ethanol, propylene glycol, acrylic polymers and polyesters (San-Martín *et al.*, 1992). Biodegradable plastic packages can be produced from lactic acid copolymers (Dillon & Martin, 1994).

The different types or qualities (or "grades") in which lactic acid is commercialized are indicated in Table 3, and Table 4 summarizes some of the most important uses of lactic acid.

Table 1: The history of commercial lactic acid production

Development	Year(s)	Reference(s)
discovered by Scheele	1780	Holten *et al.* (1971)
first industrial production	1881	Holten *et al.* (1971)
first uses in leather and textile processes	1894	Garrett, 1930
production levels $\approx 18 \times 10^4$ kg y^{-1} (100 % basis)	1897	Inskeep *et al.*, 1956
approximately 1.4×10^6 kg y^{-1} produced in U.S.: 53 % used in technical applications, 47 % in food and medicinal applications	1940	Filachione, 1952
approximately 2.9×10^6 kg y^{-1}; produced in U.S.: 40 % used in technical applications, 60 % in food and medicinal applications	1947	Filachione, 1952
used for armoured tank coolant and glycerol substitute	1939 - 1945	Vick Roy, 1985
synthetic production of lactic acid to make stearoyl-2-lactylates	early 1960's	Anon., 1963; Thorne, 1969
world production 24 - 28 $\times 10^6$ kg y^{-1}: >50 % used in food, $\approx$20 % to make stearoyl-2-lactylates, remainder in pharmaceutical and industrial applications. $\approx$50 % made by fermentation	1982	Vick Roy, 1985

Table 2: Characteristics of lactic acid [a]

Property	Characteristics
optical activity	exists as L(+), D(-) and racemic mixture
crystallization	forms colourless monoclinic crystals when highly pure
colour	none or yellowish
odour	none
consistency	syrupy liquid
solubility / miscibility / hygroscopicity	soluble in all proportions with water; miscible with water, alcohol, glycerol, furfural; insoluble in chloroform, carbon disulphide; hygroscopic
volatility	low
self-esterification	in solutions of >≈20 %, forming a cyclic dimer or a linear polymer
reactivity	versatile, *e.g.* as organic acid or organic alcohol
physical	formula weight ≈90.1; boiling point 122°C at 15 mm Hg; melting point 18°C; density 1.2; specific gravity 1.248

[a] From Lockwood *et al.* (1965); Vick Roy (1985); Sax & Lewis (1987); and Considine (1989).

In addition to the general classification of the uses of lactic acid presented in Table 3 (food, pharmaceutical and technical, the last referring to any industrial application not involved in the two previous ones), the use of lactic acid in waste treatment should be considered. Indeed, the role of lactic acid fermentation in some waste or by-products biodegradation and bioconversion operations is very important (Martin & Bemister, 1994). Many of those operations are related to the production of animal feed components. This subject is dealt with elsewhere in this book.

Lockwood (1979) reported on the economical aspects of the commercial production of lactic acids. The potential that exists for the improvement of lactic acid fermentation processes is summarized in Table 5.

One of the applications of lactic acid with a definitive environmental and biotechnological impact is its use in the production of biodegradable plastics, and this will be dealt with elsewhere in this paper.

2.1 LACTIC ACID IN NUTRITION

From the nutritional point of view, the D and L isomers of lactic acid have different effects. While L(+) lactic acid is metabolized in gluconeogenesis, which is also the fate of the L(+) lactic acid produced during human metabolism, the D(-) lactic acid is excreted or oxidized in the liver. It has been reported that the latter can produce undesirable reactions in small children (FAO/WHO Expert Committee on Food Additives, 1967). This committee also recommends that adults limit their consumption of D(-) lactic acid to 100 mg per kg body weight per day. One of the products that contributes D(-) lactic acid to the human diet is sauerkraut (Kandler & Stetter, 1977).

Table 3: Grades of commercial lactic acid [a]

Grade	Purity	% Total Acidity	Characteristics
pharmaceutical (USP)	highest	85	requires most elaborate recovery process, colourless, ≈30 % polymerized
food (FCC)	medium	≈50	pale yellow
technical	least	44 - 45	higher concentrations of contaminants, *e.g.* sugar, metals, chloride, sulphate and ash

[a] From Lockwood (1979); Vick Roy (1985).

2.2 PRODUCTION OF POLYLACTIC ACID

Waste plastic materials are considered one of the main problems in the disposal of solid wastes. It has been stated that plastics degrade very slowly and are not very suitable for landfills. It appears that biopolymers should present better characteristics than petrochemically - derived polymers. Indeed, a great advantage of natural polymers is that they are biodegradable (Pool, 1989), and emphasis is therefore being placed on the study of the production of biopolymers (Brandl *et al.*, 1990).

In microorganisms, the carbon source may be utilized for the production of both biomass and storage polymers at the same time. An alternative

Table 4: Summary of lactic acid applications

Applications	References
Food:	
acidulant (pH adjuster), flavorant, preservative and processing aid in various processed foods	Vick Roy, 1985; Anon., 1992
production of calcium and sodium stearoyl-2-lactylates (CSL and SSL) and lactylated glycerides (texture and starch conditioners, shelf-life extenders and emulsifiers)	Vick Roy, 1985; Anon., 1992
Pharmaceutical / Cosmetic:	
intermediate for pharmaceutical and cosmetic manufacture (*e.g.* of stearoyl-2-lactylate, calcium lactate, sodium lactate, ethyl lactate and aluminum salt of lactic acid)	Vick Roy, 1985; Anon., 1992
acidulant	Anon., 1992
manufacture of biodegradable polylactic acid polymers	Kulkarni *et al.*, 1971; Wehrenberg, 1981; Dillon & Martin, 1994
Industrial / Technical:	
production of stearoyl-2-lactylate, various resins, polyester plasticizers, lactylated fatty acid esters of mono- and di-glycerides, polylactic acids, biodegradable thermoplastics, humectants, rubber, leather, textiles, solvents, lubricants, herbicides, fungicides and pesticides	Lipinsky, 1981; Wehrenberg, 1981; Vick Roy, 1985; Anon., 1992
pH control, metal surface treatment, printing and dyeing, electrostatic painting, plating, soldering	Vick Roy 1985; Anon., 1992

Table 5: Potential methods for the improvement of lactic acid fermentation processes

Subject	Means
production system	microbial immobilization, continuous fermentation
materials	corrosion resistance, improved heat exchange, increased strength, decreased cost
microorganisms	genetic engineering
operational conditions	continuous controls
raw materials	use of abundant and inexpensive waste materials
recovery	use of non-toxic solvents for liquid-liquid extraction from medium in continuous fermentation and/or use of solvents with low distribution coefficient for impurities, electrodialysis

process may occur in which microorganisms growing on a carbon source, if an essential nutrient is depleted, may be induced to produce polymers by adding a polymer-forming substrate (Brandl *et al.*, 1990).

Dillon & Martin (1994) reviewed the production of some of the most promising microbially-produced biopolymers that possess potential as thermoplastics, and their applications. Polylactic acid (PLA) is a polymer that degrades in the environment and in the human body to lactic acid. Other promising biopolymers are: poly-β-hydroxybutyrates (PHB's), which are intracellular lipid storage polymers for bacteria, and pullulan, an extracellular polysaccharide synthesized by *Aureobasidium pullulans*. The same authors concluded that there is a need to find suitable and cheap fermentation substrate sources for the production of microbial polymers, and that future research will probably aim to both optimize the process and improve the production of biopolymers, and to develop new kinds, using genetic engineering techniques.

Lactic acid, with an equal number of hydroxyl and carboxyl groups, can self-condense to form a linear thermoplastic polyester, polylactic acid (Keeler, 1991). Lactic acid can be built up in molecular weight by self-condensation to a degree of polymerization of 10,000 units. Ring-opening polymerization can be applied to achieve higher molecular weights. Therefore, it is assumed that a degree of polymerization of 50,000 can be obtained by using direct condensation to splice together five 10,000-unit low-molecular-weight polymers (Keeler, 1991).

The bioconversion of waste materials to lactic acid, with the objective of producing polylactic acid by fermentation, could lower production costs. The production of polylactic acid from potato starch has been attempted in a two-step enzymatic process followed by fermentation. The starch is initially solubilized by α-amylase, followed by conversion into glucose by glucoamylase. The glucose is finally fermented by lactobacilli to produce lactic acid (Keeler, 1991).

Among the applications of PLA are its use in drug-delivery systems and sutures (costs of medical-grade polylactic acid could be considered high), and mulch films used in agriculture to reduce weed growth and to keep moisture and heat in the soil (Keeler, 1991).

3. Microorganisms

The lactic acid group of bacteria is generally loosely defined with no precise boundaries (Jay, 1992). It has been accepted that all of its members have the characteristic of producing lactic acid from hexoses. Based on the end products of glucose metabolism, those microorganisms that produce lactic acid as the only or major product are called homofermentatives, and those producing equal amounts of ethanol, carbon dioxide and lactic acid are called heterofermentative. Homofermentatives are capable of extracting, from a given amount of glucose, twice the amount of energy as heterofermentatives. If the substrate is a pentose, the homofermentative characteristic of a microorganism does not necessarily hold (Jay, 1992). Only the homofermentative microorganisms are of commercial importance for the production of lactic acid.

It has been reported that *Streptococcus cremoris* and *Streptococcus lactis* are the most important lactic acid-producing bacteria employed in the dairy industry (Sellars & Babel, 1978). However, the preferred species from the commercial point of view for the production of lactic acid belong to the genus *Lactobacillus*. Among them are *Lactobacillus bulgaricus* and *Lactobacillus delbrueckii*, which produce lactic acid from glucose and whey, respectively. Tyagi *et al.* (1991) reported the advantages of employing *Lactobacillus helveticus* in the production of lactic acid. The taxonomy and occurrence of lactic acid bacteria were discussed by Teuber (1991), and Hayakawa (1992) described the classification and actions of lactic acid bacteria. Table 6 presents some of the characteristics of lactic acid-producing bacteria.

Prescott & Dunn (1959) presented an historical overview of, and technological data for, the production of lactic acid by fungi. Soccol *et al.* (1994) studied 19 species of *Rhizopus* in shake flask culture. Only four of them produced L-lactic acid aerobically at a pH of 6, with *Rhizopus orizae* NRRL 395 producing the highest concentration (65 g/l).

The stabilisation of the production of lactic acid in resting or non-growing microbial cultures has been reported (Rees & Pirt, 1979). By imposing non-growing conditions on the process, the authors studied the biochemical controlling factors of the resting cells' metabolism in *L. delbru*eckii NCIB 8130. This species is commercially used for lactic acid production by fermentation.

Axelsson (1993) made a comprehensive study of the classification and physiology of lactic acid bacteria. Many microorganisms used in the food industry (therefore being safe, food-grade), including those that produce lactic acid, generate antimicrobial metabolites that can be used for food preservation (Ray & Daeschel, 1992). The increased commercial significance of the role of lactic acid bacteria in the production of fermented foods has boosted the interest in their biochemical and microbiological properties, including their genetics (Gasson & de Vos, 1993).

Table 6: Characteristics of some genera of lactic acid bacteria [a]

Genera	Characteristics
Lactobacillus, *Pediococcus*, *Streptococcus*	Homofermentative, optimal growth temperature ≥40°C (growth range <5 - 45°C), optimal pH between 5 and 7 (growth range 3.2 - 9.6), facultatively anaerobic, typical lactic acid yield from glucose >90 % (w/w), can produce lactic acid concentrations in media up to 3 %, may produce either or both isomers, weakly proteolytic and lipolytic, very limited synthetic abilities, all require B vitamins, preformed amino acids, and purine and pyrimidine bases

[a] From Marth (1974); Stamer (1976); Stanier *et al.* (1976); Vick Roy (1985); and Jay (1992).

3.1 MIXED CULTURE

The development of industrial-scale productions using mixed cultures of microorganisms has been hindered by the difficulties of establishing optimum conditions in the same fermenter for more than one microbial species (Kurosawa *et al.*, 1988).

A set of modified logistic and Luedeking-Piret equations were used to model the growth and lactic acid production of a mixed culture of *L. bulgaricus* and *Streptococcus thermophilus* in order to study the effect of substrate concentration (Özen & Özilgen, 1992).

Lactic acid production in fermentation processes can be an intermediary step in the production of other organic acids. Tyree *et al.* (1991) utilized mixed cultures of *Lactobacillus xylosus* ATCC 15577 and *Propionibacterium shermanii* ATCC 13673 to convert xylose and glucose to lactate, and the latter to propionic acid. Viniegra-González & Gómez (1984) discussed the lactic acid fermentation by mixed bacterial cultures.

4. Fermentation Processes

4.1 SUBSTRATES

A wide range of substrates can be utilized for the submerged production of lactic acid in fermentation processes. Prescott & Dunn (1959) and Vick Roy (1985) presented comprehensive references to raw materials studied for this process. Most of them are organic wastes and by-products of the food industry. Table 7 summarizes them.

In addition to the basic carbon and energy source, sources of nitrogen and minerals are required as nutrient supplements for the fermentation media, although their amounts should be kept to the minimum required levels in order to not interfere with the subsequent recovery processes after the fermentation operation. A study on substrates for lactic acid bacteria was presented by Salminen *et al.* (1993).

4.2 KINETICS

The study and application of kinetic models is important in the design and control of chemical and biological reactors alike. Research, such as that of Luedeking & Piret (1959*a,b*), has been conducted on the kinetics of batch and continuous lactic acid fermentation processes. Hanson & Tsao (1972) presented a study on the kinetics of *L. delbrueckii* batch and continuous fermentation on a glucose-yeast extract medium at controlled pH levels. Stieber *et al.* (1977) reported on the experimental tests of a dialysis continuous fermentation process for the production of ammonium lactate.

Unstructured models for fermentations assume that the biomass compositions remain constant during the processes, and this limits their applicability to experiments with different operating conditions. Nielsen *et al.* (1991*a,b*) presented a theoretical study of lactic acid fermentation based on a simple structured model of *S. cremoris*, in which the cell physiology is taken into consideration. These kinds of models and their modifications can be used to acquire better characterization of lactic acid fermentation in situations such as non-growth conditions. Nielsen *et al.* (1991*c*) also reported that their model described such experimental observations as *S. cremoris* carbohydrate adaptation and substrate preference on mixtures of galactose, glucose and lactose.

Ammonium lactate is a non-protein nitrogen source produced by lactic acid fermentation. This material is of industrial importance and can be produced from corn in a fermentation process (Mercier *et al.*, 1992). The kinetics of the biosynthesis of ammonium lactate were described using a simple mathematical logistic model by Mercier *et al.* (1992), who indicated that the same model can be employed in describing the kinetics of lactic acid production from glucose and corn. Stieber and Gerhard (1979*a,b*) developed a mathematical model for the continuous process, including dialysis, for ammonium lactate production from deproteinized whey. The authors modified their model and applied it to systems with dialysis and

Table 7: A summary of the fermentation substrates for lactic acid fermentation [a]

Principal substrate	Source	References
lactose	casein whey	Burton (1937); Olive (1936)
	cheese whey	Campbell (1953)
	sweet whey	Whittier & Rogers (1931)
corn sugar (dextrose)	corn	Inskeep *et al.* (1956); Machell (1959)
sucrose	molasses	Needle & Aries (1949)
	cane and beet sugar	Vick Roy (1985)
various carbohydrates	potatoes	Cordon *et al.* (1950)
	sulphite waste liquor	Leonard *et al.* (1948)
	Jerusalem artichokes	Andersen & Greaves (1942); Schopmeyer (1954)
	cellulosic materials	Schopmeyer (1954)
	sorghum extract	Samuel *et al.* (1980)

[a] From Vick Roy (1985); Prescott & Dunn (1959).

cell recycling (Stieber and Gerhard, 1981*a*), and dialysis and immobilized cell systems (Stieber and Gerhard, 1981*b*).

4.3 FERMENTATION EQUIPMENT AND TECHNOLOGY

Throughout, this article concentrates on the overall study of lactic acid fermentation, and not on the specific fermentation processes for the production of lactic acid culture concentrates. The latter, with the objective of producing active microbial biomass of lactic acid organisms, usually follows basic established technology (Porubcan & Sellers, 1979). The production and use of lactic acid bacteria in industry has been discussed by Mäyrä-Mäkinen and Bigret (1993).

In general, the production of lactic acid by fermentation can utilize standard fermentation technology. Although batch processes have been the method commonly used in industry, studies have been conducted on continuous fermentation processes. Tyagi *et al.* (1991) discussed several batch and continuous fermentation systems for the production of lactic acid, and presented a comparison of different types of bioreactors (fermenters) employed. Some of the characteristics of the fermentation process for the production of lactic acid are presented in Table 8, and Tables 9 and 10 outline some of the continuous and immobilized cell fermentation systems reported, respectively.

Conventional fermentation methods, using free microbial cells in batch or continuous fermenters, require the separation of cells from the medium at the end of each process. On the other hand, immobilization of cells on solid supports, or entrapment in a gel matrix, introduce diffusional resistances and the additional expenses associated with the immobilization step. To overcome those limitations, the use of membrane recycle bioreactors (Mehaia & Cheryan, 1986) and hollow-fibre membrane bioreactors has been studied (Vick Roy *et al.*, 1982; Mehaia & Cheryan, 1987).

Product inhibition in lactic acid fermentation processes, and its effect on lactic acid productivity, has been studied by various authors (Keller & Gerhardt, 1975; Hongo *et al.*, 1986). It has been stated that due to this inhibition, the advantages of continuous fermentation are not fully realized in lactic acid processes, and that with the exception of very low substrate concentrations, the specific growth rate of the microorganism becomes more dependent on the concentration of the product than on that of the substrate (Keller & Gerhardt, 1975). Therefore, processes that remove the product from the fermentation medium could improve the efficiency of the system. Friedman & Gaden (1970) used a dialysis culture system which removed lactic acid from the medium. The system maintained a low lactate concentration after the log phase, and this resulted in enhanced specific microbial growth rates and lactic acid production. The authors claim that these results confirmed the product inhibition effect in lactic acid fermentation. Stieber *et al.* (1977) reported that, relative to nondialysis continuous or batch processes, the dialysis continuous fermentation for the production of lactic acid allowed the use of more concentrated substrate and increased the efficiency of substrate conversion.

Monitoring of process variables is of paramount importance in a fermentation process. A semi-on-line monitoring system to study the lactic acid fermentation was developed by Nielsen *et al.* (1989). Glucose, lactic acid, protein, and optical density were measured in a computer-controlled fermenter. The authors reported that the response of the system was fast and reliable, and that it can be used for the study of mathematical fermentation models.

Table 8: Fermentation process characteristics for the production of lactic acid

Characteristics	Values or attributes	References
Temperature	43°C	Burton (1937)
	≈49°C	Inskeep *et al.* (1956)
	≈30°C, 45 - 50°C	Prescott & Dunn (1959)
	30 - 50°C	Snell & Lowery (1964)
pH	6 - 7	Burton (1937)
	5.8 - 6.0	Inskeep *et al.* (1956)
	<6	Snell & Lowery (1964)
Neutralizing agent	NH_3, K_2CO_3, Na_2CO_3	Kempe *et al.* (1950)
	$Ca(OH)_2$, $CaCO_3$	Prescott & Dunn (1959); Lockwood (1979); Vick Roy (1985)
	$Zn(OH)_2$, $ZnCO_3$	Prescott & Dunn (1959)
Initial sugar conc.	5 - 20 %	Prescott & Dunn (1959)
Final product conc.	<12 - 15 % (v/v)	Peckham (1944)
Yield	85 - 90 % (w/w)	Prescott & Dunn (1959)
	90 - 95 % (w/w)	Vick Roy (1985)
Residual sugar conc.	< 0.1 %	Daly *et al.* (1939); Vick Roy (1985)
Inoculum rate	5 - 10 % (v/v)	Vick Roy (1985)
Fermentation time	1 - 6 days	Vick Roy (1985)
Reactor productivity	1 - 3 kg m^{-3} h^{-1}	Vick Roy (1985)
Reactor material	wood	Lockwood (1979)
	316 stainless steel	Vick Roy (1985)

Table 9: Continuous fermentation studies for the production of lactic acid

System	References
L. bulgaricus, Lactobacillus casei, sweet whey	Whittier & Rogers (1931)
L. delbrueckii, glucose, yeast extract and mineral salts	Luedeking & Piret (1959*b*)
L. delbrueckii, hydrolysed corn starch	Childs & Welsby (1966)
L. delbrueckii, glucose, yeast extract medium	Hanson & Tsao (1972)
L. bulgaricus, two fermenters in series, non-aseptic, whey	Keller & Gerhardt (1975)
L. bulgaricus, dialysis fermenter, non-aseptic, whey	Stieber *et al.* (1977)
L. delbrueckii immobilized in calcium alginate, flow column using glucose and yeast extract	Stenroos *et al.* (1982)
L. delbrueckii, stirred tank with cell recycle, glucose	Vick Roy *et al.* (1983)
L. bulgaricus, membrane recycle bioreactor, whey permeate and yeast extract	Mehaia & Cheryan (1986)
L. bulgaricus, hollow-fiber bioreactor, whey permeate and yeast extract	Mehaia & Cheryan (1987)

Table 10: Fermentation systems using immobilized cells

Systems	References
L. delbrueckii and *L. bulgaricus* in dialysis culture	Friedman & Gaden (1970)
fixed-film anaerobic, upflow bioreactor with kefir mixed culture in gelatin-coated packing cross-linked with glutaraldehyde	Compere & Griffith (1975)
L. casei in polyacrylamide gel lattice, using glucose	Divies & Siess (1976)
L. bulgaricus on polyacrylamide gel beads	Ohmiya *et al.* (1977)
L. bulgaricus on polyacrylamide and alginate beads	Lee (1981)
L. delbrueckii, hollow fibre (batch) fermenter	Vick Roy *et al.* (1982)
L. casei in agar and in polyacrylamide gel	Tuli *et al.* (1985)
Sporolactobacillus inulins in porous cellulose acetate particles	Nagai *et al.* (1986)
L. bulgaricus, hollow fibre (continuous) fermenter	Mehaia & Cheryan (1987)
L. helveticus on calcium alginate beads in a continuous packed bed reactor	Boyaval & Goulet (1988)
L. bulgaricus, L. helveticus, in continuous culture	Tyagi *et al.* (1991)

Lactic acid produced in fermentation processes needs to be recovered or extracted, and purified. Some fermentation problems make this difficult. For example, in order to control the pH of the lactic acid fermentation, chemicals such as ammonia, calcium carbonate, or sodium hydroxide are added. Calcium carbonate causes precipitation of calcium lactate, which could hinder the production of a polymer-grade material from lactic acid. The removal of lactic acid from the fermentation medium can be a solution to this problem, by avoiding the lowering of the pH of the broth. A system known as reactive liquid-liquid extraction (RLLE), using amines, can accomplish this (San-Martín *et al.*, 1992). Yabannavar & Wang (1991*a*) studied extractive fermentation production of lactic acid by *L. delbrueckii* utilising a tertiary amine (Alamine 336) and oleyl alcohol, at acidic pH.

In general, processes for the recovery of lactic acid or lactate salts from the fermentation medium is a significant part of the total cost of the process (Vick Roy, 1985).

Figure 1 presents a diagram of the recovery operations of lactic acid produced in a fermentation process. Some further operations are generally required to produce an acceptable product for many applications, as has been summarized by Vick Roy (1985). Ward (1989) presented a process outline for the recovery of lactic acid as a non-volatile microbial metabolite from a fermentation process.

4.3.1 *Immobilization.* Immobilization of lactic acid bacteria may facilitate the recovery of starter bacterial cultures for reuse in fermentation processes. The need to optimize the properties of the carrier for microorganism entrapment has been acknowledged (Arnaud *et al.*, 1989).

As noted above, the use of mixed cultures of microorganisms presents the problem of establishing optimum conditions for more than one

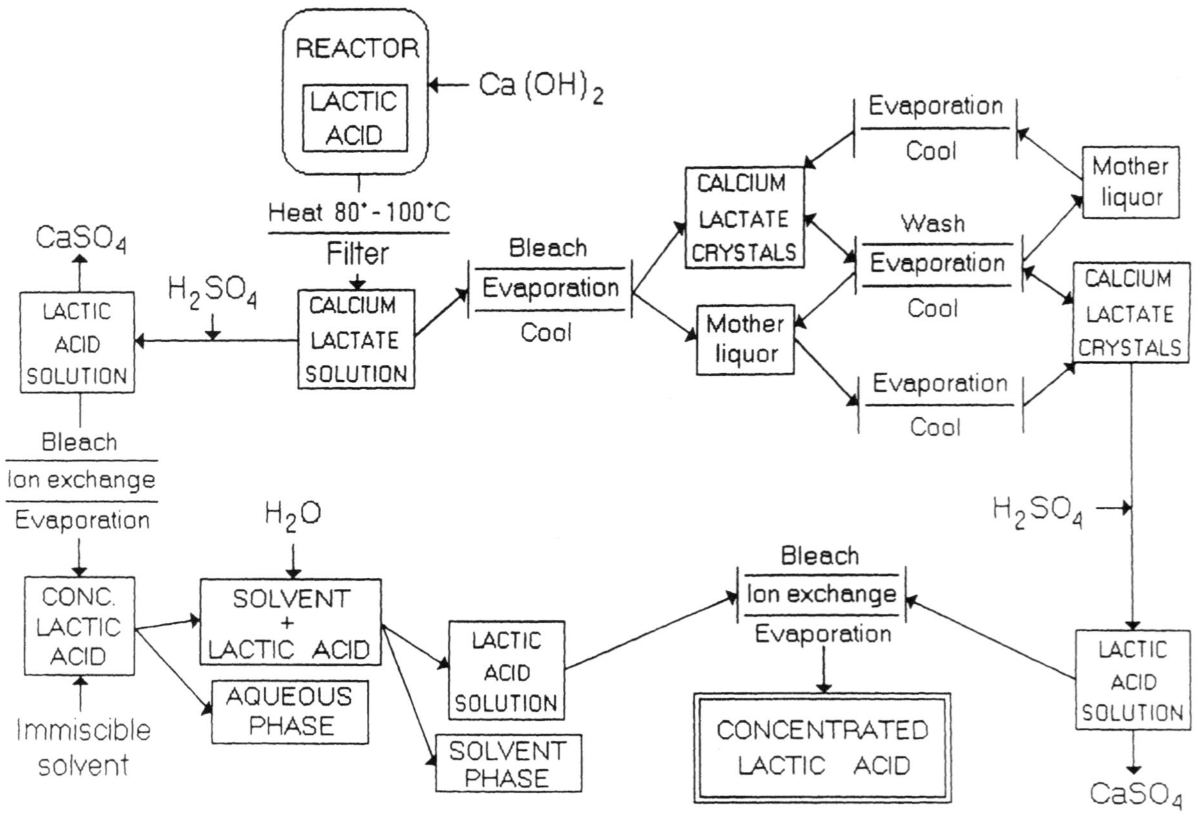

Figure 1: Recovery and purification of lactic acid produced in a fermentation process

microbial species in the same process. Kurosawa *et al.* (1988) coimmobilized in Ca-alginate beads the amylase-producing aerobic mould *Aspergillus awamori* Nakasawa IFO 4033 and *S. lactis* IFO 12007, an aerotolerant anaerobic bacterium. The mould grew on or near the surfaces of the beads, which was an aerobic zone. The bacteria grew in the oxygen-limited central part of the beads. By using starch as the main substrate source, the final concentration of lactic acid produced was 25 g/L, with a yield for lactic acid production of 0.66.

Nomura *et al.* (1987) used immobilized growing microorganisms and electrodialysis in the production of lactic acid. Lacroix *et al.* (1990) studied the rheological properties of a batch fermentation with entrapped *L. casei*. Groboillot *et al.* (1993) immobilized *Lactococcus lactis* by microencapsulation using interfacial cross-linking of chitosan. Toxicity of the cross-linking agent was observed during microencapsulation and loss of cell activity occurred afterwards. The authors suggested the need for a more compatible cross-linking agent to improve the microbial viability during the process. Mehaia & Cheryan (1987) immobilized *L. bulgaricus* in a hollow-fibre bioreactor with the objective of producing lactic acid from acid whey permeate in continuous fermentation. The authors indicated that diffusion of substrate and products separation are hindered by the physical barrier between substrate and cells. Better performance was reported with the use of a membrane bioreactor with cell recycling (Mehaia & Cheryan, 1986).

Büyükgüngör (1992) immobilized *L. bulgaricus* (ATCC 11842) in k-carrageenan, indicating that this support resulted in better mechanical and chemical stability than that offered by alginate gels. Bassi *et al.* (1987) measured the diffusivities of lactic acid in an agarose gel membrane.

Tyagi *et al.* (1991) reviewed extensively the immobilized cell processes for the production of lactic acid. An analysis of mass transfer in lactic acid fermentation for *L. delbrueckii* immobilized cells in k-carrageenan was presented by Yabannavar & Wang (1991*b*).

5. Current Developments

Development in the production of lactic acid is linked to the present general trends in biotechnology, its main factors being: existence of markets for the product, availability of appropriate substrate sources, and progress in related scientific and technological areas. Some of the relevant reported work is summarized in Table 11.

Table 11: Summary of work done on the production of lactic acid

Topics	References
kinetics of batch and continuous cultures	Hanson & Tsao (1972)
production stability in resting suspensions of *L. delbrueckii*	Rees & Pirt (1979)
immobilized *L. delbrueckii*	Stenroos *et al.* (1982)
diffusivities of lactose and lactic acid in a gel membrane	Bassi *et al.* (1987)
electrodialysis fermentation using immobilized cells	Nomura *et al.* (1987)
coimmobilized mixed culture of *A. awamori* and *S. lactis*	Kurosawa *et al.* (1988)
gel-immobilized *S. thermophilus*	Arnaud *et al.* (1989)
on-line monitoring of fermentation variables	Nielsen *et al.* (1989)
microbial system modelling	Nielsen *et al.* (1991*a,b,c*)
lactic acid fermentation as intermediate to propionic acid	Tyree *et al.* (1991)
kinetics of glucose and corn fermentation by *Lactobacillus amylophilus*	Mercier *et al.* (1992)
mixed cultures of *L. bulgaricus* and *S. thermophilus*	Özen & Özilgen (1992)
lactic acid reactive extraction with Alamine 336	San-Martín *et al.* (1992)
chitosan microencapsulation of *Lc. lactis*	Groboillot *et al.* (1993)

6. References

Andersen AA, Greaves JE (1942) d-Lactic acid fermentation of Jerusalem artichokes. Ind Eng Chem 34: 1522-1526. Quoted in Prescott SC, Dunn CG (1959) The production of lactic acid by fermentation. In: Industrial Microbiology, 3rd ed, McGraw - Hill, New York, NY, pp 304-331.

Anonymous (1963) Monsanto starts new lactic acid plant. Chem Eng News 41(49): 44-46.

Anonymous (1992) Eco-Lac Lactic Acid (Product applications information brochures, No 150 - 152). Ecological Chemical Products Company, Adell, Wisconsin, USA.

Arnaud JP, Lacroix C, Chopin L (1989) Effect of lactic fermentation on the rheological properties of k-carrageenan/locust bean gum mixed gels inoculated with *S. thermophilus*. Biotechnol Bioeng 34: 1403-1408.

Axelsson LT (1993) Lactic acid bacteria; classification and physiology. In: Salminen S, von Wright A (eds) Lactic Acid Bacteria, Marcel Dekker, New York, NY, pp 1-63.

Bassi AS, Rohani S, Macdonald DG (1987) Measurements of effective diffusivities of lactose and lactic acid in 3% agarose gel membrane. Biotechnol Bioeng 30: 794-797.

Boyaval P, Goulet J (1988) Optimal conditions for production of lactic acid from cheese whey permeate by Ca-alginate-entrapped *Lactobacillus helveticus*. Enzyme Microbiol Technol 10: 725-728.

Brandl H, Gross RA, Lenz RW, Fuller RC (1990) Plastics from bacteria and for bacteria: poly (β-hydroxyalkanoates) as natural, biocompatible, and biodegradable polyesters. Adv Biochem Eng / Biotechnol 41: 77-93.

Buchta K (1981) Lactic acid. In: Rehm HJ, Reed G (eds) Biotechnology, Verlag Chemie, Weinheim, Germany, pp 409-417.

Burton LV (1937) By-products of milk. Food Ind 9: 571-575, 617-618, 634-636. Quoted in Vick Roy TB (1985) Lactic acid. In: Moo-Young M (ed) Comprehensive Biotechnology, vol 3, Pergamon Press, Oxford, pp 761- 776.

Büyükgüngör H (1992) Stability of *Lactobacillus bulgaricus* immobilized in k-carrageenan gels. J Chem Technol Biotechnol 53: 173-175.

Campbell LA (1953) Production of calcium lactate and lactic acid from cheese whey. Can Dairy Ice Cream J 32(3): 29-31. Quoted in Prescott SC, Dunn CG (1959) The production of lactic acid by fermentation. In: Industrial Microbiology, 3rd ed, McGraw - Hill, New York, NY, pp 304-331.

Childs CG, Welsby B (1966) Continuous lactic fermentation. Process Biochem 1: 441-444.

Compere AL, Griffith WL (1975) Fermentation of waste materials to produce industrial intermediates. Dev Ind Microbiol 17: 247-252.

Considine DM (ed) (1989) Van Nostrand's Scientific Encyclopedia, 7th ed, vol 2, Van Nostrand Reinhold, New York, NY, pp 1663-1664.

Cordon TC, Treadway RH, Walsh MD, Osborne MF (1950) Lactic acid from potatoes. Ind Eng Chem 42: 1833-1836. Quoted in Prescott SC, Dunn CG (1959) The production of lactic acid by fermentation. In: Industrial Microbiology, 3rd ed, McGraw - Hill, New York, NY, pp 304-331.

Daly RE, Walsh JF, Needle HC (1939) Fermentation process of making white calcium lactate. US Patent 2,143,359, Jan 10. Quoted in Prescott SC, Dunn CG (1959) The production of lactic acid by fermentation. In: Industrial Microbiology, 3rd ed, McGraw - Hill, New York, NY, pp 304-331.

Dillon R, Martin AM (1994) Production of polymers by microorganisms. A review. Agro-Industry Hi-Tech 5(4): 27-30.

Divies C, Siess MH (1976) Étude du catabolisme de lacide l-malique par lactobacillus casei emprisonné dans un gel de polyacrylamide. Annales de Microbiologie (Inst Pasteur) 127 B: 525-539.

Erickson LE, Fung DYC, Tuitemwong P (1991) Anaerobic fermentations. In: Rehm HJ, Reed G (eds) Biotechnology, 2nd ed, vol 3, VCH, Weinheim, pp 295-318.

FAO/WHO Expert Committee on Food Additives (1967) Toxicological evaluation of some antimicrobials, antioxidants, emulsifiers, stabilizers, flour-treatment agents, acids and bases. In: Lactic Acid, FAO/WHO, Rome, Italy. Quoted in Steinkraus KH (1983) Handbook of Indigenous Fermented Foods, Marcel Dekker, New York, NY, pp 108.

Filachione EM (1952) In: Kirk RE, Othmer DF (eds) Encyclopedia of Chemical Technology, 1st ed, vol 8, Wiley, New York, NY, pp 167-180.

Friedman MR, Gaden EL (1970) Growth and acid production by *Lactobacillus* (L.) *delbrueckii* in a dialysis culture system. Biotechnol Bioeng 12: 961-974.

Garrett JF (1930) Lactic acid. Ind Eng Chem 22: 1153-1154.

Gasson M, de Vos W (eds) (1993) Genetics and Biotechnology of Lactic Acid Bacteria, Blackie Academic & Professional, London.

Groboillot AF, Champagne CP, Darling GD, Poncelet D, Neufeld RJ (1993) Membrane formation by interfacial cross-linking of chitosan for microencapsulation of *Lactococcus lactis.* Biotechnol Bioeng 42: 1157-1163.

Hanson TP, Tsao GT (1972) Kinetic studies of the lactic acid fermentation in batch and continuous culture. Biotechnol Bioeng 14: 233-252.

Hayakawa K (1992) Classification and actions of food microorganisms with particular reference to fermented foods and lactic acid bacteria. In: Nakazawa Y, Hosono A (eds) Functions of Fermented Milk, Challenges for the Health Sciences, Elsevier Applied Sciences, London, pp 127-164.

Holten CM, Müller A, Rehbinder D (eds) (1971) Lactic Acid, Properties and Chemistry of Lactic Acid and Derivatives, Verlag Chemie, Weinheim, Germany.

Hongo M, Nomura Y, Iwahara M (1986) Novel method of lactic acid production by electrodialysis fermentation. Appl Environ Microbiol 52: 314-319.

Inskeep GC, Taylor GG, Breitzke WC (1956) Lactic acid from corn sugar. Modern Chemical Processes 3: 96-107.

Jay JM (1992) Modern Food Microbiology, 4th ed, Van Nostrand Reinhold, New York, NY.

Kandler O, Stetter KO (1977) Sauerkraut. In: Symposium on Indigenous Fermented Foods, Bangkok, Thailand. Quoted in Steinkraus KH (1983) Handbook of Indigenous Fermented Foods, Marcel Dekker, New York, p 108.

Keeler R (1991) Don't let food go to waste - make plastic out of it. R&D Magazine Feb: 52-57.

Keller AK, Gerhardt P (1975) Continuous lactic acid fermentation of whey to produce a ruminant feed supplement high in crude protein. Biotechnol Bioeng 17: 997-1018.

Kempe LL, Halvorson HO, Piret EL (1950) Effect of continuously controlled pH on lactic acid fermentation. Ind Eng Chem 42: 1852-1857. Quoted in Prescott SC, Dunn CG (1959) The production of lactic acid by fermentation. In: Industrial Microbiology, 3rd ed, McGraw - Hill, New York, NY, pp 304-331.

Kulkarni RK, Moore EG, Hegyeli AF, Leonard F (1971) Biodegradable poly(lactic acid) polymers. J Biomed Mater Res 5: 169-181.

Kurosawa H, Ishikawa H, Tanaka H (1988) L-lactic acid production from starch by coimmobilized mixed culture system of *Aspergillus awamori* and *Streptococcus lactis*. Biotechnol Bioeng 31: 183-187.

Lacroix C, Paquin C, Arnauld JP (1990) Batch fermentation with entrapped growing cells of *Lactobacillus casei*. Optimization of the rheological properties of the entrapment gel matrix. Appl Microbiol Biotechnol 32: 403-408.

Lee KH (1981) Hanguk Nonghwa Hakhoe Chi (Korea) 24: 149. Quoted in Tyagi RD, Kluepfel D, Couillard D (1991) Bioconversion of cheese whey to organic acids. In: Martin AM (ed) Bioconversion of Waste Materials to Industrial Products, Elsevier Applied Science, London, pp 313-333.

Leonard RH, Peterson WH, Johnson MJ (1948) Lactic acid from fermentation of sulphite waste liquor. Ind Eng Chem 40; 57-67. Quoted in Prescott SC, Dunn CG (1959) The production of lactic acid by fermentation. In: Industrial Microbiology, 3rd ed, McGraw - Hill, New York, NY, pp 304-331.

Lipinsky ES (1981) Chemicals from biomass: petrochemical substitution options. Science 212: 1465-1471.

Lockwood LB (1979) Production of organic acids by fermentation. In: Peppler HJ, Perlman D (eds), Microbial Technology, 2nd ed, vol 1, Academic Press, New York, NY, pp 355-387.

Lockwood LB, Yoder DE, Zienty M (1965) Lactic Acid. Ann NY Acad Sci 119: 854-867.

Luedeking R, Piret EL (1959*a*) A kinetic study of the lactic acid fermentation. Batch process at controlled pH. J Biochem Microb Tech Eng 1: 393-412.

Luedeking R, Piret EL (1959*b*) Transient and steady states in continuous fermentation. Theory and experiment. J Biochem Microb Tech Eng 1: 431-459.

Machell G (1959) Production and applications of lactic acid. Ind Eng Chem 35: 283-290. Quoted in Vick Roy TB (1985) Lactic acid. In: Moo-Young M (ed) Comprehensive Biotechnology, vol 3, Pergamon Press, Oxford, pp 761-776.

Marth EH (1974) Fermentations. In: Webb BH *et al* (eds), Fundamentals of Dairy Chemistry, Ch 13, AVI, Westport, CT. Quoted in Jay JM (1992) Modern Food Microbiology, 4th ed, Van Nostrand Reinhold, New York, NY, p 375.

Martin AM, Bemister P (1994) Use of peat extract in the ensiling of fisheries wastes. Waste Management and Research 12: 467-479.

Mäyrä-Mäkinen A, Bigret M (1993) Industrial use and production of lactic acid bacteria. In: Salminen S, von Wright A (eds) Lactic Acid Bacteria, Marcel Dekker, New York, NY, pp 65-95.

Mehaia MA, Cheryan M (1986) Lactic acid from acid whey permeate in a membrane recycle bioreactor. Enzyme Microb Technol 8: 289-292.

Mehaia MA, Cheryan M (1987) Immobilization of *Lactobacillus bulgaricus* in a hollow-fibre bioreactor for production of lactic acid from acid whey permeate. Appl Biochem Biotechnol 14: 21- 27.

Mercier P, Yerushalmi L, Rouleau D, Dochain D (1992) Kinetics of lactic acid fermentation on glucose and corn by *Lactobacillus amylophilus*. J Chem Technol Biotechnol 55: 111-121.

Nagai S, Ozaki M, Fukunishi K, Yamazaki K (1986) Japanese patent 6158588, 25 March. Quoted in Tyagi RD, Kluepfel D, Couillard D (1991) Bioconversion of cheese whey to organic acids. In: Martin AM (ed) Bioconversion of Waste Materials to Industrial Products, Elsevier Applied Science, London, pp 313-333.

Needle MC, Aries RS (1949) Lactic acid and lactates. Sugar 44: 32-36. Quoted in Vick Roy TB (1985) Lactic acid. In: Moo-Young M (ed) Comprehensive Biotechnology, vol 3, Pergamon Press, Oxford, pp 761-776.

Nielsen J, Nikolajsen K, Villadsen J (1989) FIA for on-line monitoring of important lactic acid fermentation variables. Biotechnol Bioeng 33: 1127-1134.

Nielsen J, Nikolajsen K, Villadsen J (1991*a*) Structured modelling of a microbial system: I. A theoretical study of lactic acid fermentation. Biotechnol Bioeng 38: 1-10.

Nielsen J, Nikolajsen K, Villadsen J (1991*b*) Structured modelling of a microbial system: II. Experimental verification of a structured lactic acid fermentation model. Biotechnol Bioeng 38: 11-23.

Nielsen J, Nikolajsen K, Villadsen J (1991*c*) Structured modelling of a microbial system: III. Growth on mixed substrates. Biotechnol Bioeng 38: 24-29.

Nomura Y, Iwahara M, Hongo M (1987) Lactic acid production by electrodialysis fermentation using immobilized growing cells. Biotechnol Bioeng 30: 788-793.

Ohmiya K, Ohashi H, Kobayashi T, Shimizu S (1977) Hydrolysis of lactose by immobilized microorganisms. Appl Environ Microbiol 33: 137-146.

Olive, TR (1936) Waste lactose is a raw material for a new lactic acid process. Chem Met Eng 43: 480-483. Quoted in Prescott SC, Dunn CG (1959) The production of lactic acid by fermentation. In: Industrial Microbiology, 3rd ed, McGraw - Hill, New York, NY, pp 304-331.

Özen S, Özilgen M (1992) Effects of substrate concentration on growth and lactic acid production by mixed cultures of *Lactobacillus bulgaricus* and *Streptococcus thermophilus*. J Chem Technol Biotechnol 54: 57-61.

Peckham GT (1944) The commercial manufacture of lactic acid. Chem Eng News 22: 440. Quoted in Vick Roy TB (1985) Lactic acid. In: Moo-Young M (ed) Comprehensive Biotechnology, vol 3, Pergamon Press, Oxford, pp 761-776.

Pool R (1989) In search of the plastic potato. Science 245: 1187-1189.

Porubcan RS, Sellars RL (1979) Lactic starter culture concentrates. In: Peppler HJ, Perlman D (eds), Microbial Technology, 2nd ed, pp 59-92.

Prescott SC, Dunn CG (1959) The production of lactic acid by fermentation. In: Industrial Microbiology, 3rd ed, McGraw - Hill, New York, NY, pp 304-331.

Ray B, Daeschel M (eds) (1992) Food Biopreservatives of Microbial Origin, CRC Press, Boca Raton, FL.

Rees JF, Pirt J (1979) The stability of lactic acid production in resting suspensions of *Lactobacillus delbrueckii*. J Chem Technol Biotechnol 29: 591-602.

Salminen S, Ramos P, Fonden R (1993) Substrates and lactic acid bacteria. In: Salminen S, von Wright A (eds) Lactic Acid Bacteria, Marcel Dekker, New York, NY, pp 295-306.

Samuel WA, Lee YY, Anthony WB (1980) Lactic acid fermentation of crude sorghum extract. Biotechnol Bioeng 22: 757-777.

San-Martín M, Pazos C, Coca J (1992) Reactive extraction of lactic acid with Alamine 336 in the presence of salts and lactose. J Chem Technol Biotechnol 54: 1-6.

Sax NI, Lewis RJ Sr (revisors) (1987) Hawley's Condensed Chemical Dictionary, 11th ed, Van Nostrand Reinhold, New York, NY, p 680.

Schopmeyer HH (1954) Lactic acid. In: Underkofler LA, Hickey RJ (eds) Industrial Fermentations, Chemical Publishing Co., New York, NY, pp 391-419.

Sellars RL, Babel FJ (1978) Cultures for the Manufacture of Dairy Products (revised edition), CHR Hansen's Laboratory, Inc, Milwaukee, WI, pp 6-7.

Snell RL, Lowery CE (1964) Calcium (L+) lactic acid production. US Patent 3,125,494, March 17. Quoted in Vick Roy TB (1985) Lactic acid. In: Moo-Young M (ed) Comprehensive Biotechnology, vol 3, Pergamon Press, Oxford, pp 761-776.

Soccol CR, Stonoga VI, Raimbault M (1994) Production of l-lactic acid by *Rhizopus* species. World J Microbiol Biotechnol 10: 433-435.

Stamer JR (1976) Lactic acid bacteria. In: de Figueiredo MP, Splittstoesser DF (eds) Food Microbiology: Public Health and Spoilage Aspects, AVI, Westport, CT, pp 404-426.

Stanier RY, Adleburg EA, Ingraham JL (1976) The Microbial World, Prentice Hall, Englewood Cliffs, NJ, pp 496- 501.

Stenroos SL, Linko YY, Linko P (1982) Production of L-lactic acid with immobilized *Lactibacillus delbrueckii*. Biotechnol Lett 4: 159-164.

Stieber RW, Coulman GA, Gerhardt P (1977) Dialysis continuous process for ammonium-lactate fermentation of whey: experimental tests. Appl Environ Microbiol 34: 733-739.

Stieber RW, Gerhard P (1979*a*) Dialysis continuous process for ammonium-lactate fermentation of whey: improved mathematical model

and use of deproteinized whey. Appl Environ Microbiol 37: 487-495.
Stieber RW, Gerhard P (1979*b*) Continuous process for ammonium lactate fermentation of deproteinized whey. J Dairy Sci 62: 1558-1566.
Stieber RW, Gerhard P (1981*a*) Dialysis continuous process for ammonium-lactate fermentation: simulated prefermentor and cell-recycling systems. Biotechnol Bioeng 23: 523-534.
Stieber RW, Gerhard P (1981*b*) Dialysis continuous process for ammonium-lactate fermentation: simulated and experimental dialysate-free, immobilized-cell system. Biotechnol Bioeng 23: 535-549.
Teuber M (1991) Lactic acid bacteria. In: Rehm HJ, Reed G (eds) Biotechnology, 2nd ed, vol 1, VCH, Weinheim, pp 325-366.
Thorne JGM (1969) Synthetic lactic acid. Chem Processing (January): 8-9.
Tuli A, Sethi RP, Khanna PK, Marwaha SS, Kennedy JF (1985) Lactic acid production from whey permeate by immobilized *Lactobacillus casei*. Enzyme Microbiol Technol 7: 164-168.
Tyagi RD, Kluepfel D, Couillard D (1991) Bioconversion of cheese whey to organic acids. In: Martin AM (ed) Bioconversion of Waste Materials to Industrial Products, Elsevier Applied Science, London, pp 313-333.
Tyree RW, Clausen EC, Gaddy JL (1991) The production of propionic acid from sugars by fermentation through lactic acid as an intermediary. J Chem Technol Biotechnol 50: 157-166.
Van Ness JH (1981) Hydroxycarboxylic acids, lactic acid. In: Mark HF, Othmer DE, Overberger CG, Seaborg GT (eds) Encyclopedia of Chemical Technology, 3rd ed, vol 13, Wiley, New York, NY, pp 80-103.
Vick Roy TB (1985) Lactic acid. In: Moo-Young M (ed) Comprehensive Biotechnology, vol 3, Pergamon Press, Oxford, pp 761-776.
Vick Roy TB, Blanch HW, Wilke CR (1982) Lactic acid production by *Lactobacillus delbrueckii* in a hollow fiber fermenter. Biotechnol Lett 4: 483-488.
Vick Roy TB, Mandel DK, Dea DK, Blanch HW, Wilke CR (1983) The application of cell recycle to continuous fermentative lactic acid production. Biotechnol Lett 5: 665-670.
Viniegra-González G, Gómez J (1984) Lactic acid production by pure and mixed bacterial cultures. In: Wise DL (ed) Bioconversion Systems, CRC Press, Boca Raton, Florida, pp 17-39.
Ward OP (1989) Fermentation Biotechnology. Prentice Hall, Englewood Cliffs, NJ, pp 85, 139.
Wehrenberg RH II (1981) Lactic acid polymers: strong, degradable thermoplastics. Mater Eng 94: 63-66.
Whittier EO, Rogers LA (1931) Continuous fermentation in the production of lactic acid. Ind Eng Chem 23: 532-534. Quoted in Prescott SC,

Dunn CG (1959) The production of lactic acid by fermentation. In: Industrial Microbiology, 3rd ed, McGraw - Hill, New York, NY, pp 304-331.
Yabannavar VM, Wang DIC (1991*a*) Extractive fermentation for lactic acid production. Biotechnol Bioeng 37: 1095-1100.
Yabannavar VM, Wang DIC (1991*b*) Analysis of mass transfer for immobilized cells in an extractive lactic acid fermentation. Biotechnol Bioeng 37: 544-550.

Casein-breakdown by *Lactococcus lactis*

Bert Poolman, Vincent Juillard, Edmund R.S. Kunji, Anja Hagting and Wil N. Konings*

Department of Microbiology

University of Groningen

Kerklaan 30

9751 NN Haren

The Netherlands

This chapter is an updated version of Poolman *et al.*, Journal of Applied Bacteriology Symposium Supplement 1995, 79, 65S-75S
*** corresponding author**

NATO ASI Series, Vol. H 98
Lactic Acid Bacteria:
Current Advances in Metabolism, Genetics and Applications
Edited by T. Faruk Bozoğlu and Bibek Ray

1. INTRODUCTION

Mixed starter cultures used to manufacture Dutch cheeses are composed of lactic acid bacteria of which *Lactococcus* spp. are the dominant organisms (Hugenholtz 1986). Several metabolic properties of lactococci serve special functions which directly or indirectly have impact on processes such as flavour development and ripening of the cheese (Olson 1990). These functions are (i) fermentation and depletion of the milk sugar lactose, (ii) reduction of the redox potential, (iii) citrate fermentation, and (iv) degradation of casein. The degradation of casein by lactococci yields peptides and amino acids that are the sources of essential and growth-stimulating amino acids for *Lactococcus lactis* (Thomas and Pritchard 1987; Chopin 1993). In addition to being an important nutritional source for the starter culture bacteria, the casein degradation products also play a crucial role in the development of flavour in cheese. Certain peptides contribute to the formation of a typical cheese flavour, whereas others, undesirable bitter-tasting peptides, can result in an off-flavour. The need for a better understanding of the processes leading to the formation of these flavour peptides has prompted various research groups to study the components of the proteolytic pathway. Research on the proteolytic pathway of *L.lactis* is the most advanced. Most, if not all, of the components of this pathway have been identified, the majority of the enzymes purified and biochemically characterized, and the genetics of the corresponding genes studied (Table I).

An analysis of the ß-casein breakdown products has indicated that the peptides liberated by the proteinases are, with a few exceptions, too long to be taken up by *L.lactis* (Smid *et al.* 1991). Consequently, breakdown of these peptide fragments by extracellular peptidases is required to fulfil the needs of *L.lactis* for essential and growth-stimulating amino acids. On the basis of the product formation resulting from the action of purified P_I proteinases from different *L.lactis* strains on ß-casein and the knowledge at that time about peptide transport systems, the suggestion was made that at least a number of aminopeptidases and/or endopeptidases would be located extracellularly (Smid *et al.* 1991).

Table I. Well-characterized Enzymes of the Proteolytic Pathway of *Lactococcus lactis*

Enzyme	Name[1]	Type[2]	Substrates[3]	Characterization[4]	Reference[5]
Proteinase	PrtP	Serine (P_I)[6]	β-, κ-casein	P, S, Ext	Kok *et al.* (1988)
	PrtP	Serine (P_{III})[6]	α_{S1}-, β-, κ-casein	P, S, Ext	Vos *et al.* (1989)
Di-tripeptide transporter	DtpT	Δp-driven	Di/tripeptides	S, M	Hagting *et al.* (1994)
Oligopeptide transporter	Opp[7]	ATP-driven	Oligopeptides	S, M	Tynkkynen *et al.* (1993)
General Aminopeptidases	PepN	Metallo	Lys-↓pNA	P, S, Int	Tan *et al.* (1992b)
	PepC	Thiol	His-↓ßNA	P, S, Int	Chapot-Chartier *et al.* (1993)
	PepP	Metallo	X-↓Pro-Pro-$(X)_N$	P	Monnet *et al.* (1993)
Glutamyl aminopeptidase	PepA	Metallo	Glu-↓pNA	P, S, Int	M. Gasson (pers. comm.)
X-pro-dipeptidyl Aminopeptidase	PepX	Serine	X-Pro-↓pNA	P, S, Int	Mayo *et al.* (1991)
Prolidase	PRD	Metallo	X-↓Pro	P	Kaminogawa *et al.* (1984)
Proline Iminopeptidase	PIP	Metallo	Pro-↓X-(Y)	P	Baankreis and Exterkate(1991)
Pyrrolidonyl carboxylyl peptidase	PCP	Serine	PyroGlu-↓pNA	S, Int	A. Haandrikman (pers. comm.)
Dipeptidase	PepD	Metallo	Leu-↓Leu	P	Van Boven *et al.* (1988)
Tripeptidase	PepT	Metallo	Leu-↓Gly-Gly	P, S, Int	Mierau *et al.* (1994)
	TRP	Metallo	Leu-↓Leu-Leu	P, Ext?	Salhstrøm *et al.* (1993)
Endopeptidases	PepO	Metallo	$(X)_N$-↓$(Y)_M$	P, S, Int	Mierau *et al.* (1993b)
	$PepO_2$	Metallo	$(X)_N$-↓$(Y)_M$	P, S, Int	M. Hellendoorn *et al.* (1996)
	PepF	Metallo	$(X)_N$-↓$(Y)_M$	P, S, Int	V. Monnet (pers. comm.)
	$PepF_2$	Metallo	$(X)_N$-↓$(Y)_M$	P, S, Int	V. Monnet *et al.* (1996)

[1] See Tan et al. (1993) for details of the nomenclature.
[2]The enzyme type is based on inhibitor studies and/or sequence information from which the active site region can be inferred. An indication is given of whether the proton motive force (Δp) or ATP provides the driving force for transport.
[3]Typical substrates and the peptide cleavage sites (↓) are indicated.
[4]Characterisation: P, enzyme has been purified and biochemically characterised; S, gene has been isolated and sequenced; Ext, enzyme is located extracellularly; Int, enzyme is located intracellularly (may or may not be associated with the membrane); M, membrane protein (complex).
[5]The references cited refer to the publication in which the gene sequence was reported. If sequence data are not available the reference given is to the publication in which purification of the enzyme is described.
[6]Only one particular proteinase is usually present in a strain; the best studied enzymes are those from L.lactis Wg2 (P_I) and SK11 (P_{III}) (Kok et al. 1988; Vos et al. 1989). Upstream of and in opposite orientation to prtP, a gene is present (prtM) that encodes a protein essential for PrtP activation (maturation).
[7]The oligopeptide transport system is composed of 5 subunits that are encoded by OppDFBCA.

Studies in recent years have indicated that the proteinase PrtP is synthesized with a typical signal peptide sequence, but that this property is not shared by any of the peptidases analyzed so far (Kok and De Vos 1993). These findings are consistent with biochemical and immunological data which indicate that PrtP is present outside the cell (cell wall-associated), whereas the peptidases are found in the cytoplasm (Tan *et al.* 1992a; Baankreis 1992). The apparent discrepancy between the need for extracellular peptidases and the experimental data supporting an intracellular location of the enzymes has led us to investigate the hydrolysis of ß-casein *in vitro* and *in vivo*, and to follow the fate of the breakdown products *in vivo* using genetically well-defined peptide transport mutants. The outcome of these studies leads to the remarkable conclusion that all amino acids required for growth of *L.lactis* are released from β-casein by the action of PrtP in a form that can be transported by the oligopeptide transport system. Only one or two amino acids are accumulated in amounts insufficient to allow maximal growth rates of *L.lactis* on β-casein. These experiments and the recent advances in the understanding of the molecular genetics and biochemistry of the components of the proteolytic pathway of *L.lactis* are described.

2. PROTEINASE ACTIVITY *IN VITRO*

Caseins constitute *ca* 80% of all proteins present in bovine milk. The four different types of caseins found in milk, α_{S1}-, α_{S2}-, β-, and κ-casein, are organized in micelles to form soluble complexes (Schmidt 1982). In free solution, the caseins behave as non-compact and largely flexible molecules with a high proportion of residues accessible to the solvent, i.e. like random coil-type proteins (Holt and Sawyer 1988). On the basis of the patterns of breakdown of α_{S1}-, β-, and κ-caseins, two protease specificity-classes have been defined (Visser *et al.* 1986). These proteinase activities are generally referred to as P_I and P_{III}. The primary substrate of P_I-type enzymes is ß-casein although κ-casein is also degraded, while P_{III}-type enzymes degrade α_{S1} -, β- and κ-caseins (Pritchard and Coolbear 1993). The breakdown of ß-casein by P_I - and P_{III}-type proteinases is different. The genes encoding several P_I - and P_{III} -type proteinases have been cloned and sequenced, and the primary amino acid sequences of the proteins appear to be more than 98% identical (Kok and De Vos 1993). From the analysis of hybrid proteins composed of domains of the P_I-type proteinase of *L.lactis* Wg2 and the P_{III} -type enzyme of strain SK11, the differences in the proteinase specificities have been confined to two regions i.e. one in the catalytic domain (residues 131-166 of the mature protein) and another in which the amino acid residues 747-748 are particularly important (Vos *et al.* 1991). It is worthwhile emphasizing that PrtP activity is essential for degradation of caseins, and that this enzyme allows the bacteria to grow in casein-containing environments such as milk.

The products resulting from the action of proteinases on ß-casein have been analyzed *in vitro* using purified P_I and P_{III} enzymes of different *L.lactis* strains (Monnet *et al.* 1989; Visser *et al.* 1988; Reid *et al.* 1991; Pritchard and Coolbear 1993). These studies indicate that only part of ß-casein is degraded and that relatively large fragments - only a few contain less than 8 amino acid residues - are formed (cleavage sites are indicated in Fig.1). These analyses, however, are hampered by a relatively low resolution in the detection of the peptide fragments, and only the peptide composition of the major peaks has been determined. Using chromatography techniques with a highly improved resolution in combination with mass spectrometry, more than 95% of the peptides formed by the action of the P_I proteinase of *L.lactis* Wg2 on ß-casein have been recovered (Juillard *et al.* 1994). The results indicated that ß-casein is degraded to peptide fragments of 4 to 30 residues, of which a major fraction falls in the range of 4-10 residues (Fig.2). The proteinase activity did not yield detectable amounts of di- and tripeptides.

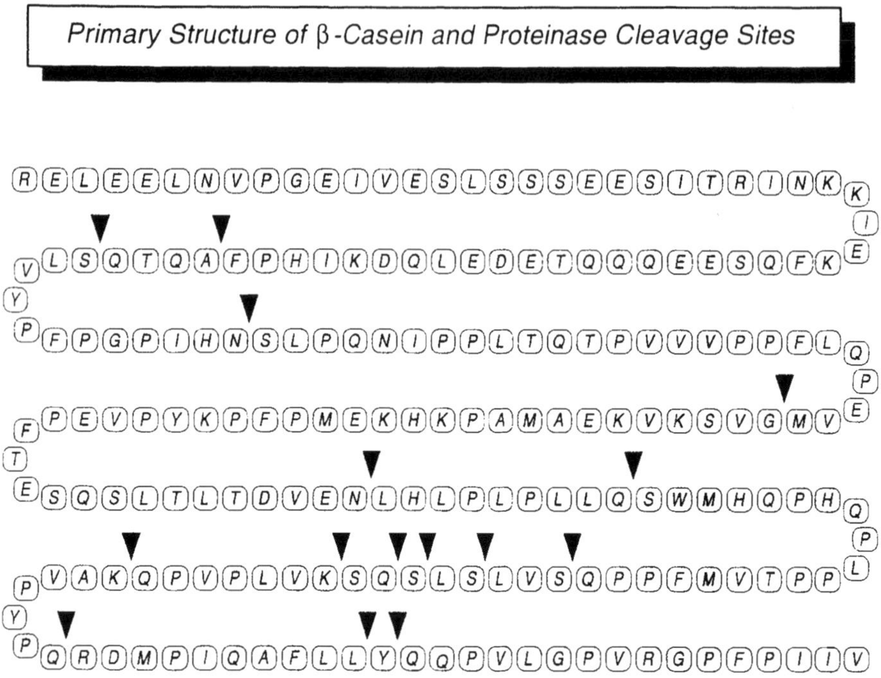

Fig. 1. Primary structure of β-casein and P_I proteinase cleavage sites. The cleavage sites were determined for the proteinases of Lactococcus lactis NCDO763, AC1, HP and H2 (see Pritchard and Coolbear 1993). Cleavage sites are indicated by arrows; note that not every site has been observed for each of the enzymes.

It seems, therefore, that many more peptide fragments are formed by the proteinase than orginally thought, and that the size of many of these peptides is such that uptake by the oligopeptide transport system of *L.lactis* is possible (Kunji *et al.* 1993; Tynkkynen *et al.* 1993). Thus, the question remains whether these peptides are hydrolyzed further by extracellular peptidases or whether they are transported into the cell directly. To answer this question it is important to have a collection of genetically well-defined mutants that are deficient in one or more peptidases and/or peptide transport systems. The role of a particular enzyme in casein degradation can then be addressed by analysing the degradation of casein *in vivo*.

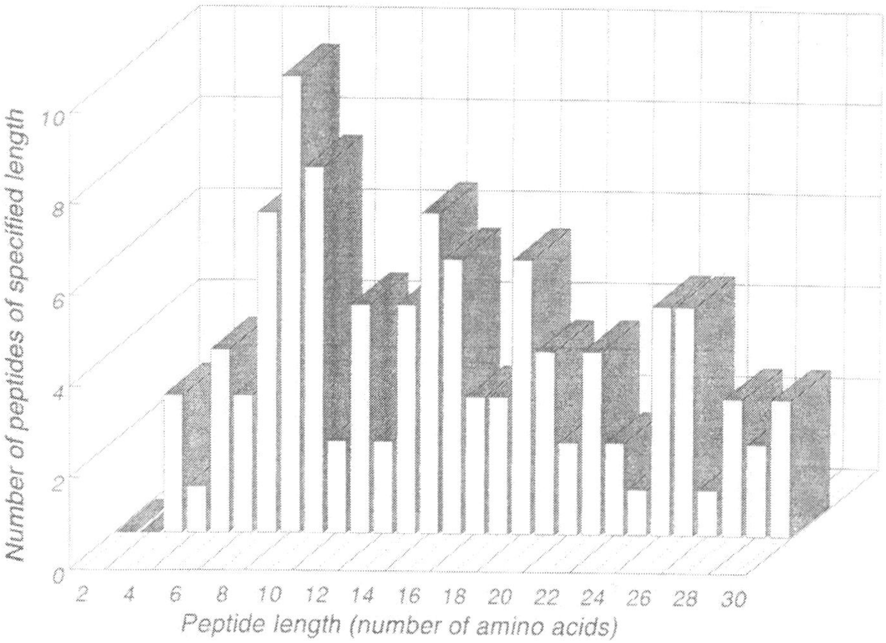

Fig.2. Size-distribution of the peptides released from β-casein by the P_I proteinase of Lactococcus lactis Wg2. Data from Juillard et al. (1994), using mass spectrometry detection of peptide fragments. Note the difference in size-distribution when the data are compared with those of Fig.1.

Before describing these experiments current information on lactococcal peptidases and peptide transport systems pertinent to the analyses will be summarized. Information on specificities, enzyme effectors, primary sequences etc. are given in a number of recent reviews (Tan *et al.* 1993; Kok and DeVos 1993; Pritchard and Coolbear 1993) and in the chapters of J. Kok and W. De Vos

3. PEPTIDASES

Table I summarizes the components of the proteolytic pathway that have been characterized in great detail. Most of the peptidases identified in *L.lactis* have been isolated and purified to homogeneity, and the corresponding genes cloned and sequenced. Inspection of amino-terminal sequences obtained from Edman degradation of purified enzymes and of those deduced from nucleotide sequences indicates that the N-terminal end of all the peptidases investigated is unprocessed. The localization of two aminopeptidases (PepN and PepC), an X-prolyl-dipeptidyl aminopeptidase (PepX), an endopeptidase (PepO), a tripeptidase (PepT) and a glutamyl aminopeptidase (PepA) has been addressed by cell fractionation studies and electron microscopy of immunogold-labeled peptidases (Tan *et al.* 1992a; Baankreis 1992). Although a location at the periphery of the cell has been suggested for some of the enzymes, each of the peptidases could clearly be demonstrated intracellularly. Some association of PepX, PepO, PepT and PepA with membrane fractions has been observed, however. This location has been observed also for other typical cytoplasmic enzymes such as ß-galactosidase (Foucaud and Poolman 1992) and malolactic enzyme (Poolman *et al.* 1991). The recent claim for the isolation of a di-tripeptidase from the cell wall fraction of a *L.lactis* strain (Sahlstrøm *et al.* 1993) awaits confirmation by further biochemical and genetic studies. Despite earlier suggestions to the contrary it can be concluded that, on the basis of the available experimental data, there is no good evidence that peptidases of *L.lactis* are located extracellularly.

Using a variety of genetic approaches the genes encoding inactivation of PepX, PepN, PepT and PepO, and combinations of genes (*pepX pepO, pepN pepO, pepN pepX and pepT pepX*) have been inactivated (Mayo *et al.* 1991; Tynkkynen *et al.* 1993; Mierau *et al.* 1993a; 1994). Although the hydrolysis patterns of selected peptides were altered, the single as well as the double mutants retained their ability to grow in milk. This is not very surprising because, in general, lactic acid bacteria contain more than one enzyme able to hydrolyze a particular peptide or to produce a specific amino acid (Tan *et al.* 1993). Also, as is to be expected, the specific effects of inactivation of peptidase genes on the utilization of caseins are difficult to prove since individual activities (not necessarily a single enzyme) can have essential housekeeping functions in the cell - e.g. turnover of endogenous proteins (Goldberg and St. John 1976). Beyond a critical activity, inactivation of peptidase genes may be lethal irrespective of whether amino acids, peptides or caseins are used as sources of nitrogen for growth.

4. PEPTIDE TRANSPORT

4.1. *Transport of di- and tripeptides*

An essential role for oligopeptide peptide transport system(s) in the utilization of casein is to be expected if extracellular casein hydrolysis is catalyzed by the action of PrtP only, i.e. when peptidases are not involved in the extracellular degradation. To establish whether peptide transport plays a role in the utilization of casein by *L.lactis* and to determine how many distinct transport activities are present, mutants have been selected for resistance towards toxic peptides (Smid *et al.* 1989b; Kunji *et al.* 1993). Using the toxic dipeptide L-alanyl-ß-chloro-L-alanine, transport mutants of *L. lactis* have been isolated. These peptide-transport mutants (DtpT^{-}) are unable to grow in a chemically defined medium supplemented with a mixture of caseins as sole source of nitrogen. Their growth is similar to that of the wild-type strain when the medium contains a mixture of amino acids (Smid *et al.* 1989b). These experiments indicate that, in addition to the proteinase PrtP, a di-tripeptide transport activity forms an essential part of the proteolytic pathway of *L.lactis.* It also suggests that at least a single amino acid is taken up in the form of a di- or tripeptide although such small peptides have not been detected in the degradation of ß-casein by PrtP (Fig.2; Juillard *et al.* 1994). It is possible that these di- and tripeptides are obtained from casein species other than the -form (possibly -casein). Control experiments have verified that proteinase, peptidase and oligopeptide transport activities are not lowered by the mutation(s) that result in the DtpT^{-} phenotype (Smid *et al.* 1989b; Kunji *et al* 1993). In fact, the proteinase activity is somewhat increased in the DtpT^{-} strain.

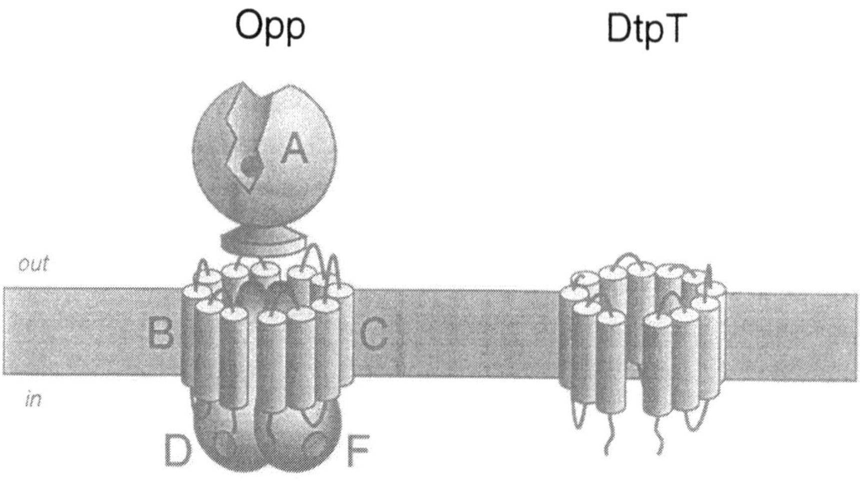

Fig.3. The oligopeptide and di-tripeptide transport systems of Lactococcus lactis.

Transport studies in peptidase-free membrane vesicles of *L.lactis* have shown that relatively hydrophilic dipeptides such as L-alanyl-L-glutamate, L-alanyl-L-alanine, and L-glutamyl-L-glutamate are transported by a proton motive force-driven transport mechanism (Smid *et al.* 1989a). The gene (*dtpT*) encoding this di-tripeptide transport protein has been cloned by complementation and functionally expressed in *E.coli* and *L.lactis* using plasmid constructs (Hagting *et al.* 1994). The amino acid sequence deduced from the nucleotide sequence of *dtpT* shows no significant similarity to that of other proteins, but the secondary structure of DtpT is similar to that of other proton-linked transport proteins (Fig.3).

In fact, DtpT represents the first example of a bacterial transport system that transports peptides in symport with proton(s). Other known bacterial peptide transport systems belong to the superfamily of ABC transporters that couple transport to the hydrolysis of ATP (Higgins 1992). Using flanking regions of the di-tripeptide transport gene, *dtpT* has

been deleted from the chromosome via homologous recombination. Characterisation of the Δ*dtpT* strain indicates that DtpT is the only transport protein for hydrophilic di- and tripeptides in *L.lactis*. Recently a third peptide transport system has been found that facilitates the uptake of more for hydrophobic substrates (Foucaud *et al.*,).

4.2. ***Transport of oligopeptides***

From mutant analysis it has become apparent that *L.lactis* possesses, in addition to a di-tripeptide transport system(s), a transporter that is specific for peptides of four to at least six residues (oligopeptides) (Kunji *et al.* 1993). Growth experiments in chemically defined media containing peptides of varying length have suggested that the oligopeptide transport system may also transport peptides up to lengths of 7 and 8 residues (Tynkkynen *et al.* 1993), and possibly even longer. On the basis of metabolic inhibitor studies it has been concluded that this oligopeptide transport is driven by ATP rather than the proton motive force (Kunji *et al.* 1993).

The genes encoding the oligopeptide transport protein were cloned fortuitously. A spontaneous mutant (MG1614V) of *L.lactis* MG1614 was isolated that did not grow in milk, despite the presence of a functional proteinase gene. When a particular chromosomal DNA fragment of *L.lactis* SSL135 was introduced into MG1614V, growth in milk was restored (Von Wright *et al.* 1987). Nucleotide sequencing of part of this chromosomal fragment has revealed the presence of six open reading frames (Tynkkynen *et al.*1993). Five of these frames encode polypeptides that are typical components of binding protein-dependent transport systems (OppDFBCA) (Fig.3), and the system has been classified as a member of the ATP-binding cassette (ABC) superfamily (Higgins 1992). The sixth open reading frame encodes the endopeptidase PepO (Table I). The five subunits of the oligopeptide transport system include a peptide binding protein (OppA), two transmembrane domains (OppB and OppC), and two ATP-binding proteins (OppD and OppF). The translated sequence of the *oppA* gene has the consensus prolipoprotein cleavage site Leu-Ser-Ala-↓Cys between residues 22 and 23 (Tynkkynen *et al.* 1993). This suggests that the mature protein is anchored to the

outer surface of the cytoplasmic membrane, via an amino-terminal lipo-cysteine anchor (Von Heijne 1989; Poolman 1993). OppA serves as the receptor protein that delivers the peptides to the membrane-bound translocator complex. OppB and OppC are highly hydrophobic proteins that, on the basis of hydropathy profiling, are able to span the cytoplasmic membrane in α-helical configuration six times. These proteins are likely to constitute the pathway that facilitates the translocation of oligopeptides across the membrane. OppD and OppF are homologous to the ATP binding protein(s) (domains) of the ABC-transporter superfamily (Higgins 1992). These proteins most likely couple the hydrolysis of ATP to conformational changes in OppB/C that allow passage of the peptides across the membrane.

To discriminate between OppDFBCA and PepO as essential components of the proteolytic pathway of *L.lactis*, two integration mutants have been constructed, one defective in OppA and the other in PepO. Growth of these mutants in milk and in a chemically defined medium with oligopeptides has shown that the OppDFBCA system, but not the endopeptidase, is essential for the utilization of casein - and oligopeptides (Tynkkynen *et al.* 1993).

5. CASEIN DEGRADATION *IN VIVO*

Gene inactivation experiments have demonstrated that PrtP, Opp and DtpT are essential for growth of *L.lactis* on media containing casein(s) as sole source of amino acids. Such a crucial role in casein hydrolysis has not been observed for any of the peptidases investigated to date. Moreover, various lines of evidence indicate that the peptidases are located intracellularly and, consequently, cannot play a role in the extracellular hydrolysis of caseins or casein-derived peptides, unless cell lysis occurs. These observations together with the tentative evidence that the Opp system may transport peptides with a length of 4 up to at least 8 residues, suggest that a major fraction of the essential and growth-stimulating amino acids enter the cell in the form of oligopeptides. To prove that the Opp system can transport various oligopeptides with a chain length of 8 or longer is not easy. Such peptides are not readily available and/or do

not mimic the casein hydrolysis fragments. An *in vivo* approach was adopted therefore, to study the role of peptide uptake in the degradation of β-casein, by the use of genetically well-defined peptide transport mutants that expressed PrtP. The strains used for the analysis were of the following phenotypes: PrtP^{+} DtpT^{+} Opp^{+} (= wild-type MG1363/pLP712), PrtP^{+} DtpT^{-} Opp^{+}, PrtP^{+} DtpT^{+} Opp^{-}, and PrtP^{+} DtpT^{-} Opp^{-} (Table 2).

Table 2

Characterisation of *Lactococcus lactis* Peptide Transport Mutants

Properties	Wild-type	DtpT^{-}Opp^{+}	DtpT^{+}Opp^{-}	DtpT^{-}Opp^{-}
Proteinase Activity[1]	1.9	3.6	2.1	4.8
μ_{MAX} (h^{-1})[2]				
Amino acids	0.92	0.99	0.94	0.96
L-Ala-L-Glu	0.76	0.00	0.72	0.02
Leu-enkephalin	0.76	0.69	0.00	0.03
Milk	0.52	0.52	0.07	0.06
Alanine Uptake[3]				
L-Alanine	55	65	56	75
Di-L-alanine	108	1	120	1
Tetra-L-alanine	21	25	2	1
L-Ala-Cl-Ala[4]	Sensitive	Resistant	Sensitive	Resistant

[1]*Proteinase activity was estimated from the hydrolysis of MeO-Suc-Arg-Pro-Tyr-p-nitroanilide-HCl (activity in nmol.min^{-1} x mg of protein).*
[2]*Maximal growth rates (μ_{MAX}) were calculated from the increase in A_{660} in chemically-defined media containing all amino acids; all amino acids minus L-glutamate plus L-Ala-L-Glu; all amino acids minus L-leucine plus Leu-enkephalin or milk as indicated.*
[3]*Alanine uptake refers to the initial rates of accumulation of alanine with L-alanine, L-alanyl-L-alanine (Di-L-alanine) or L-alanyl-L-alanyl-L-alanyl-L-alanine (tetra-L-alanine) as substrate.*
[4]*L-Ala-Cl-Ala refers to sensitivity (growth inhibition) of the cells to L-alanyl-β-chloro-alanine when present in a chemically-defined medium.*

The amino acids and peptides formed both extra- and intracellularly from the degradation of ß-casein by these strains have been analyzed (Kunji *et al.* 1994).

Using MG1363 as parent strain (PrtP^{-} DtpT^{+} Opp^{+}), *dtpT* was deleted from the chromosome with or without subsequent disruption of the *oppA* gene (*oppA* encodes the binding protein of the oligopeptide transport system). The genes essential for expression of a functional proteinase (*prtP* and *prtM*) were then introduced into the wild-type and mutant strains by conjugation with plasmid pLP712 using *L.lactis* NCDO712 as the donor strain. The various mutations were confirmed by Southern hybridization analyses. The phenotypes of the strains were checked by various criteria (Table 2). (i) Proteinase activity was assayed by monitoring the hydrolysis of the chromogenic substrate MeO-Suc-Arg-Pro-Tyr-pNA. The increased proteinase activity in the DtpT^{-} Opp^{+} and DtpT^{-}Opp^{-} strains is caused by an elevated expression of the proteinase; lack of a functional di-tripeptide transport system relieved inhibition of expression by dipeptides (unpublished results). (ii) The presence or absence of a functional di-tripeptide transport system was inferred by the sensitivity of the strains to the toxic dipeptide L-alanyl-ß-chloro-L-alanine; DtpT^{+} strains were sensitive whereas DtpT^{-} strains were resistant. (iii) The presence or absence of a particular transport system was inferred from the uptake of L-alanine, L-alanyl-L-alanine (di-L-alanine) and L-alanyl-L-alanine-L-alanine-L-alanine (tetra-L-alanine) which are substrates specific for the L-alanine/glycine, di-tripeptide and oligopeptide transport systems, respectively. The results indicated clearly that di-L-alanine uptake was lacking in the DtpT^{-} strains, whereas tetra-L-alanine was absent in the Opp^{-} strains. (iv) Growth of the bacteria was

monitored in milk and in a chemically defined medium containing all essential amino acids except for one which was supplied in the form of a di- or pentapeptide. The results of these experiments, summarized in Table 2, indicate that the phenotypes were consistent with the mutations made.

The effect of the various mutations on the utilization of ß-casein was studied *in vivo* by incubating chloramphenicol-treated cells of *L.lactis* (wild-type and peptide transport mutants) with ß-casein in the presence of glucose as a source of metabolic energy.

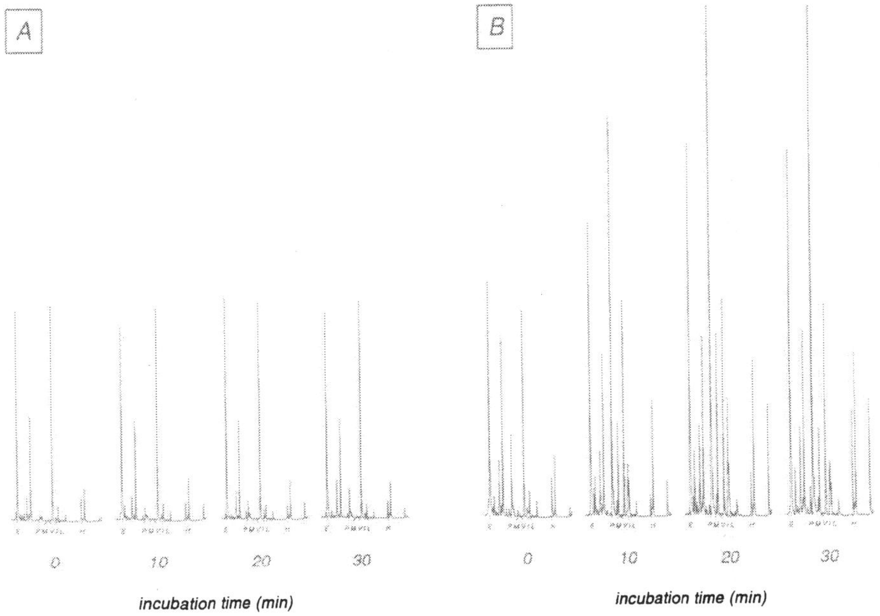

Fig.4. Elution profiles of the intracellular amino acids of MG1363 $prtP^{+}\Delta dtpT$ opp^{+} (right panel) and MG1363 $prtP^{+}dtpT^{+}$ $oppDFCB^{+}$ oppA::pLS19A (left panel) after incubation of chloramphenicol-treated cells with ß-casein in the presence of glucose for varying periods of time. E, glutamate; P, proline; M, methionine; V, valine; I, isoleucine; H, histidine.

At given time intervals the cells were separated from the medium by rapid filtration, the cells extracted with perchloric acid, and both the extra- and intracellular fractions analyzed for the presence of amino acids and/or peptides after derivatisation with

dansylchloride. Typical reversed-phase HPLC elution profiles are shown in Fig.4. From these data it is clear that the pools of various amino acids increased in time in the PrtP⁺ DtpT⁻ Opp⁺ strain whereas no significant changes in the pool sizes were observed in PrtP⁺ DtpT⁺ Opp⁻. All the peptides taken up by the cells were rapidly hydrolysed intracellularly. Consequently, significant amounts of peptides were never detected in the intracellular fraction (see also Kunji *et al.* 1993; Tynkkynen *et al.* 1993; Hagting *et al.* 1994). In fact, all the peaks can be attributed to a particular amino acid.

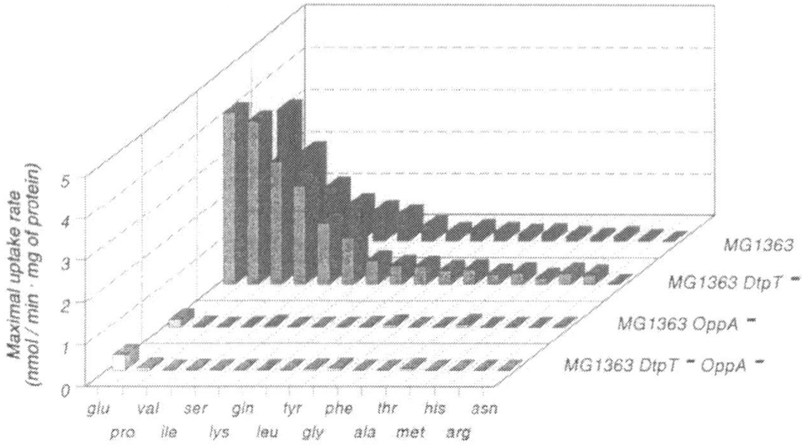

Fig.5. Maximal rates of appearance of amino acids in the cytoplasm of chloramphenicol-treated cells incubated with β-casein in the presence of glucose. The phenotype of the various strains is indicated. Since the intracellular glutamate pools were very high, aspartate could not be separated from glutamate in the HPLC analysis (the sum of both pools is indicated under Glu).

The data for all four strains are summarized in Fig.5. In this Figure the rates at which the amino acids appear in the cytoplasm are compared.

The results show that the wild-type MG1363/pLP712 and the PrtP⁺ DtpT⁻ Opp⁺ strain accumulate all amino acids that have been analyzed[1]. Growth experiments indicate that, with the exception of histidine and leucine, the amino acids are accumulated sufficiently to support maximal growth of *L.lactis* (E.R.S. Kunji, unpublished results). Remarkably, none of the amino acids is accumulated significantly in the strains lacking a functional oligopeptide transport system (Opp⁻), irrespective of whether or not DtpT is present (Fig.6). Control experiments in which the cells were fed with a mixture of amino acids mimicing the composition of ß-casein, rather than ß-casein itself, showed that the pool sizes of individual amino acids increased to the same extent in all four strains (data not shown). These experiments reveal a number of important properties of the proteolytic pathway of *L.lactis.* First, most, if not all, of the essential and growth-stimulating amino acids of *L.lactis* can be released from ß-casein by the action of the proteinase PrtP in a form that can be transported by the cells. Second, these peptides are taken up by the oligopeptide transport system. Third, consistent with the observation that PrtP does not release significant amounts of di- and tripeptides from ß-casein (Juillard *et al.* 1994), inactivation of the di-tripeptide transport system has no effect on the utilization of the protein. On the other hand di-tripeptide transport mutants selected on the basis of resistance towards L-alanine-ß-chloro-L-alanine have a decreased ability to grow on caseins (Smid *et al.* 1989b). It is possible that this is due to an inability to release essential amino acids (probabely histidine and/or leucine) in the form of small peptides from proteins other than ß-casein (e.g. κ-casein). Fourth, if peptidases had been present extracellularly, it is likely that amino acids as well as di- and tripeptides would have been released. The observation that a single mutation, that abolishes oligopeptide transport activity, results in an inability to accumulate amino acids argues strongly against the presence of extracellular peptidases. Fifth, the observation that *L.lactis* can grow well on a chemically defined medium supplemented with ß-casein, histidine plus leucine as sole source of amino acids indicates that uptake of the oligopeptides occurs at

---

ä1. A single tryptophan is present in ß-casein whereas cysteine residues are absent. These amino acids have not been detected inside the cell. In *L.lactis* intracellular asparagine is converted largely into aspartate and glutamine into glutamate (Poolman *et al.* 1987), consequently, the uptake of these amino acids cannot be inferred from the individual concentration

rates that are sufficient to meet most of the (essential) amino acids required for the growth of the organism. (Glu/Gln, Leu, Val, Ile, Met, His) are essential for *L.lactis* MG1363, whereas another four (Asn, Pro, Phe and Ala) are needed to support This is quite remarkable given the competition between the peptides for a single binding protein. This is especially so since at least six amino acids reasonable rates of growth (unpublished results).

Finally, the extracellular medium was analyzed with respect to the appearance/accumulation of amino acids and peptides. In the case of the PrtP$^+$ DtpT$^+$ Opp$^-$ and PrtP$^+$ DtpT$^-$ Opp$^-$ strains the accumulation of various peptides in the medium reflected the degradation of ß-casein by the proteinase PrtP. A much smaller number of peptides accumulated in the medium with the wild-type and PrtP$^+$ DtpT$^-$ Opp$^+$ strains. The observation that some peptides do accumulate in the medium despite a functional Opp system may be a consequence of the size exclusion limits of the oligopeptide transporter (peptides up to a length of 30 amino acids are formed; Fig.2; Juillard *et al.* 1994). Furthermore, although the lactococcal oligopeptide transport system must have a broad substrate specificity, certain peptides may not be transported due to competitition of the peptides for entry via a single oligopeptide transport system. Part of the peptide fraction may also be taken up at a rate that is lower than that of their production by the proteinase. As a result a significant fraction of peptides generated by PrtP will accumulate in the medium. Only minor amounts of amino acids appear in the extracellular medium. Since peptides taken up by *L.lactis* are rapidly hydrolyzed by intracellular peptidases, large outwardly directed amino acid concentration gradients are formed. These are the most likely explanation for leakage of some amino acids from the cells.

6. CONCLUSIONS

Lactic acid bacteria that are used in the dairy industry for the production of food products have an active proteolytic system that is involved in the degradation of milk proteins (caseins).

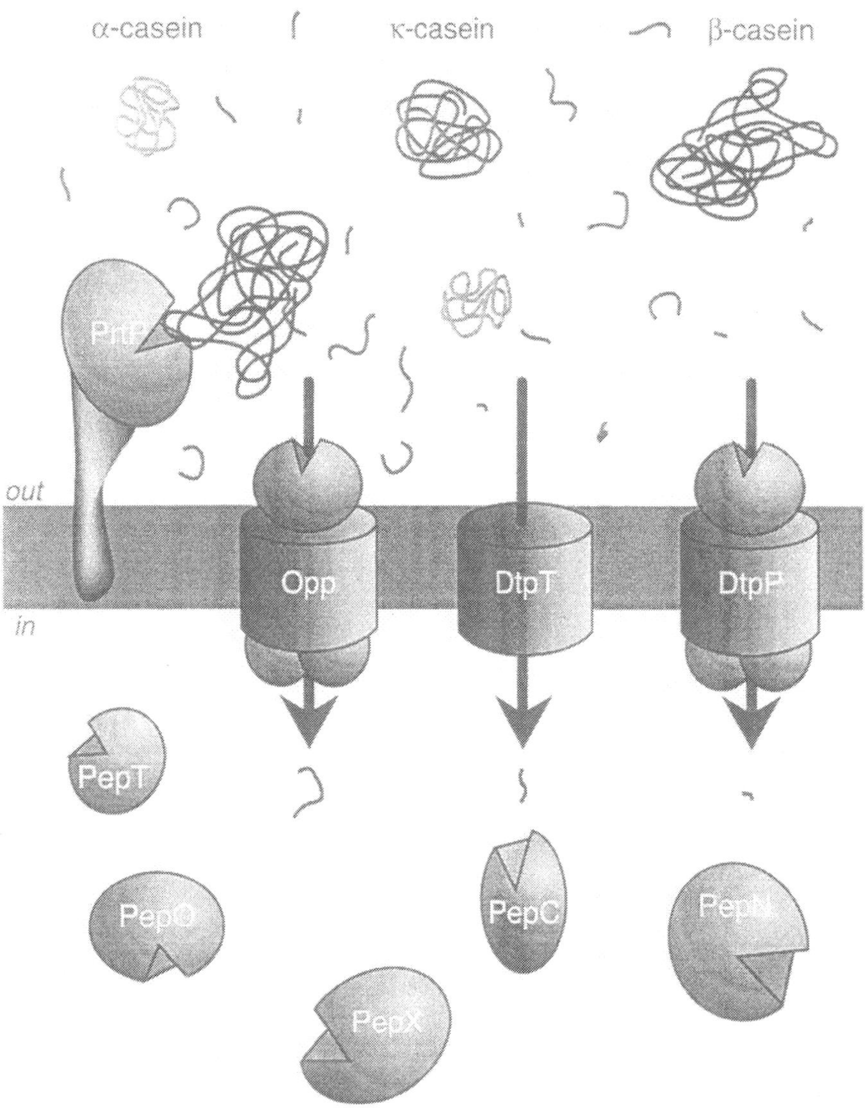

Fig. 6. The proteolytic pathway of Lactococcus lactis.

In the proteolytic pathway of *L.lactis* (Fig.6), casein is hydrolyzed extracellularly to (oligo)peptides by a protease (PrtP). A large fraction of the peptides formed is taken up

by the cells via the oligopeptide system (Opp); other peptides accumulate in the medium. A minor role in the uptake of essential amino acids in the form of peptides is played by the di-tripeptide transport system (DtpT). Although several peptidases have been identified in *L.lactis,* biochemical and genetic studies indicate that the enzymes are located intracellularly and play no role in the extracellular degradation of casein-derived peptides. Since peptides are known to contribute to flavour development in cheese, engineering of the specificity of the (oligo)peptide transport systems, to produce a different spectrum of peptides, could result in different and perhaps improved flavour. Additionally, the growth of the organisms on casein-based media could be accelerated.

7. REFERENCES

Baankreis, R. and Exterkate, F.A. (1991) Characterization of a peptidase from *Lactococcus lactis* ssp. *cremoris* HP that hydrolyses di- and tripeptides containing proline or hydrophobic residues as the amino-terminal amino acid. *Systematic Applied Microbiology* 14, 317-323.

Baankreis, R. (1992) The role of lactococcal peptidases in cheese ripening. Ph.D. thesis, University of Amsterdam, The Netherlands.

Chapot-Chartier, M.P., Nardi, M., Chopin, M.-C., Chopin, A. and Gripon, J.-C. (1993) Cloning and sequencing of *pepC*, a cysteine aminopeptidase gene from *Lactococcus lactis* subsp. *cremoris* AM2. *Applied and Environmental Microbiology* 59, 330-333.

Chopin, A. (1993) Organization and regulation of genes for amino acid biosynthesis in lactic acid bacteria. *FEMS Microbiology Reviews* 12, 21-37.

Foucaud, C. and Poolman, B. (1992) Lactose transport protein of *Streptococcus thermophilus*. Functional reconstitution of the protein and characterization of the kinetic mechanism of transport. *Journal of Biological Chemistry* 267, 22087-22094.

Foucaud, C., Kunji, E.R.S., Hagting, A., Richard, J., Konings, W.N. Desmazeaud, M. and Poolman, B. Specificity of peptide transport systems in *Lactococus lactis*; evidence for a third system which transports hydrophobic di- and tripeptides. *Journal of Bacteriology* 177, 4652-4657.

Goldberg, A.L. and St. John, A.C. (1976) Intracellular protein degradation in mammalian and bacterial cells. *Annual Review of Biochemistry* 45, 747-803.

Hagting, A., Kunji, E.R.S, Leenhouts, K.J., Poolman, B. and Konings, W.N. (1994) The di- and tripeptide transport protein of *Lactococcus lactis*. A new type of bacterial peptide transporter. *Journal of Biological Chemistry* 269, 11391-11399

Higgins, C.F. (1992) ABC transporters: from microorganisms to man. *Annual Review of Cellular Biology* 8, 67-113.

Holt, C. and Sawyer, L. (1988) Primary and predicted secondary structures of the caseins in relation to their biological functions. *Protein Engineering* 2, 251-259.

Hugenholtz, J. (1986) Population dynamics of mixed starter cultures. *Netherlands Milk Dairy Journal* 40, 129-140.

Juillard, V., Laan, H., Kunji, E.R.S., Heronimus-Stratingh, C.M., Bruins, A.P. and Konings, W.N. (1995) The extracellular P_1-type proteinase of *Lactococcus lactis* hydrolyzes ß-casein into more than one hundred oligopeptides. *Journal of Bacteriology* 177, 3472-3478

Kaminogawa, S., Azuma, N., Hwang, I.K., Susuki, Y. and Yamauchi, K. (1984) Isolation and characterization of a prolidase from *Streptococcus cremoris* H61. *Agricultural and Biological Chemistry* 48, 3035-3040.

Kok, J. and De Vos, W.M. (1993) The proteolytic system of lactic acid bacteria. In *'Genetics and Biotechnology of Lactic Acid Bacteria'* (Eds. M.J. Gasson and W.M. De Vos), pp 169-210, Blackie Academic & Professional, London, U.K.

Kok, J., Leenhouts, K., Haandrikman, A.J., Ledeboer, A.M. and Venema, G. (1988) Nucleotide sequence of the cell wall proteinase gene of *Streptococcus cremoris* Wg2. *Applied and Environmental Microbiology* 54, 231-238.

Konings, W.N., Poolman, B. and Driessen, A.J.M. (1989) Bioenergetics and solute transport in lactococci. *Critical Reviews in Microbiology* 16, 419-476.

Kunji, E.R.S., Smid, E.J., Plapp, R., Poolman, B. and Konings, W.N. (1993) Di-tripeptides and oligopeptides are taken up via distinct transport mechanisms in *Lactococcus lactis*. *Journal of Bacteriology* 175, 2052-2059.

Kunji, E.R.S., Hagting, A., De Vries, C.J., Juillard, V., Haandrikman, A.J., Poolman, B. and Konings, W.N. (1994) Transport of β-casein-derived peptides by the oligopeptide transport system is a crucial step in the proteolytic pathway of Lactococcus lactis. *Journal of Biological Chemistry* 270, 1569-1574.

Mayo, B., Kok, J., Bockelman, W., Haandrikman, A., Leenhouts, K.J. and Venema, G. (1993) Effect of X-prolyl dipeptidyl aminopeptidase deficiency on *Lactococcus lactis*. *Applied and Environmental Microbiology* 59, 2049-2055.

Mierau, I., Haandrikman, A.J., Leenhouts, K.J., Kok, J. and Venema, G. (1993a) Analysis of lactococcal strains mutated in peptidase genes. *FEMS Microbiology Reviews* 12, p83.

Mierau, I., Tan, P.S.T., Haandrikman, A.J., Kok, J., Leenhouts, K.J., Konings, W.N. and Venema, G. (1993b) Cloning and sequencing of the gene for a lactococcal endopeptidase, an enzyme with sequence similarity to mammalian enkephalinase. *Journal of Bacteriology* 175, 2087-2096.

Mierau, I., Haandrikman, A.J., Velterop, O., Tan, P.S.T., Leenhouts, K.J., Konings, W.N., Venema, G. and Kok, J. (1994) Tripeptidase gene (*pepT*) of *Lactococcus lactis*: molecular cloning and nucleotide sequencing of *pepT* and construction of a chromosomal deletion mutant. *Journal of Bacteriology* 176, 2854-2861.

Monnet, V., Bockelman, W., Gripon, J.-C. and Teuber, M. (1989) Comparison of cell wall proteinases from *Lactococcus lactis* subsp. *cremoris* AC1 and *Lactococcus lactis* subsp. *lactis* NCDO 763. *Applied Microbiology and Biotechnology* 31, 112-118.

Monnet, V., Mars, I. and Gripon, J.-C. (1993) Purification and specificity of a lactococcal aminopeptidase P. *FEMS Microbiology Reviews* 12, p80.

Olson, N.F. (1990) The impact of lactic acid bacteria on cheese flavour. *FEMS Microbiology Reviews* 87, 131-148.

Poolman, B. (1993) Energy transduction in lactic acid bacteria. *FEMS Microbiology Reviews* 12, 125-148.

Poolman, B., Kunji, E.R.S., Hagting, A., Juillard, V., Konings, W.N. (1995) The proteolytic pathway of *Lactococcus lactis*,

Poolman, B., Molenaar, D., Smid, E.J., Ubbink, T., Abee, T., Renault, P.P. and Konings, W.N. (1991) Malolactic fermentation: electrogenic malate uptake and malate/lactate antiport generate metabolic energy. *Journal of Bacteriology* 173, 6030-6037.

Poolman, B., Smid, E.J. and Konings, W.N. (1987) Kinetic properties of a phosphate-bond-driven glutamate-glutamine transport system in *Streptococcus lactis* and *Streptococcus cremoris*. *Journal of Bacteriology* 269, 2755-2761.

Pritchard, G.G. and Coolbear, T. (1993) The physiology and biochemistry of the proteolytic system in lactic acid bacteria. *FEMS Microbiology Reviews* 12, 179-206.

Reid, J.R., Ng, K.H., Moore, C.H., Coolbear, T. and Pritchard, G.G. (1991) Comparison of bovine β-casein hydrolysis by P_I and P_{III}-type proteinases from *Lactococcus lactis* subsp. *cremoris*. *Applied Microbiology and Biotechnology* 35, 477-483.

Salhlstrøm, S., Chrzanowska, J. and Sørhaug, T. (1993) Purification and characterization of a cell wall peptidase from *Lactococcus lactis* subsp. *cremoris* IMN-C12. *Applied and Environmental Microbiology* 59, 3076-3082.

Schmidt, D.G. (1982) Association of caseins and casein micelle structure. In *'Developments in dairy chemistry'* (Ed. P.F. Fox), vol.1, pp. 61-68, Elsevier, London, U.K.

Smid, E.J., Driessen, A.J.M. and Konings, W.N. (1989a) Mechanism and energetics of dipeptide transport in membrane vesicles of *Lactococcus lactis. Journal of Bacteriology* 171, 292-298.

Smid, E.J., Plapp, R. and Konings, W.N. (1989b) Peptide uptake is essential for growth of *Lactococcus lactis* on the milk protein casein. *Journal of Bacteriology* 171, 6135-6140.

Smid, E.R.S., Poolman, B. and Konings, W.N. 1991 Casein utilization by lactococci. *Applied and Environmental Microbiology* 57, 2447-2452.

Tan, P.S.T., Chapot-Chartier, M.-P., Pos, K.M., Rousseau, M., Boquien, C.-Y., Gripon, J.-C. and Konings, W.N. (1992a) Localization of peptidases in lactococci. *Applied and Environmental Microbiology* 58, 285-290.

Tan, P.S.T., Poolman, B. and Konings, W.N. (1993) Proteolytic enzymes of *Lactococcus lactis. Journal of Dairy Research* 60, 269-286.

Tan, P.S.T., Van Alen-Boerrigter, I.J., Poolman, B., Siezen, R.J., De Vos, W.N. and Konings, W.N. (1992b) Characterization of the *Lactococcus lactis pepN* gene encoding an aminopeptidase homologous to mammalian aminopeptidase N. *FEBS Letters* 306, 9-16.

Thomas, T.D. and Pritchard G.G. (1987) Proteolytic enzymes of dairy starter cultures. *FEMS Microbiology Reviews* 46, 245-268.

Tynkkynen, S., Buist, G., Kunji, E.R.S., Haandrikman, A., Kok, J., Poolman, B. and Venema, G. (1993) Genetic and biochemical characterization of the oligopeptide transport system of *Lactococcus lactis. Journal of Bacteriology* 175, 7523-7532.

Van Boven, A., Tan, P.S.T. and Konings, W.N. (1988) Purification and characterization of a dipeptidase from *Streptococcus cremoris* Wg2. *Applied and Environmental Microbiology* 54, 43-49.

Visser, S., Exterkate, F.A., Slangen, C.J. and De Veer, G.J.C.M. (1986) Comparative study of action of cell wall proteinases from various strains of *Streptococcus cremoris* on bovine α_{S1}-, β-, and κ-casein. *Applied and Environmental Microbiology* 52, 1162-1166.

Visser, S., Slangen, C.J., Exterkate, F.A. and De Veer, G.J.C.M. (1988) Action of a cell wall proteinase (P_I) from *Streptococcus cremoris* HP on bovine β-casein. *Applied Microbiology and Biotechnology* 29, 61-66.

Von Heijne, G. (1989) The structure of signal peptides from bacterial lipoproteins. *Protein Engineering* 2, 531-534.

Von Wright, A., Tynkkynen, S. and Suominen, M. (1987) Cloning of a *Streptococcus lactis* subsp. *lactis* chromosomal fragment associated with the ability to grow in milk. *Applied and Environmental Microbiology* 53, 1584-1588.

Vos, P., Boerrigter, I.J., Buist, G., Haandrikman, A.J., Nijhuis, M., De Reuver, M.J., Sieze, R.J., Venema, G., De Vos, W.M. and Kok, J. (1991) Engineering of *Lactococcus lactis* serine proteinase by construction of hybrid enzymes. *Protein Engineering* 4, 479-484.

Vos, P., Simons, G., Siezen, R.J. and De Vos, W.M. (1989) Primary structure and organization of the gene for a prokaryotic cell evelope-located serine proteinase. *Journal of Biological Chemistry* 264, 13579-13585.

Mechanism of Nisin-induced Pore-Formation

Gert N. Moll · Wil N. Konings* & Arnold J.M. Driessen
Department of Microbiology
University of Groningen
Kerklaan 30
9751 NN Haren
The Netherlands

Introduction

Bacteriocins were originally defined as peptides/proteins that exert exclusively antimicrobial activity against bacteria which are related to the producer bacteria. This definition still fits some of these competitor-killing proteins. Recently, an increasing number of antimicrobial peptides, among which lantibiotics, have been identified with a *broader* spectrum. Such a broader range of antimicrobial activity makes these peptides more interesting for existing applications like for instance food additives. In addition, among the bacteriocins produced by bacteria, lantibiotics are the *most efficient* antimicrobial peptides (Klaenhammer 1993). During their synthesis lantibiotics are subject to a series of posttranslational modifications which give rise to the presence of special dehydrated amino acid residues. Lantibiotics owe their name to the presence of intramolecular thioether rings, known as lanthionines. Figure 1 shows the structure of nisin with five intramolecular rings. Some lantibiotics such as Pep5 (Sahl et al. 1981; Sahl et al. 1985), nisin and epidermin induce voltage-dependent ion current in membranes. It has been proposed to classify these elongated cationic pore forming lantibiotics as

abbreviations: $\Delta\psi$ transmembrane electrical potential · Δp protonmotive-force
*corresponding author

NATO ASI Series, Vol. H 98
Lactic Acid Bacteria:
Current Advances in Metabolism, Genetics and Applications
Edited by T. Faruk Bozoğlu and Bibek Ray

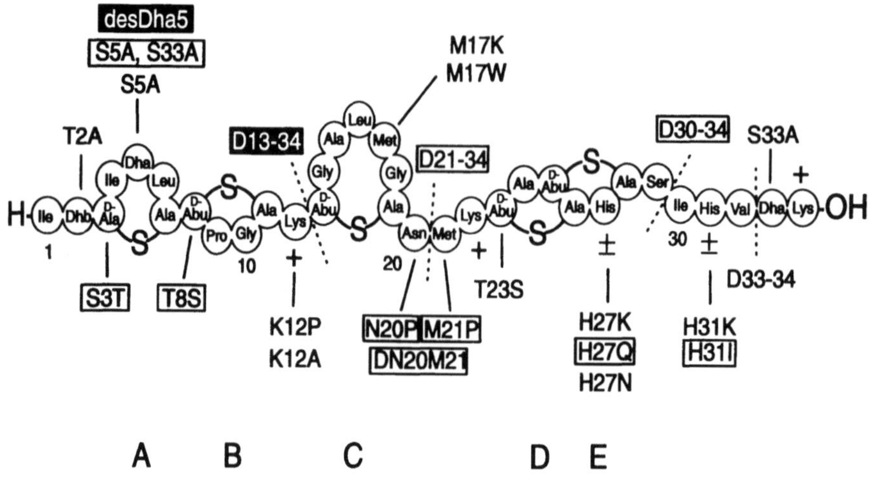

Fig 1: Schematic structure of nisin A showing the position of modification and site-directed mutants. When indicated boxed, alteration of the molecule resulted in a dramatic loss of activity. Modications indicated in black boxes caused an almost complete loss of antimicrobial activity. Fragmentation of the nisin molecule by proteolysis is indicated by dotted lines. Dhb, dehydrobutyrine; Dha, dehydroalanine; D-Ala$_s$, alanine moiety of lanthionine or 3-methyllanthionine; D-Abu, α-aminobutyric acid moiety of 3-methyllanthionine.

"type A" lantibiotics (Jung 1991). Other lantibiotics, "type B", usually with a rather globular structure, are enzyme inhibitors and immunologically active peptides, like ancovenin, duramycin, mersacidin and actagardine (Jung 1991).

Here, we concentrate only on the pore-forming lantibiotics and especially on nisin, the best-studied lantibiotic. Nisin is composed of 34 amino acids (Fig 1) and is able to kill a broad spectrum of Gram-positive bacteria. Gram-negative bacteria, such as for instance *Escherichia coli*, are only affected by nisin when the outer membrane is disrupted (Stevens et al. 1991).

In addition to the pore forming activity, nisin is known to inhibit the outgrowth of bacterial spores (Gross and Morell 1971; Morris et al. 1984) and it might interfere with some enzymatic step in the biosynthesis of the cell wall

(Reisinger et al. 1980; Sahl 1991). Nisin also affects the activity of autolytic enzymes (Bierbaum and Sahl 1985; 1987; 1988; 1991).

When the dehydroalanine residue at position 5 is replaced by alanine, full bactericidal activity remains but the capacity to inhibit the outgrowth of spores of *Bacillus* species is lost (Liu and Hansen 1993; Chan et al. 1995a). Hence the mechanism of pore formation differs from the mechanism of inhibiting the outgrowth of spores.

Nisin is soluble in aqueous solution at pH 2, but at high pH values it forms oligomers and is inactivated (Liu & Hansen 1990).

Pore formation by nisin in the cytoplasmic target membrane results in the efflux of ions, amino acids and ATP, and a collapse of the protonmotive-force (Δp) (Bruno et al. 1992; Gao et al. 1991; García Garcerá et al. 1993; Okereke & Montville 1992; Ruhr & Sahl 1985; Sahl 1991; Sahl et al. 1995). Nisin and other similar lantibiotics are cationic, and require a trans-negative electrical potential ($\Delta\psi$) for pore forming action (Sahl et al. 1987; Sahl 1991; Kordel & Sahl 1986).

Nisin displays also activities on liposomes. It forms pores in anionic phospholipid containing liposomes and has anion carrier activity in liposomes which do not contain these lipids. Since liposomes can be used to study pore formation by nisin, it seems that *in vivo* no binding to a putative target cell *protein*-receptor is necessary. On the other hand the possibility remains, that anionic phospholipids act as "nisin receptors" and that binding is required for pore formation.

Here we summarize a variety of work on the action of cationic lantibiotics on intact cells, proteoliposomes, liposomes and black lipid membranes, converging to the current model(s) for the molecular mechanism(s) of pore formation by nisin and possibly also that by other lantibiotics.

Specific anion efflux from PC liposomes

Nisin and epilancin K7 (Driessen et al. 1995) induce the release of the fluorophores carboxyfluorescein and fluorescein from phosphatidylcholine liposomes. More bulky fluorescein derivatives are less or not susceptible to lantibiotic induced efflux. Neither any influx of a *neutral* fluorescein-ester, nor any efflux of the *cationic* fluorescent indicator *N*-(ethoxycarbonylmethyl)-6-methoxyquinolium ion (MQAE) was mediated by nisin or epilancin K7. Under conditions that nisin elicits carboxyfluorescein release from phosphatidylcholine liposomes, nisin had little effect on the $\Delta\psi$. Nisin mediated carboxyfluorescein efflux is relatively slow and occurs only in phosphatidylcholine liposomes and not at all in phosphatidylglycerol liposomes. (García Garcerá et al. 1993). The ratio of the number of molecules nisin/liposomes determines the speed with which carboxyfluorescein is released by the liposomes; saturation occurs at approximately 100 molecules nisin/liposome. No threshold $\Delta\psi$ (membrane potential) is necessary for the nisin carrier activity, but a $\Delta\psi$ inside negative enhances (García Garcerá et al. 1993), and a $\Delta\psi$ inside positive inhibits (Driessen et al. 1995).

Taken together it appears that, in the absence of stable interactions with anionic phospholipids, the cationic nisin acts as an anion-carrier and functions according to the following mechanism. When added from the outside, nisin binds to the lipid surface and transverses the membrane in a $\Delta\psi$ (inside negative) stimulated manner. It then binds the anion (such as carboxyfluorescein) on the inside, and subsequently transverses the membrane as a binary complex with the anion. On the outer surface, the anion is released and nisin recycles to bind another anion on the inside. Penetration of the lantibiotic into the membrane is likely to be rate-limiting, and for epilancin K7 direct evidence has been obtained, that this step is promoted by $\Delta\psi$ (Driessen et al. 1995).

Why do anionic phospholipids inhibit the carrier activity of nisin? Nisin may

remain bound to the negative surface charge of anionic phospholipid as a partially neutralized complex at the lipid surface or interface. The "sites" on nisin that might bind anions in phosphatidylcholine liposomes may be already occupied by the anionic lipids and would thus not be available to bind other negatively charged groups. Specific interaction of lantibiotics, nisin and epilancin K7, with negatively charged phospholipids has been well documented (Driessen et al. 1995).
In the case of phosphatidylglycerol membranes no carrier activity at all is measured. Since target bacterial membranes are rich in anionic phospholipid, low anion carrier activity is to be expected *in vivo*.

Binding to anionic phospholipids

Fluorescence studies show that nisin strongly binds to phosphatidyl glycerol liposomes but not (even in the presence of a $\Delta\psi$) to phosphatidylcholine liposomes (Driessen et al. 1995). Once nisin is bound to the lipid membrane it disturbs the lipid bilayer structure as evidenced by NMR and fluorescence studies. Nisin seems to restrict the fluidity of phosphatidylglycerol membranes near the interface between the aqueous phase and the lipid head group (Driessen et al. 1995). Interestingly, nisin causes aggregation of phosphatidyl glycerol liposomes and promotes the transfer of octadecylrhodamine B between these liposomes.

Further studies with a novel cationic lantibiotic, epilancin K7, indicate that this molecule, or at least the tyrosine residue located in the middle of the molecule, penetrates into the hydrophobic carbon region of the lipid bilayer upon the imposition of a $\Delta\psi$.

NMR data indicate that nisin in solution adopts a rather flexible structure with an amphiphilic character (van de Ven et al. 1991). Nisin binds to micelles of zwitterionic dodecylphosphocholine or of anionic sodiumdodecylsulphate (van den

Hooven et al. 1993; Lian et al. 1991; Lian et al. 1992) in approximately similar conformations with the charged side exposed to the aqueous phase.

Energetics of pore-formation

The effect of temperature on pore formation by nisin Z in *Listeria monocytogenes* target cells has been investigated (Abee et al. 1995). The rate of nisin Z induced K^+ efflux decreased sharply with temperature. Nisin Z was unable to produce K^+ efflux below 7°C if the cells were grown at 30°C. Similarly, the lag-time with respect to K^+ efflux increased very sharply at lower temperatures for 30°C grown cells. Abee and co-workers suggested that the ordering of the lipid hydrocarbon chains, which takes place when the temperature is decreased, and consequently the decrease of membrane fluidity might be responsible for the lowered efficiency of nisin Z action at low temperatures.

Nisin pores cause a rapid dissipation of $\Delta\psi$ (Ruhr and Sahl 1985; Gao et al. 1991; Okereke and Montville 1992) and ΔpH (Gao et al. 1991) in intact cells, cytoplasmic membrane vesicles, cytochrome c oxidase proteoliposomes and liposomes. Nisin induced pores have a low specificity (Sahl 1991). In cytochrome c oxidase proteoliposomes a threshold $\Delta\psi$ of 100 mV was required for pore formation (Driessen et al. 1995). In contrast, at relatively very high concentrations of nisin, no $\Delta\psi$ threshold level seems to be necessary for *E. coli* liposomes (Gao et al. 1991).

The rate with which nisin dissipates the -artificially imposed- $\Delta\psi$ in liposomes depends strongly on the magnitude of the $\Delta\psi$, *i.e.*, the higher the $\Delta\psi$, the faster it is dissipated by nisin (Abee et al. 1991). Nisin dissipates ΔpH, trans-alkaline, in liposomes both in the absence and presence of a $\Delta\psi$. Again, the dissipation rate increases with the magnitude of the ΔpH. This strongly suggests that the ΔpH promotes pore formation. Higher concentrations of nisin are needed for the Δp dissipation in cytochrome c oxidase vesicles as compared to liposomes. To all likelihood this is due to the pumping activity of the cytochrome c oxidase.

When intact cells are treated with Pep5, nisin, subtilin, epidermin and gallidermin, threshold values between 50 and 100 mV are measured. These peptides cause rapid efflux of K^+, amino acids and ATP from the cytoplasm of various Gram-positive test bacteria, with a concomitant decrease in Δψ. In line with the existence of a threshold potential it was furthermore observed that efflux and depolarization are much faster with energized cells than with starved cells. In addition, cells that have been de-energized with uncouplers of the oxidative phosphorylation are completely insensitive.

Most interestingly, data are also available on the effect of lantibiotics on the Δψ across black lipid membranes (Benz et al. 1991). At a low Δψ within the range of 10 to 40 mV the lantibiotics nisin, Pep5, subtilin, gallidermin and epidermin do not induce any permeability in artificial membranes. At threshold values between 50 and 80 mV, negative at the trans-side, nisin and Pep5 cause an increase in permeability of several orders of magnitude. The membrane conductance is an exponential function of the voltage applied. When the trans-side is positive, no conductance increase is brought about neither by nisin nor by Pep5. This is in agreement with the observations that also in whole cells and liposomes, a threshold potential, inside negative, is needed for nisin pore formation. Threshold values may reflect the energetically rate limiting step of the pore formation, that is the insertion of the peptide(s) into the membrane.

Gallidermin and epidermin, on the other hand, increase the conductivity irrespective of the polarity of the applied voltage. In addition, gallidermin and epidermin channels are far more stable than those of other lantibiotics. This can be explained by the observation that threshold values are lower for these two lantibiotics as compared with the others. Pore formation in the case of gallidermin and epidermin is less voltage dependent.

According to black lipid membrane studies lantibiotics act by the formation of transient channels with highly variable single-channel conductances. The channels lifetime is in the order of milliseconds to seconds, depending on the applied voltage, the composition of the lipid membrane and the type of lantibiotic. Pores that have been formed remain stable for a short period of time even when the electrical potential across the black lipid bilayer is lowered below the threshold value of pore-formation. The channel diameter is estimated to be around 1 nm for low (100-500 nM) concentrations of nisin and Pep5, and around 2 nm for subtilin (Benz et al. 1991).

Atomic structure

For a full elucidation of the mechanism of pore formation knowledge on the atomic structure(s) of nisin is essential. The 3D structure of nisin on the basis of NMR data has been first determined in aqueous solution (van de Ven et al. 1991). The main part of the molecule consists of the lanthionine rings A, B, C (residues 3-19) linked by a “hinge” region to two intertwined double rings (D, E), residues 23-28. The proximal N-terminal and C-terminal parts are both quite flexible.

Interestingly, rings A, B and C have in common that the hydrophobic part is located opposite to the thioether bonds. The hydrophobic face is composed of the residues Ile4, Leu6, Pro9, Leu16 and Met17 and the hydrophilic Lys12 is located at the opposite side. Also the region from residue 21-28 is amphipathic: Lys22 and His27 are on one side and Met21 and His21 on the opposite side. The amphipathicity of the residues Ser29-Lys34 is not yet clear. However, in a mixed water-trifluoroethanol solvent the α-helical character of the residues 23-28 seems to extend to the C-terminus. In the latter case the helix is amphipathic with His27,

His 31 and Lys34 on one side. The C terminal half of the molecule contains four positively charged side chains of Lys22, His27, His31 and Lys34. Hence the molecule is amphipatic in two ways: First, the majority of the residues in the N-terminal half is hydrophobic and only a single charged residue is present, Lys12, whereas the charged and hydrophilic residues are mainly located in the C-terminal half of the molecule. Secondly, both the first domain, containing ring A, B and C as well as the second domain, containing ring D and E, have a hydrophobic and a hydrophilic side. Strikingly the peptide-bonds between the residues 23-24, 25-26 and 27-28, which form the intertwined D and E ring, are parallel. The formation of these D and E rings might follow an eventual α helical structure of the 23-28 region of prenisin. The resulting structure looks like a somewhat "overwound α-helix".

The structure of nisin has also been determined in membrane mimicking models: 1) dodecylphosphocholine micelles (DPC), 2) sodium dodecylsulphate (SDS) micelles and 3) a mixture of trifluoroethanol (TFE) and water (van den Hooven et al. 1993).

Nisin binds to micelles of zwitterionic DPC and anionic SDS. The largest differences in chemical shift between nisin in solution and nisin complexed to DPC micelles concern the amide protons of Dha5, Leu6 and Abu8 and H^{β}protons of Dha5. These observations may have various causes, such as for instance a varied interaction with the micelles, an altered intrinsic susceptibility to changes in chemical surrounding or/and altered hydrogen-bridging. In the structure of nisin in aqueous solution a γ-turn is observed around Dha5. In contrast, for nisin complexed to both micelles no typical β- or γ-turns are present.

In this respect it is of interest to note studies with a Dha5Dhb mutant (Kuipers et al.1992) and with [α-OH-Ala^5]nisin. Both the bactericidal activity as well as the extent of structural change following interaction with micelles decrease in the following order: nisin A > Dha5Dhb mutant > [α-OH-Ala^5]nisin (van den Hooven 1995). This suggests that the micelle bound conformation of ring A might correctly mimick the membrane bound nisin and that this conformation is essential for activity.

Nisin complexed to DPC as well as nisin in TFE/water (3/1) is in the monomeric form; nisin complexed to SDS micelles is mostly in the monomeric form (>90%). Following an increasing addition of TFE or DPC micelles, the conformation of nisin changes gradually, until 70% TFE or a DPC/nisin ratio of 30. No titration is possible with SDS micelles.

The above NMR data were complemented with CD measurements. The CD spectra of nisin in the three model systems, TFE/water 3/1, SDS micelles, DPC micelles, differ significantly from the spectrum of nisin in aqueous solution, which suggests a structural change (van den Hooven et al. 1993).

Since both micelles seem to be usefull models for membrane binding, the interaction of nisin with the DPC and SDS micelles and in particular the surface location and orientation has been further studied. The hydrophobic residues are slightly immersed into the micelles and oriented towards the center, whereas the more polar or charged residues have an outward orientation.

The whole nisin molecule itself does not seem to be embedded into the micelles. However, on the basis of the temperature coefficients of the amide protons, both dehydroalanines (residues 5 and 33) as well as the residues Ile30-Lys34, which surround the second dehydroalanine, seem to be shielded to some extent from the solvent. Spin label experiments indicate that the interactions of the residues Ser29-Lys34 with the micelles are the strongest (van den Hooven 1995). It should be taken in mind that, since nisin$^{1-32}$ has full activity, the membrane interaction of the last two residues can not be essential for activity.

Taken together the main result of the structural analyses indicates a structural change in the A ring upon the membrane binding of nisin.

Model of pore formation

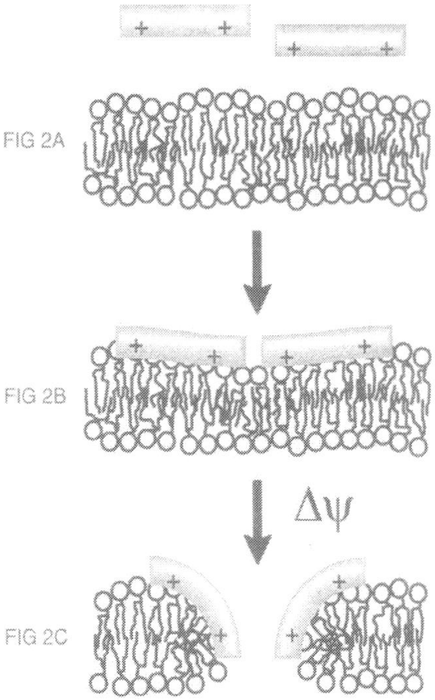

Fig 2: Wedge-model of pore-formation. At first nisin is in solution (A). Then bound nisin disturbs the lipid dynamics (B). Above the $\Delta\psi$ threshold level, nisin inserts. Co-insertion of bound, anionic phospholipids results in bending of the lipid surface giving rise to a wedge-like, nonspecific pore.

In view of the small size of nisin and other lantibiotics one has to assume that several molecules participate in channel formation. Artificial membranes also give some information on the manner of channel formation. Single channel measurements do not reveal distinct steps but a variety of fast current fluctuations on the millisecond scale, including bursts and pulse like spikes. In the case of nisin, Pep5 and subtilin, at higher voltages both the lifetime and the fluctuations increase (Benz et al. 1991). This indicates very unstable oligomers with a highly fluctuating number of lantibiotic molecules. Presently data on pore formation by lantibiotics seem to suggest the following models. Lantibiotics may initially be bound at the lipid surface and subsequently, in the presence of a $\Delta\psi$, insert into the membrane. The membrane-inserted molecules may form a cluster around a central pore as in the "barrel-stave model". To all likelihood the order of events will be first insertion and subsequently aggregation. Alternatively a “wedge-model" (Fig 2) can be envisioned where the amphiphilic molecules adhere to the surface of the membrane causing destabilisation of the bilayer structure thereby promoting pore formation, such as proposed for annexins (Demange et al.1994). Binding of nisin to anionic phospholipids indeed results in a local disorder of thephospholipid headgroups. Nisin has a strong effect on the lipid dynamics, (dehydration of the lipid surface, non-bilayer intermediate structures, stimulation of lipid transfer between liposomes, *i.e.*, bilayer-bilayer contacts) when incubated with liposomes at high concentrations. However, binding of the lantibiotic to the membrane by itself does not induce pores. Pore-formation requires the presence of a high $\Delta\psi$. If the membrane potential is sufficient high,the electrical force exerted on the cationic nisin may cause the insertion of at least part of the lantibiotic, forming a wedge (Fig 2). By carrying the anionic lipids along it may induce a local bending of the membrane surface. This will result in the formation of a metastable non-selective water-filled pore. As suggested by the NMR data, nisin may expose its charged side to the aqueous phase under these conditions. An important aspect of this model is that the actual pore is formed by an array of nisin molecules that temporarily force the lipids in a thermodynamically unfavourable non-bilayer conformation.Such pores are intrinsically unstable, as the lipid will try to rearrange into a bilayer structure. Moreover, since the association and dissociation of nisin oligomers is likely to be a dynamic event, lowering of the magnitude of $\Delta\psi$ below the threshold of insertion will result in a relaxation of the nisin molecules to the surface-bound state and a disassembly of the pore.

Engineering and chemical modications

How do the site directed mutagenesis studies and results of engineering of the nisin molecule (Summarized in Fig 1) contribute to pore-formation-models? Data are available on both nisin A and Z. Nisin Z is a natural occurring variant of nisin A with the histidine at position 27 replaced by an asparagine.

Kuipers and co-workers (1991; 1992; 1994) obtained several nisin Z mutants by site directed mutagenesis. Residue Ile1 can be replaced without significant loss of activity. Mutant T2S gives rise to a dehydroalanine (Dha) residue at the dehydrobutyrine (Dhb2) position and this mutant has twofold higher activity against two target organisms. Replacement of AlaS3 by AbuS leads to a dramatic loss of activity, probably due to a conformational change in the first lanthionine ring. Replacement of the Ser-5 codon by a Thr codon led to a mutant with a dehydrobutyrine (Dhb) instead of a dehydroalanine (Dha) residue at position 5. Its antimicrobial activity was 2-10 fold lower relative to wild-type nisin Z depending on the indicator strain used. It has been speculated that either the additional methyl-group of the Dhb residue might affect the interaction with the target cell membrane or that the reactivity of the Dha-5 double bond is crucial for nisin action, since the less reactive double bond of Dhb would then indeed be expected to yield a lower bactericidal capacity. In line with the latter hypothesis is the drastic reduction of activity following saturation of the double bond in Dha-5 (Chan et al. 1989a; Rollema et al. 1991; Lian et al. 1992).

As already shortly indicated under the heading "spatial structure" an intact ring A is important for activity. This is shown, for example, by the activities of the two major breakdown products of nisin, formed by acid-catalysed hydrolysis of the two dehydroalanine residues. Nisin$^{1\text{-}32}$, lacking the two C-terminal residues, has full bacteriocidal activity, but (*des*-ΔAla$_5$)-nisin$^{1\text{-}32}$, in which dehydroalanine 5 is also missing, is inactive (Chan et al.1989). The observation that the ΔAla$_5$ -Ala mutant retains full activity (Chan et al.1995a) suggests that it is the opening of the ring rather than the absence of the dehydro residue which leads to the loss of activity in (*des*-ΔAla$_5$)-nisin$^{1\text{-}32}$. Indeed, even a distortion of ring A produced by converting D-Ala$_s$ to D-Abu at position 3 (achieved by a serine to threonine substitution in the structural gene) results in a complete loss of activity.

Mutations have also been made in the C-ring. Replacing simultaneously the codons for Met-17 and Gly-18 by codons for Gln and Thr, respectively, led to two mutant species, one containing a Thr residue and the other containing a dehydrobutyrine residue at position 18. The Dhb-18, Gln-17 mutant showed similar activities as nisin Z. On the other hand, the Thr-18, Gln-17 mutant is about 4-fold less active against two indicator strains but twice as active against *Micrococcus flavus* than the wild-type nisin Z. The substitutions at the positions 17 and 18 appear to influence the global structure of nisin which might cause the observed alteration in activity spectrum of the Thr-18, Gln-17 mutant.

Above the relevance of an intact ring A has been described. The integrity of ring C is similarly essential for activity: proteolytic cleavage of either the Ala_{15}-Leu_{16} bond (Chan et al. 1993; 1995b) or of the Lys_{17}-Gly_{18} bond in a Met_{17} to Lys mutant leads to complete loss of activity. It remains to be established whether the additional charges introduced by proteolysis contribute to this loss of activity. Also other data point at the importance of an intact ring C for activity. If Thr-13 is replaced by Cys-13, a disulphide bond between residues 13 and 19 is formed and antimicrobial activity is low. Destruction of this bond by DTT leads to virtually complete loss of activity. This indicates that integrity of the third is relevant for antimicrobial activity (Kuipers et al. 1994).

NMR studies (Lian et al. 1991; 1992) indicated flexibility of the "hinge" between rings A and B and between ring C and D. In solution the flexibility of this first hinge (B/C) was found nearly as high as the second hinge (C/D). Also other NMR studies indicated a highly flexible hinge region in nisin connecting rings ABC with DE (van den Hooven et al. 1993; van de Ven et al. 1991). According to the wedge-model of pore-formation, this region may act as a twist that allows nisin to bend the lipid surface (Driessen et al. 1995). Attempts to immobilize these rings by replacing residues 20 and 21 by proline residues resulted in a decrease of the activity. Deletion of residues 20 and 21, in order to change the orientation of rings 4 and 5 with respect to the rings ABC, led even to a more dramatic loss of activity (Kuipers et al. 1994).

Studies of proteolytic fragments of the molecule have identified other residues which are important for bacteriocidal activity. For example, while nisin 1-32 has full activity, the further removal of Ile_{30}-His_{31}-Val_{32} to give nisin 1-29 leads to a 30 fold drop in activity (Chan et al. 1993; 1995b). The fragment $nisin^{1-12}$ is a particularly interesting one since, while it is itself inactive, it antagonizes the bacteriocidal effects of nisin (Chan et al. 1995b) and the nisin-induced $\Delta\psi$ dissipation in cytochrome c oxidase proteoliposomes (J Clark and G Moll, unpublished work). This suggests that rings A, B and C are required for binding, but that more C-terminal portions of the molecule are required in addition for pore formation. It remains to be established whether the rings A, B and C are involved in the oligomerisation of nisin and/or its binding to phospholipids in the membrane.

According to the wedge model, an electrostatic interaction between the cation side-chains of nisin and anionic lipids is responsible for tight surface binding. Mutagenesis studies demonstrate that the positive charge on position 12 is not essential for the antimicrobial activity. In contrast, for optimal activity the imidazole groups of histidine at position 27 and 31 need to be protonated.

Although there is no direct evidence supporting this view at this stage, it seems conceivable that rings ABC or parts thereof are involved in oligomerization of nisin, while the positively charged carboxyl-terminal moiety of the molecule inserts in the membrane as depicted in Figure 2. Further studies on the mode of action of nisin, nisin-derivatives and site-directed mutants will reveal the intimate features of pore-formation.

References

Abee T, Gao FH & Konings WN (1991) The mechanism of action of the lantibiotic nisin in artificial membranes. in: Nisin and novel lantibiotics. Jung G & Sahl H-G eds) pp 373-386 ESCOM Leiden

Abee T, Rombouts FM, Hugenholz J, Guihard G & Letellier L (1994) Mode of action of nisin Z against *Listera monocytogenes* Scott A grown at high and low temperatures. Appl and Environm Microbiol 60:962-1978

Benz R, Jung G & Sahl H-G (1991) Mechanism of channel formation by lantibiotics in black lipid membranes. in: Nisin and novel lantibiotics. (Jung G & Sahl H-G eds) pp 359-372 ESCOM Leiden

Bierman G & Sahl H-G (1991) Induction of autolysis of *Staphylococcus simulans* 22 by Pep5 and nisin and influence of the cationic peptides on the activity of the autolytic enzymes.in: Nisin and novel lantibiotics. (Jung G & Sahl H-G, eds) pp 386-396 ESCOM Leiden

Bierman G & Sahl H-G (1985) Induction of autolysis of *Staphylococci* by the basic peptide antibiotics Pep5 and nisin and their influence on the activity of autolytic enzymes. Arch Microbiol 141:249-254

Bierman G & Sahl H-G (1987) Autolytic system of *Staphylococcus simulans* 22: influence of cationic peptides on activity of N-acetylmuramoyl-L-alanine amidase. J Bact 169:5452-5458

Bierman G & Sahl H-G (1988) Influence of cationic peptides on the activity of the autolytic endo-β-N-acetylglucosamidase of *Staphylococcus simulans* 22. FEMS Microbiol Lett 58:223-228

Bruno MEC, Kaiser A & Montville TJ (1992) Depletion of proton motive force by nisin in *Listera monocytogenes* cells. Appl Environ Microbiol 58:2255-2259

Chan WC, Roberts GCK & Bycroft BW (1993) Structure-activity relationships of nisin and subtilin. In: Peptides 1992. Schneider CH & Eberle AN (eds) Leiden: ESCOM Science Publishers pp 770-771

Chan WC, Dodd HM, Maclean K, Lian LY, Bycroft BW, Gassen MJ & Roberts GCK (1995a) Structure-activity relationships in the peptide antibiotic nisin: the role of dehydroalanine 5. Molec Microbiol submitted

Chan WC, Leyland M, Clarke J, Lian LY, Gassen MJ, Bycroft BW & Roberts GCK (1995b) Structure-activity relationships in the peptide antibiotic nisin: biological activity of proteolytic fragments of the molecule. In prep

Chan WC, Bycroft BW, Lian LY & Roberts GCK (1989) Isolation and characterisation of two degradation products derived from the peptide antibiotic nisin. FEBS Lett 252:29-36

Demange P, Voges D, Benz J, Liemann S, Göttig P, Berendes R, Burger A & Huber R (1994) Annexin V: the key to understanding ion selectivity and voltage regulation. Trends Biochem Sci 19:272-276

Driessen AJM, van den Hooven HW, Kuiper W, van de Kamp M, Sahl H-G, Konings RNH & Konings WN (1995) Mechanistic studies of lantibiotic-induced permeabilization of phospholipid vesicles. Biochemistry 34:1606-1614

Gao FH, Abee T & Konings WN (1991) Mechanism of action of the peptide antibiotic nisin in liposomes and cytochrome c oxidase-containing proteoliposomes. Appl Environ Microbiol 57:2164-2170

García Garcerá MJ, Elferink MGL, Driessen AJM & Konings WN (1993) In vitro pore-forming activity of the lantibiotic nisin. Eur J Biochem 212:417-422

Gross E & Morell JL (1971) The presence of dehydroalanine in the lantibiotic nisin and its relationship to activity. J Am Chem Soc 93:4634-4635

van den Hooven HW, Fogolari F, Rollema HS, Konings RNH, Hilbers CW, van de Ven FJM (1993) NMR and circular dichroism studies of nisin in non-aqueous environments. FEBS Lett 319:189-194

van den Hooven HW (1995) Structure of the lantibiotic nisin in aqueous solution and in membrane-like environments. PhD Thesis University of Nijmegen the Netherlands

Van den Hooven HW, Fogolari F, Rollema HS, Konings RNH, Hilbers CW & Van de Ven FJM (1993) NMR and circular dichroism studies of the lantibiotic nisin in non aqueous environments. FEBS Lett 319:189-194

Jung G (1991) Lantibiotics: a survey. in: Nisin and novel lantibiotics. Jung G & Sahl H-G (eds) pp1-30 ESCOM Leiden

Van de Kamp M, Horstink LM, van den Hooven HW, Konings RNH, Hilbers CW, Frey A, Sahl H-G, Metzger JW, Jung G, & van de Ven FJM (1995) Sequence analysis by NMR spectroscopy of the peptide lantibiotic epilancin K7 from *Staphylococcus epidermis* K7. Eur J Biochem 227:757-771

Klaenhammer TR (1993) Genetics of bacteriocins produced by lactic acid bacteria. FEMS Microbiol Rev 12:39-86

Kordel M & Sahl H-G (1986) Susceptibility of bacterial eukaryotic and artificial membranes to the disruptive action of the cationic peptides Pep5 and nisin. FEMS Microbiol Lett 34:139-144

Kuipers OP, Yap WMGJ, Rollema HS, Beerthuizen MM, Siezen RJ & de Vos WM (1991) Expression of wild-type and mutant nisin genes in *Lactococcus lactis*. in: nisin and novel lantibiotics, Jung G & Sahl H-G (eds) Leiden: ESCOM Science Publishers pp 250-259

Kuipers OP, Rollema HS, Yap WMGJ, Boot HJ, Siezen RJ & de Vos WM (1992) Engineering dehydrated amino acid residues in the antimicrobial peptide nisin. J Biol Chem 267:24340-24346

Kuipers OP, Rollema HS, Bongers R, van den Bogaard P, Kosters H, de Vos WM & Siezen RJ (1994) Structure-function relationships of nisin studied by protein engineering. 2nd International Workshop on Lantibiotics, November 20-24 1994 Arnhem the Netherlands

Lian LY, Chan WC, Morley SD, Roberts GCK, Bycroft BW & Jackson D (1991) NMR studies of the solution structure of nisin A and related peptides. In: Nisin and novel lantibiotics. Jung G & Sahl H-G (eds) Leiden: ESCOM Science Publishers pp 43-58

Lian LY, Chan WC, Morley SD, Roberts GCK, Bycroft BW & Jackson D (1992) Solution structures of nisin A and its two major degradation products determined by NMR. Biochem J 283:413-420

Liu W & Hansen JN (1990) Some chemical and physical properties of nisin, a small protein antibiotic produced by *Lactococcis lactis*. Appl Environ Microbiol 56:2551-2558

Liu W & Hansen JN (1993) The antimicrobial effect of a structural variant of subtilin against outgrowing *Bacillus cereus* T spores and vegetative cells occurs by different mechanisms. Appl Environm Microbiol 59:648-651

Morris SL, Walsh RC & Hansen JN (1984) Identification and characterization of some bacterial membrane sulfhydryl groups which are targets of bacteriostatic and antibiotic action. J Biol Chem 259:13590-13594

Okereke A & Montville TJ (1992) Nisin dissipates the proton motive force of the obligate anaerobe *Clostridium sporogenes* PA 3679. Appl Environ Microbiol 58:2463-2467

Reisinger P, Seidel H, Tschesche H & Hammes WP (1980) The effect of nisin on murein synthesis. Archiv Microbiol 127:187-193

Rollema HS, Both P & Siezen RJ (1991) in: Nisin and novel lantibiotics. Jung G & Sahl H-G (eds) Leiden: ESCOM Science Publishers pp 123-130

Ruhr E & Sahl H-G (1985) Mode of action of the peptide antibiotic nisin and influence on the membrane potential of whole cells and on cytoplasmic and artificial membrane vesicles. Antimicrob Agents Chemother 27:841-845

Sahl H-G & Brandis H (1981) Production, purification and chemical properties of an antistaphylococcal agent produced by *Staphylococcus epidermis*. J Gen Microbiol 127:377-383

Sahl H-G, Groβgarten M, Widger WR, Cramer WA & Brandis H (1985) Structural similarities of the Staphylococcin-like peptide Pep-5 to the peptide antibiotic nisin. Antimicrobial Agents and Chemother 27:836-840

Sahl H-G, Jack RW, & Bierbaum G (1995) Lantibiotics:biosynthesis and biological activities of peptides with unique post-translational modifications. Eur J Biochem 230:827-853

Sahl H-G, Kordel M & Benz R (1987) Voltage dependent depolarization of bacterial membranes and artificial lipid bilayers by the peptide antibiotic nisin. Arch Microbiol 149:120-124

Sahl H-G (1991) Pore formation in bacterial membranes by cationic lantibiotics. in: Nisin and novel lantibiotics. Jung G & Sahl H-G (eds) pp 347-358 ESCOM Leiden

Stevens KA, Sheldon BW, Klapes NA & Klaenhammer TR (1991) Nisin treatment for inactivation of *Salmonella* species and other Gram-negative bacteria. Appl Environ Microbiol 57:3613-3615

van de Ven FJM, van den Hooven HW, Konings RNH, Hilbers CW (1991) The structure of nisin in aqueous solution. Eur J Biochem 202:1181-1188

INDUCTION OF PENICILLIN RESISTANCE IN *S.THERMOPHILUS*

Faruk Bozoglu, Laura-Christina Chirica
Osman Yaman and Candan Gürakan
Department of Food Engineering,
Middle East Technical University
06531 Ankara - Turkey

β-lactams

Availability of antibiotics for the treatment of bacterial infections goes back over sixty years. Use of cell wall synthesis inhibitors are shown to be an effective means of aiding host defenses in cases of bacterial infection in man and animals. Generally four groups of inhibitors of cell wall synthesis are commonly utilized and categorized under β-lactams. These antibiotics are among the most useful chemotheropeutic tools available to medicine. All cell wall inhibitors need to reach at least cytoplasmic membrane to interact with their target. For gram negative organisms the other membrane affords a potential barrier to the cell wall inhibitors to interact with their targets. Other potential cellular barriers include: capsular materials, peptidoglycan and possibly the cytoplasmic membrane. Once past the permeability barriers this group will interact to inhibit the formation of peptidoglycan. To be effective all of cell wall inhibitors need first reach their targets. Most of the clinically important cell wall inhibitors have targets either on the surface of the cytoplasmic membrane for β- lactams, within this membrane for bacitracin and vancomycin , or in the cytoplasm of the cell for cycloserine.

The molecular weight of cell wall inhibitors is not big enough to be sieved by peptidoglycan layer, therefore this cell wall sieving ability seems unlikely to inhibit the access of such agents in gram positive cells (Bryan,1984). However the sieving effect of gram negative outer membranes prevents adequate concentration of many cell wall inhibitors limiting the spectrum of these drugs. Also the chemical structure of β-lactams greatly influences the permeation of the drugs across the outer membrane to their targets at the surface of the cytoplasmic membrane.

NATO ASI Series, Vol. H 98
Lactic Acid Bacteria:
Current Advances in Metabolism, Genetics and Applications
Edited by T. Faruk Bozoğlu and Bibek Ray

Resistance to cell wall synthesis inhibitors

Decreased permeability

There is a big correlation between decreased permeability and increased resistance to β-lactam antibiotics and when there is a decrease in permeability barrier, the susceptibility increases. Studies on some gram negative bacterial species has shown that, induced mutants or isolates from clinical materials which are resistant to β-lactam is due to the decreased ability of such antibiotics to penetrate the outer membrane of the cell. Change in permeability is attributed to the structural changes and/or alteration in lipopolysaccharite content of the cell wall. As the oligosaccharide O antigens are hydrophobic their elimination will result an increase in the penetration of the outer membrane by those antibiotics. The treatment of cells by EDTA that removes LPS stabilizing divalent cations increase the sensitivity of cells to many drugs. Mutants without noticeable changes in their outer membrane profile, but with alterations in their LPS has been shown to have increased susceptibility to some antibiotics.

Decreased affinity for targets

Cell wall examination of all bacteria has revealed to contain penicillin binding proteins (PBS). Studies on PBS profiles of taxonomically related bacteria tend to be the same, and membrane fractions of various species contain compounds that have high affinity to β-lactams. This proteins have been thoroughly studied in *E. coli* and they have been associated with cell elongation., septum formation and maintenance of rod or bacillus shape. If sublethal concentrations of β-lactams are administrated , morphological changes are observed due to the association of PBSs with antibiotics. Studies with mutants showed a decrease affinity of PBS for penicillin and result of this alteration was an increase in the resistance of the mutant to the same antibiotic tested.

β-lactamases

β-lactamases are those enzymes that inactivate β-lactam antibiotics and function to protect the cells against β-lactam attack. This type of activity has been known more than 50 years, and interaction between substrate and enzyme at both the cellular and molecular level have been studied intensively. β-lactamases from gram positive bacteria dominated early studies and bacilli and staphylococci are the main genera in those studies. In presence of β-lactams these organisms produce intra or extra cellular β-lactamases that are mostly active against penicillin. Unlike gram positive bacteria, the outer membrane that is present in gram negative bacteria acts as a barrier to the permeability of hydrophobic molecules like β-lactams both into and out of the cell. Thus, the majority of β-lactamases in these cells are periplasmic. Therefore accessibility of β-lactams in gram negative cells depends on the penetration characteristics of β-lactam molecules. Though, majority of the β-lactamases produced by gram positive organisms are inducible and exocellular, those produced by gram negative organisms may be inducible or constitutive and with few exceptions, periplasmic (Pham and Bell, 1992).

In case of gram positive bacteria extracellular nature of β-lactamases provides a strong defense against β-lactame molecules, and they have the ability to deactivate β-lactame molecules at low concentrations before gaining access to the cell. However, the dilution factor is a disadvantage for such enzymes and it is perhaps for this reason that gram positive organisms also contain low level of cell wall associated β-lactamases (Fig.1). Although the concentration of cell wall associated β-lactamase is low, their strategic location close to the target proteins provide a second line of defense for gram positive bacteria. Therefore, diffusion of unscavenged β-lactam molecules to the cell wall may still be eliminated by the cell wall associated β-lactamases before the target proteins are affected. In case of gram negative bacteria, the specific location of periplasmic β-lactamases results in a major defensive advantage to the organism in presence of β-lactams and only small amounts of enzyme may be required for effective

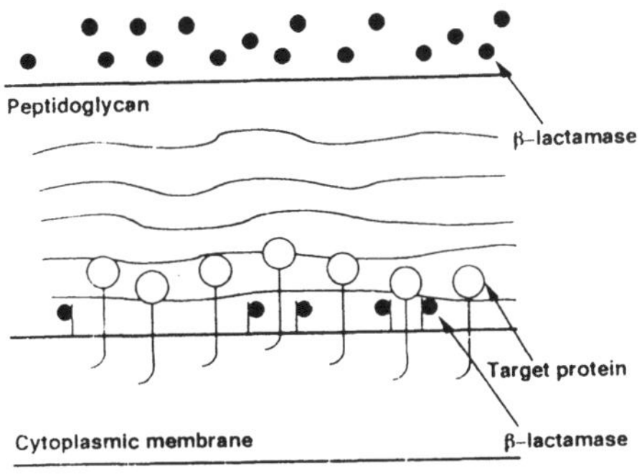

Figure 1. Location of β-lactamase in gram positive bacteria

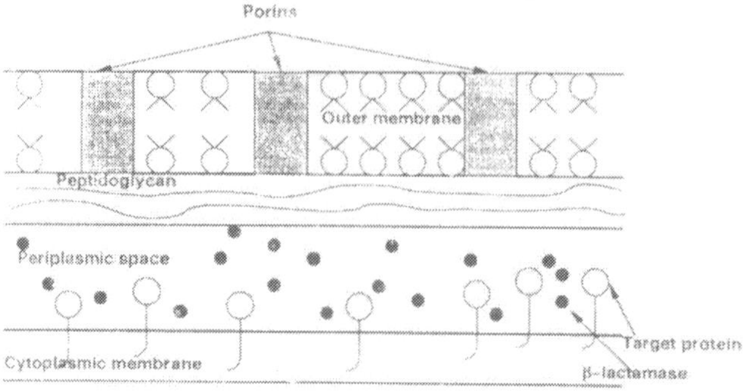

Figure 2. Location of β-lactamase in gram negative bacteria

activity because of the resultant clustering of strategic placement of molecules within the periplasmic space (Fig.2)

Genetics of resistance to cell wall synthesis antibiotics

Existence of antibiotic resistant bacteria was realized even during the initial period surrounding the commercial development of penicillin. Today two broad categories of antibiotic resistance are recognized: intrinsic and acquired. The term intrinsic is used to imply that inherent features of the cell are responsible for preventing antibiotic action and usually resistant strains emerge from previously sensitive bacterial populations upon the exposure of the antibiotic concerned. Intrinsic resistance is usually expressed by chromosomal genes , whereas acquired resistance may result from mutation in genetic frame work or by acquisition of plasmid transposons (Rice, 1991, Podbielski 1991). Not all chromosomally-specified antibiotic resistance is due to the mutation. For instance, the intrinsic resistance of the majority of gram negative bacteria to hydrophobic compounds is due to the expression of normal, resident genes involved in the synthesis and assembly of lipopolysaccharide in the outer membrane and β-lactamases that confer resistance to β-lactam antibiotics. Plasmid determined antibiotic resistance is more common than chromosal resistance in many important bacteria. This could be mainly due, 1) in the absence of antibiotic pressure, the majority of cell in a particular population need not maintain plasmids, thereby reducing the biochemical stress on the cell that would otherwise be required for replication and expression of DNA. Therefore, it is not necessarily essential for every member of population to maintain particular resistance plasmid. In the selective pressure plasmid-carrying progeny will survive and replicate, 2) being more mobile, plasmids can be acquired form other bacteria in addition to inheritance from mother cells, keeping in mind that plasmids can be transferred both within and between certain species, 3) plasmids are frequent vectors of transposons that accounts for the emergence of multiple-resistance. Since tranposons do not require extensive DNA homology to insert in to the recipient molecule, it is possible for bacteria to acquire several different resistance genes by a series of transposon insertion in resident plasmid.

Starter cultures

The microbial ecology of food and related fermentation has been studied for many years. Many of these, including such foods as ripened cheese, fermented sausages, sauerkraut and yogurt are preserved products that have been fermented by suitable starter cultures. The group of bacteria used in such fermentation processes is comprised of at least eight genera. While the lactic group is loosely defined with no precise boundaries, all members share the property of producing lactic acid from hexoses. Fermented foods, the bacterial strains used and products are given in Table I.

Factors that interfere with the development of the cultures

Factors that can interfere with the growth of cultures in milk are varied. They are partly due to the quality of the milk, i.e., to the standards of hygiene and the criteria followed for milking and storing . They are partly due to particular contamination and mutation of the culture themselves.

Product	Starter species	Associated flora
Cheese	S.lactis S.cremoris L.cremoris	S.lactissubp. diacetylactis
Yogurt	S.thermophilus L.bulgaricus	
Kefir	S.lactis L.bulgaricus S.fragili S.cereviciae	L.dextranicum T.kefir

Table I. Examples of some fermented milk products

Infection by bacteriophage

Lactic cultures used in the production of fermented milk can be infected by phages, generally resulting in lysis of the bacterial cell. Possible precautions against infection by bacteriophages are as those adopted in production of cheeses with mesophilic lactic microflora, such as alternation of the strains and enforcing high standards of hygiene with a wide use of chlorine based compounds and use of strains that posses bacteriophage stability. In recent years, systems in which cell protect themselves from phage attack were well studied and reviewed by Hill (1993).

Presence of antibiotics

The presence of antibiotics due to the treatment of mastitic animals that results in milk, is a real problem in modern production particularly in fermented milk products, because the lactic acid bacteria used for the fermentation are among the most sensitive. With regard to penicillin the critical levels in milk for the bacteria in interest are given in Table II. There are a few measures that can be taken to reduce the effect of penicillin. Commercial sterilization of milk inactivates 50% of the penicillin, however use of penicillinase would be too expensive.

	IU/mL
S.thermophilus	0.004-0.05
S.cremoris	0.05- 0.1
S.lactis	0.1-0.3
L.cremoris	0.05-0.1
L.acidophilus	0.3-0.6
L.casei	0.3-0.6

Table II. Critical levels of penicillin

Induction of penicilline resistance in *S.thermophilus*

S.thermophilus is one of the essential species in yogurt production. Its activity is correlated with its fermentation capacity that is acid production.

Presence of residual penicillin even in the levels of 0.5 units/ml is quite inhibitory for the activity of *S.thermophilus*. However, when grown by several passages, starting from subbacteriocidal penicillin concentration (% 0.1 U/ml), it is possible to isolate *S.thermophilus* cells that are resistant even up to 45 U/ml of penicillin. Need for multiple passages under penicillin stress indicates the presence of resistant cells that are inducible within the population. Emerging of those resistant cells mostly results from the synthesis of β-lactamases (Kitano and Tomasz,1979). β-lactamase is the most common mechanism which accounts for the bacterial resistance to various penicillin and cephalosporin (Bush and Singer, 1989). Studies on *S.thermophilus* resistant cells, has shown the presence of extracellular heat labile proteiniecous substances that could be categorized as β-lactamase. Two different extracellular fractions with β-lactamase activities were determined with different molecular weights (Chirica, 1995). Activities of both fractions were specific to penicillin but not to cephaloridin (Table III).

Substrate	Enzyme activity (U/mg)	
	FI	FII
Penicillin G	8.10	3.11
Ampicilin	6.42	2.40
Oxacillin	5.78	2.02
6-APA	0.92	0.24
Cefuroxine	0.00	0.00

Table III. Activities of fractions FI and FII to different antibiotics

These two fractions of β-lactamase differ only in their kinetic constants and the enzyme specifities tested with different penicillin, but there are a lot of similarities between the two fractions such as optimal pH, optimal temperature and effect of inhibitors (Table IV).

Studies on the crushed saline washed cells also showed β-lactamase activity, indicating possible cell wall associated enzyme system (Yaman, 1994). Similar activity was reported for *E.corodence* indicating a cell-bound chromosome encoded β-lactamase (Bush and Singer, 1989).

Gram staining of penicillin sensitive *S.thermophilus* cells gives gram positive reaction, however the penicillin resistant cells give gram negative reaction that indicates the loss of some cell wall characteristics. Similar morphological changes under β-lactam stress has been reported by Kitano and Tomasz (1979) indicating an autolytic activity in the presence of penicillin resulting cell wall degradation. The efficiency to trigger autolytic cell wall degradation in many cells reported earlier and shown in our studies

	FI	FII
opt. pH	7.00	7.00
opt. Temp. (^{o}C)	41	41
Inhibitors (% inhibition)		
Iodine (0.006mM)	82	100
$CuSO_4$ (0.04mM)	65	89
$FeSO_4$ (0.05mM)	88	95
EDTA (5mM)	0	0

Table IV.Characteristics of fractions FI and FII. Penicillin G is used as the substrate.

may account for another mechanism for the resistance to penicillin. It is obvious that penicillin resistant *S.thermophilus* cells posses different mechanisms of resistance to penicillin. The primary one is the synthesis of exocellular β-lactamases and the second one is the alteration of the cell wall structure that possibly changes the lipid content that decrease the succeptibility of the cell to penicillin (Kiteno and Tomasz, 1979) and synthesis of cell wall associated β-lactamase molecules

Studies on the growth and acid production of penicillin resistant and sensitive cells of *S.thermophilus* were compared. Elongation of the lag phase and decrease in acid production rate were observed for penicillin resistant cells (Yaman,1994). The mechanism by which penicillin resistant bacterial enzymes bring about on the growth of cells is not clearly understood. However effect of chemically different β-lactam antibiotics that elicit a variety of different morphological changes and distinct growth inhibitory mechanism in some bacteria were reported for other microorganisms (Bryan,1984).

References:

Bryan, L.E., 1989, Antimicrobial drug resistance. Academic Press. Inc., Orlando- Florida

Bush, K. and Singer, S.D., 1989, New plasmid born β-lactamases. Infection, Vol. 17, pp.429-433

Chirica, L.C., 1995, Partial purification and biochemical characterization of β-lactamase from penicillin G resistant strains of S.thermophilus . Master thesis, Dept. of Biochemistry, METU, Ankara- Turkey.

Hill, C., 1993, Bacteriophage and bacteriophage resistance in lactic acid bacteria. FEMS Microbiology Reviews, Noordwijkerhout, the Netherlands

Kitano, K., and Tomasz, A. 1979, E.coli mutants tolerant to β-lactam antibiotics. J. of Bacteriol. Vol. 140, No: 3, pp.955-963

Pham, J.N. and Bell S.M. 1992, β-lactamase induction by imipenem in Y.enterocolita . Pathology Vol. 24 pp.201-204

Rice, L.B., Eliopoudos, G.M., Wennersten, C., Goldman, D., Jacoby, G.A. and Moellering, R.C. 1991,Chromosomally mediated β-lactamase production and gentamicin resistance in E.feacalis . Antimicrobial Agents and Chemotheraphy. Vol. 35, No:2 pp.272-276

Podbielski, A., Schönling, J., Melzer, B., Warnatz, K., and Leusch, H.G. 1991, Molecular characterization of a new plasmid-encoded SHV-type β-lactamase (SHV-2 variant) conferring high-level cefotaxime resistance upon K.pneumonia. J. of Gen. Microb. Vol. 137 pp.567-578

Yaman, O., 1994, Isolation of penicillin G resistant strains of S.thermophilus. Master thesis, Dept. of Food Eng., METU, Ankara-Turkey.

Kinetics of Food Processes Involving Pure or Mixed Cultures of Lactic Acid Bacteria

Mustafa Özilgen
Food Engineering Department
Middle East Technical University
06531 Ankara, Turkey

Microbial kinetics

Proliferation of the microorganisms in a batch growth medium may include four ideal stages (Table 1). Some of these growth phases may be prevented intentionally in any process with appropriate experimental design.

Table 1. Ideal microbial growth phases in batch medium (Özilgen, 1996)

growth phase	growth rate expression	comments
lag phase	$\frac{dx}{dt} = 0$ [1]	time required for adaptation of the microorganisms to a new medium or recovery from injury
exponential phase	$\frac{dx}{dt} = \mu x$ [2]	variation of the population size is proportional with the number of microorganisms in the culture
stationary phase	$\frac{dx}{dt} = 0$ [3]	microbial growth may stop because of lack of a limited nutrient or product inhibition
death phase	$\frac{dx}{dt} = - k_d x$ [4]	microbial death may start after depletion of the cellular reserves in the stationary phase or with the effect of the unfavorable factors

Specific growth rate μ is a constant in the exponential growth phase, which implies that the growth rate is proportional with the viable microbial population, where all the members have equal potential for growth. The specific growth rate may be regarded as the frequency of producing new microorganisms by the ones already present.

NATO ASI Series, Vol. H 98
Lactic Acid Bacteria:
Current Advances in Metabolism, Genetics and Applications
Edited by T. Faruk Bozoğlu and Bibek Ray

The logistic model is frequently used to simulate microbial growth when a microbial population inhibits its own growth via depletion of a limited nutrient, product accumulation, or unidentified reason:

$$\frac{dx}{dt} = \mu x(1 - \frac{x}{x_{max}}) \qquad [5]$$

where μ is initial specific growth rate and x_{max} is the maximum attainable value of x. The logistic equation is an empirical model, it simulates the data when microbial growth curve follows a sigmoidal path to attain the stationary phase. It is mostly based only on experimental observations only. When $x \ll x_{max}$, the term in parenthesis is almost one and neglected, then the equation simulates the exponential growth, i.e., Equation [2]. When x is comparable with x_{max}, the term in parenthesis becomes important and simulates the inhibitory effect of over-crowding on microbial growth. When $x = x_{max}$ the term in parenthesis becomes zero, then the equation will predict no growth, i.e., stationary phase as described with equation [3].

The Luedeking-Piret model (1959) is among the most popular product formation models of food microbiology interest:

$$\frac{dc_{Pr}}{dt} = \alpha\, x + \beta\frac{dx}{dt} \qquad [6]$$

where c_{Pr} is product concentration, α and β are constants. The term αx represents the product formation rate by the microorganisms regardless of their growth; $\beta dx/dt$ represents the additional product formation rate during growth in proportion with the growth rate. This is an empirical equation, because it simply relates the experimental observations mostly without considerable theoretical basis. When growth-associated product formation rates are much greater than the non-growth-associated product formation rates equation [6] may be written as:

$$\frac{dc_{Pr}}{dt} = \beta\frac{dx}{dt} \qquad [7]$$

when non-growth associated product formation rates are much greater than the growth-associated product formation rates, [6] becomes:

$$\frac{dc_{Pr}}{dt} = \alpha x \qquad [8]$$

Equation [6] was originally suggested for simulating lactic acid production by ***Lactobacillus delbrueckii*** (Luedeking and Piret, 1959); it was later used for numerous other microbial products.

In a fermentation process substrate is allocated for three basic uses:

$$\begin{Bmatrix}\text{total rate of}\\ \text{substrate}\\ \text{utilization}\end{Bmatrix} = \begin{Bmatrix}\text{rate of substrate}\\ \text{utilization for}\\ \text{biomass synthesis}\end{Bmatrix} + \begin{Bmatrix}\text{rate of substrate}\\ \text{utilization for}\\ \text{maintenance}\end{Bmatrix} + \begin{Bmatrix}\text{rate of substrate}\\ \text{utilization for}\\ \text{product formation}\end{Bmatrix} \qquad [9]$$

Substrate consumption to keep the cells alive without growth or product formation is referred to as ***maintenance***. Equation [9] may be expressed in mathematical terms as:

$$-\frac{dc_S}{dt} = \frac{1}{Y_{x/s}}\frac{dx}{dt} + k_m x + \frac{1}{Y_{p/s}}\frac{dc_{Pr}}{dt} \qquad [10]$$

where $Y_{x/s}$ is the cell yield coefficient defined as grams of biomass produced per grams of substrate used, and $Y_{p/s}$ is the product yield coefficient, i.e., grams of product produced per grams of substrate used. The yield coefficients usually remains constant as long as the fermentation behavior remains unchanged, but there are also examples to the variable yield coefficients in the literature, which may indicate shift in substrate preference, availability of oxygen, etc. during the course of the process (Özilgen ***et al.***, 1988).

Microorganisms produce heat as a by-product of the energy metabolism which may be summarized as:

$$\text{Glucose} + 36\ P_i + 36\ \text{ADP} + 6\ O_2 \xrightarrow{\text{respiration}} 6\ CO_2 + 42\ H_2O + 36\ \text{ATP} + Q \qquad [11]$$

$$\text{Glucose} + 2\ P_i + 2\ \text{ADP} \xrightarrow{\text{anaerobic fermentation}} 2\ \text{lactate} + 2\ H_2O + 2\ \text{ATP} + Q \qquad [12]$$

where P_i and Q represent inorganic phosphate atoms and metabolic heat generation, respectively. ADP and ATP are abbreviations for adenosine diphosphate and adenosine triphosphate, respectively. It should be noticed that [11] and [12] do not describe the overall metabolic activity, where numerous metabolic intermediates and minor products are not accounted. ATP is the energy currency of the cell. ATP to ADP conversion is coupled with the energy consuming metabolic reactions, i.e., biosynthesis, cellular transport, etc. and make them thermodynamically feasible. ADP is converted back to ATP through equations [11] or [12]. Metabolic heat generation rate may be related with the growth rate as:

$$\begin{bmatrix} \text{metabolic heat} \\ \text{generation rate} \\ \text{per unit volume} \\ \text{of fermentor} \end{bmatrix} = \frac{1}{Y_\Delta}\frac{dx}{dt} \qquad [13]$$

where Y_Δ is the heat generation coefficient defined as grams of biomass production coupled with one unit, i.e., kJ, of heat evaluation.

Most food and beverage fermentations involve mixed culture of microorganisms. Typical examples may include yogurt cultures (***Streptococcus thermophilus*** and ***Lactobacillus bulgaricus***); wine (***Saccharomyces cerevisiae*** and wild microbial species, including the lactic acid bacteria), cheese production (mixed culture of various microorganisms, including various genera and strains of lactic acid bacteria).

Kinetics of microbial growth and lactic acid production during yogurt production (Özen and Özilgen, 1992)

Lactic acid production by mixed cultures of ***Streptococcus thermophilus*** and ***Lactobacillus bulgaricus*** is the most important chemical process involved in yogurt production. Interaction of these microorganisms in milk is favorable for both of the species, but not obligatory. Both numbers of the individual microorganisms and the total amounts of lactic acid produced by mixed cultures of these microorganisms are considerably higher than those obtained with pure cultures of each microorganism. In the mixed cultures, stimulation of growth of ***S. thermophilus*** was attributed to the production of certain amino acids, i.e., glycine, histidine, valine, leucine and isoleucine, by ***L. bulgaricus. S. thermophilus*** stimulates growth of ***L. bulgaricus*** by producing formic acid. The logistic equation was modified to simulate the mixed culture growth of the microorganisms as:

$$\frac{dx_S}{dt} = \mu_S x_S \left(1 - \frac{\tau_S x_S + \tau_L x_L}{x_{Smax} + x_{Lmax}}\right) \qquad [14]$$

and

$$\frac{dx_L}{dt} = \mu_L x_L \left(1 - \frac{\tau_S x_S + \tau_L x_L}{x_{Smax} + x_{Lmax}}\right) \qquad [15]$$

where subscript L and S refers to ***L. bulgaricus*** and ***S. thermophilus***, respectively. Some microbial species died at the end of the process, and the death rates were simulated with Equation [4]. The Luedeking-Piret equation was modified to simulate lactic acid production by the mixed culture of the microorganisms as:

$$\frac{dc_{Pr}}{dt} = \alpha_S x_S + \alpha_L x_L + \beta_S \frac{dx_S}{dt} + \beta_L \frac{dx_L}{dt} \qquad [16]$$

Comparison of the solutions of [14] - [16] with the experimental data is shown with a typical example in Figure 1. The kinetic analysis clearly showed that the contribution of each microbial species to the mixed culture growth process changed drastically when the substrate concentration was about 15 % (Figure 2). This optimum initial substrate concentration was in agreement with the result of the previous studies and the optimum find by trial and error procedure in the commercial yogurt production.

Kinetic aspects of leavening with mixed cultures of *Lactobacillus plantarum* and *Saccharomyces cerevisiae* (Yöndem *et al.*, 1992)

Sour dough is produced by leavening of dough with mixed cultures of yeast and bacteria. A typical German sour dough was reported to contain 10^7-10^{11} sour dough bacteria and 10^5-10^6 yeasts/g. Microflora of the German sourdough contains yeasts ***Saccharomyces cerevisiae, Pichia saitoi,***

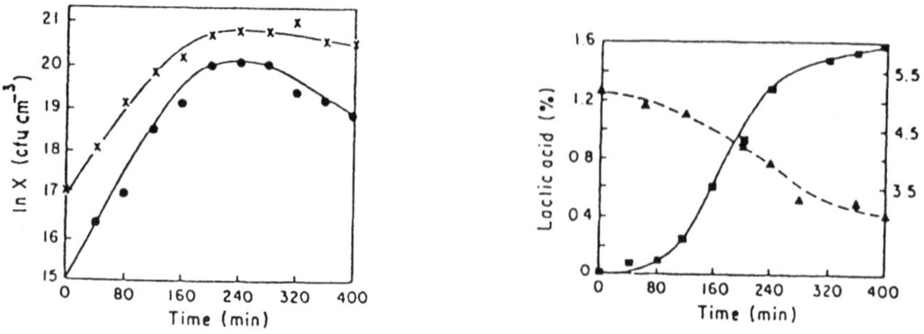

Figure 1. Variation of the concentrations of ***L. bulgaricus*** (●), *S. thermophilus* (x), lactic acid (■) and pH (▲) in 20 % non-fat dry milk medium. Data and simulations are shown in symbols and solid lines, respectively (pH values were not simulated). Constants: μ_L = 0.046 min^{-1}, μ_S = 0.032 min^{-1}, τ_L = 0.19, τ_S = 1.34, α_L = 5.1x10^{-12} % cm^{3}cfu^{-1}min^{-1}, α_S = 6.1x10^{-15} % cm^{3}cfu^{-1}min^{-1}, β_L = 1.9x10^{-10} % cm^{3}cfu^{-1}, β_S = 4.0x10^{-10} % cm^{3}cfu^{-1}, x_{Lmax} = 5.0x10^8 cfu cm^{-3}, x_{Smax} = 1.1x10^9 cfu cm^{-3}, k_L = 9.4x10^{-3} min^{-1} and k_S = 2.0x10^{-3} min^{-1} (Özen and Özilgen, 1992, © SCI. Reproduced by permission).

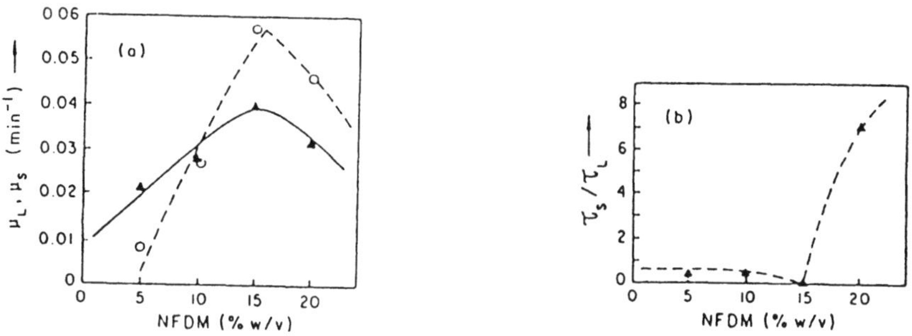

Figure 2. (a) Variation of the specific growth rates, i.e., parameters μ_L and μ_S with the initial substrate concentration. (▲) *S. thermophilus*, (o) ***L. bulgaricus.*** (b) Variation of the ratio τ_S/τ_L with the initial non-fat dry milk concentration (Özen and Özilgen, 1992; © SCI. Reproduced by permission).

Thrichosporon penicillatum, and lactic acid bacteria ***Lactobacillus acidophilus***, ***L. casei***, ***L. plantarum***, ***L. farcimins***, ***L. brevis***, ***L. buchneri***, ***L. fermentum***, ***L. fractivorans***, and ***L. alimentarius***. San Francisco (USA) sour dough contains the yeast ***Torulopsis holmii*** and ***Saccharomyces inusitatus*** and homofermentative lactobacilli, with the predominant species ***L. Sanfrancisco***. There are also numerous other yeast and lactic acid bacteria reported to be isolated from sour dough varieties as reviewed by Yöndem ***et al.***, (1992). Martinez-Anaya ***et al.*** (1990) reported that breads made with mixed cultures of yeasts and bacteria were superior to the ones produced with single strains, and that ***S. cerevisiae*** and ***L. plantarum*** were the most effective yeast and bacterium, respectively.

In a leavening process ***S. crevicae*** consumes the simple sugars available in the flour and produces carbon dioxide. After 30 minutes to 1 hour from the beginning of fermentation carbon dioxide production decreases with the depletion of the simple sugars then ***S. cerevisiae*** adapts it self to consume maltose and carbon dioxide production rate increases again. Most sour dough lactic acid bacteria can consume both mono- and di-saccharides, their major product is lactic acid (with homofermentative bacteria) or combination of lactic acid, acetic acid ethanol (with heterofermentative bacteria) plus carbon dioxide. Kinetic models may help to understand the details of the microbial processes which may not be understood by plotting the data only. Yöndem ***et al.***, (1992) performed experiments during sour dough production at 25 °C with 100 % yeast, 80 % yeast plus 20 % ***L. plantarum***, 60 % yeast plus 40 % ***L. plantarum***, 40 % yeast plus 60 % ***L. plantarum***, 20 % yeast plus 80 % ***L. plantarum*** and 100 % ***L. plantarum*** inoculum. It was observed that leavening process was pursued by multiphasic growth, which was attributed to depletion of simple sugars and adaptation of the microorganisms to the remaining complex sugars. The optimum growth was observed with 80 % yeast plus 20 % ***L. plantarum***. Comparison of the mathematical models with three sets of experimental data are depicted in Figure 3. It should be noticed that ***L. plantarum*** is a poor carbon dioxide producer (Yöndem ***et al.***, 1992). One mole of ideal gas occupies 22.4 L of volume under standard temperature and pressure, therefore a poor carbon dioxide producer makes significant contribution to the overall leavening activity.

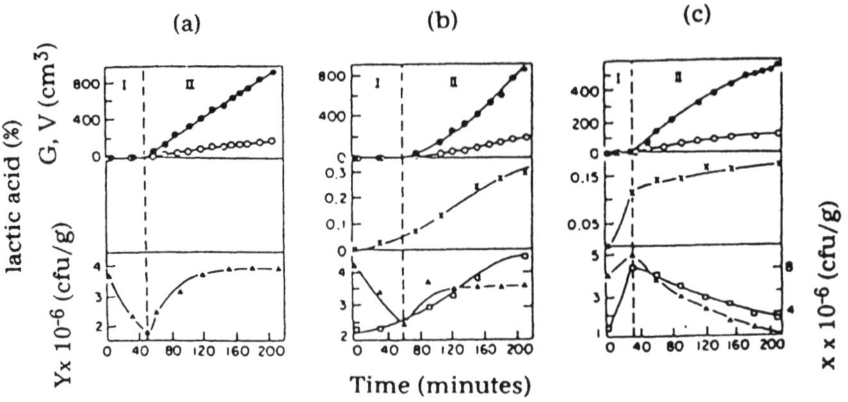

Figure 3. Comparison of the kinetic models with the experimental data at 25 °C. Simulations (—) , experimental data: (•) gas production, (o) dough volume increase, (x) lactic acid, (▲) *S. cerevisiae*, (□) *L. plantarum*. Inoculum: (a) 100 % yeast, (b) 80 % yeast plus 20 % *L. plantarum*, (c) 60 % yeast plus 40 % *L. plantarum*. Microbial counts of *S. cerevisiae* and *L. plantarum*, total volume of gas production and volume of the dough were shown with Y, X, G and V, respectively. Growth of the microorganisms were simulated with Equation [1] in Figure 3c and with Equation [5] in Figures 3a and 3b. Death of the microorganisms were simulated with Equation [4]. Gas production was expressed with Equation [16] after making the appropriate modifications in the nomenclature. Dough volume increase was expressed as $\frac{dV}{dt} = \phi\left(1 - \frac{V}{V_{max}}\right)\frac{dG}{dt}$, where V_{max} and ϕ were the maximum attainable dough volume and a constant, respectively (Yöndem *et al.*, 1992, © Academic Press. Reproduced by permission).

Kinetics of spontaneous wine production (Özilgen *et al.*, 1991)

A spontaneous wine production process is actually a mixed-culture and multi-product process, commenced by the natural microorganisms of the grapes. Natural grape microorganisms consists of various genera of molds, yeasts and lactic acid bacteria. Generally wine yeast, ***Saccharomyces cerevisiae***, is extremely low in population on the grapes. In the wine making process it multiplies with a strong fermentative capacity, excludes most of the other microorganisms from the medium and eventually invade the raw grape juice. Winemaking process may be analyzed in two phases: alcoholic fermentation and malo-lactic fermentation. During the alcoholic fermentation process wine microorganisms consume the fermentable sugars and produce ethanol. In the alcoholic fermentation process growth of the biomass may be simulated in exponential growth, stationary and death phases as described with Equations [2], [3] and [4], respectively. Total biomass concentration, denoted by x, is actually a mixed culture and the microbial species contribute to x actually change with time. In spontaneous wine fermentations, as the process proceeds, alcohol-sensitive microorganisms are inhibited and the alcohol tolerant microorganisms dominate the culture. Alcohol-tolerant microorganisms are generally better alcohol producers, therefore parameter β of the Luedeking-Piret equation may be related with the ethanol concentration in the medium:

$$\beta = \beta_0 + \beta_1 c_{pr} + \beta_2 c_{pr}^2 \qquad [17]$$

where β_0, β_1 and β_2 are constants. After substituting [17] in [6]:

$$\frac{dc_{pr}}{dt} = \alpha\, x + (\beta_0 + \beta_1 c_{pr} + \beta_2 c_{pr}^2)\frac{dx}{dt} \qquad [18]$$

Substrate utilization in the exponential growth phase was:

$$-\frac{dc_S}{dt} = \frac{1}{Y_{x/s}}\frac{dx}{dt} + \frac{1}{Y_{p/s}}\frac{dc_{pr}}{dt} \qquad [19]$$

Lactic acid bacteria are the major species contributing the total biomass concentration x in the malo-lactic fermentation phase, where lactic acid bacteria uses malic acid as a substrate to produce lactic acid. Equation [5] simulates microbial growth in the malo-lactic fermentation phase.

Temperature changes in the fermentation vessel were modeled after making thermal energy balance:

$$-\sum_{i=1}^{n} U_i A_i (T - T_{env}) + \frac{V}{Y_\Delta}\frac{dx}{dt} = \frac{d(\rho c V T)}{dt} \qquad [20]$$

where the term $-\sum_{i=1}^{n} U_i A_i (T - T_{env})$ represents the energy loss from the fermentor surfaces, and the term (V/Y_Δ) (dx/dt) represents the thermal energy generation coupled with microbial growth. This term becomes zero after the end of the exponential growth in the alcoholic fermentation. The term $d(\rho cVT)/dt$ is thermal energy accumulation. Equation [20] was rearranged after substituting [2] for dx/dt:

$$\frac{dT}{dt} = K_1 (T_{env} - T) + K_2 e^{\mu t} \qquad [21]$$

where $K_1 = \sum_{i=1}^{n} (U_i A_i/\rho c V)$ and $K_2 = (\mu x_0 / Y_\Delta \rho c V)$. Parameter K_2 became zero after the end of the exponential growth phase of the alcoholic fermentation. The model is compared with a typical set of experimental data in Figure 4.

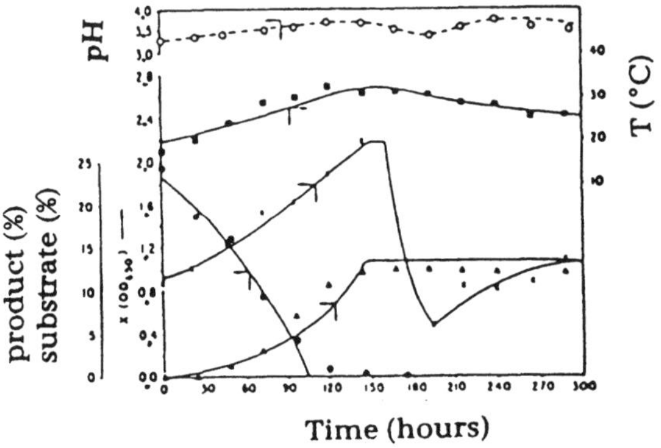

Figure 4. The course of the wine production. Experimental data and simulations are shown in symbols, and solid lines, respectively. (o) pH, (■) temperature, (●) reducing sugars, (▲) ethanol, (x) biomass. The model constants are: $\mu = 0.006\ h^{-1}$, $k_d = 0.042\ h^{-1}$, $\alpha = 0.0045$ % ethanol/OD_{650} h, $\beta = 3.6 + 1.25\ cP + 0.004c_p^2$ % ethanol/OD_{650}, $Y_{x/s} = 0.04\ OD_{650}$/% reducing sugars, $Y_{p/s} = 10$ % ethanol/% reducing sugars, $K_1 = 0.004\ h^{-1}$, $K_2 = 0.08\ °C\ h^{-1}$, $\mu_M = 0.032\ h^{-1}$, $x_{max} = 1.08\ OD_{650}$

Reprinted by permission of the publisher from Kinetics of Spontaneous Wine Production (Özilgen *et al.*, 1991) *Enzyme and Microbial Technology,* 13(3), 252-256.

Microbial death during drying and freeze-drying

Drying data is usually expressed as the amounts of water carried per unit dry mass of dry solids (Özilgen, 1996):

$$\omega = \frac{\text{mass of water}}{\text{mass of dry solids}} \qquad [22]$$

The basic model of drying is used since 1920's and discussed in detail in the literature (McCabe *et al.*, 1985; Porter *et al.*, 1973). At the beginning of a drying process when the surface is covered with a continuous film of water, it attains the wet-bulb temperature and the evaporation rate from the surface becomes the rate determining step. This is called the constant rate period, where water movement is assumed to be controlled by several factors, including capillary suction and diffusion. The mechanism of drying in slow drying non-porous materials, such as soap, gelatin, glue, and in the later stages of drying of clay, wood, textiles, leather, paper, foods and starches is liquid diffusion (Geankoplis, 1983). Diffusion of liquid water may result because of concentration gradients between the depths of the solid, where the concentration is high, and the surface, where it is low. These gradients are set up during drying from the surface (Treybal, 1980). The capillaries extend from small reservoirs of water in the solid to the drying surface. In a drying proceeds, at first moisture moves by capillary to the surface rapidly enough to maintain a uniform wetted surface and the rate of drying is constant. Surface tension of the liquid is a controlling factor on flow rate of water flow due to capillary suction (Labuza and Simon, 1970). When the drying solids attain the critical moisture content, dry spots start appearing on the surface. This is the first falling rate period and continues until all the liquid film disappears from the surface, where water supply rate to the surface is not sufficient, therefore the dry spots enlarge and the water film decrease continuously. Then the second falling rate period starts, where evaporation occurs below the surface of the biomass and diffusion of vapor occur from the place of vaporization to the surface (Treybal, 1980). The drying process stops when equilibrium is established between the biomass and the ambient air. With the depletion of the sub-surface water reservoirs the mechanical structure collapses and the biomass undergoes shrinkage. If the biomass should be subject to too high temperatures at the beginning of the second falling rate period, the entrance of the partially empty capillaries may shrink rapidly and prevent removal of the remaining water at the deeper locations. This is called ***case hardening***. This entrapped moisture may diffuse to the surface during storage and may cause microbial spoilage. Case hardening may be prevented by using moderately moist or warm air for drying.

A vapor pressure versus temperature phase diagram for free water is shown in Figure 5. The process 1 → 2 is conventional drying under constant pressure, where liquid water is evaporated. The process 3 → 4 represents freeze drying, where ice is converted directly into vapor. In a freeze drying process biomass is frozen, then water is removed by transfer from the solid state to the vapor state by sublimation. It should be noticed that the freeze drying requires vacuum and conducted at lower temperatures therefore expensive equipment and long processing times are needed, which increases the cost of the process.

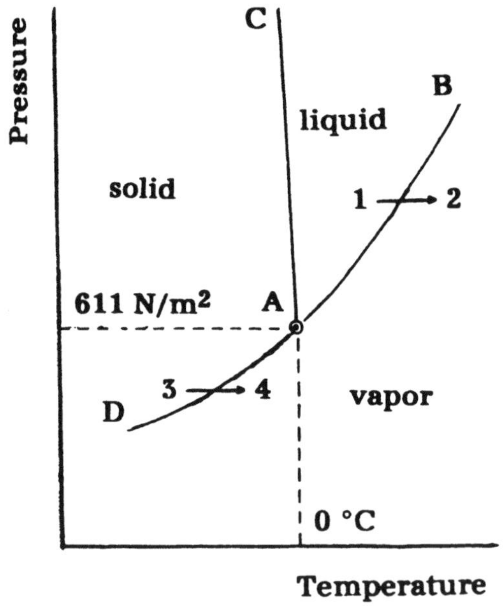

Figure 5. A qualitative vapor pressure versus temperature phase diagram for free water. $\overline{AC}$: melting line, $\overline{AD}$: sublimation line, $\overline{AB}$: vaporization line.

Baker's yeast (Alpas, *et al.*, 1996) and lactic acid bacteria (Johnson and Etzel, 1993; Fu and Etzel, 1995) are subjected to drying with commercial purposes. Production of the baker's yeast has been reviewed in numerous references (Ponte and Tsen, 1978; Peppler, 1979; Reid, 1982; Reed and Nagodawithana, 1991). In a typical process yeast is produced with fermentation, centrifuged, washed and concentrated into yeast cream of approximately 18 to 21 % solids, then filtered in filtered press or vacuum filters. The filter cake is mixed with small amounts of water emulsifiers and oils and converted into compressed yeast pellets via extrusion. The compressed yeast is perishable needs to be kept in closed packages in the refrigerator. Active dry yeast is manufactured by a similar process as the compressed yeast, but the moisture content of the extrudate is substantially reduced. The yeast pellets can be tunnel dried to obtain a product containing less than 8 % moisture without substantial loss of viability.

Starter culture of yogurt and other fermented milk products are usually freeze-dried lactic acid bacteria. Spray drying of the lactic acid bacteria is being studied as an alternative to the freeze drying to reduce the cost of the drying process (Fu and Etzel, 1995). The spray-dried, non-viable lactic acid bacteria may also be used as source of active enzymes in the dairy products to accelerate the ripening process (Johnson and Etzel, 1993). Drying of baker's yeast causes some damage to the cellular structure, including rupture of the cytoplasmic membrane, changes in the composition of the nucleic acids, proteins, lipids, carbohydrates and polyphosphates (Bekker and Rapoport, 1987), the damage may also cause the death of the cells; therefore an active dry yeast preparation can not be better than the compressed yeast that it is made from. The rate at which the compressed yeast may be dried depend on the temperature, velocity and humidity of the drying air stream, as well as the size and shape of the granulated yeast. It is generally believed that in a yeast drying process at moisture levels above 20 % the changes in cell viability and cell wall permeability are minor; at moisture levels below 20 %, respiration stops and cell constituents are leached if the yeast is rehydrated in water (Reed and Nagodawithana, 1991).

Fu and Etzel (1995) while studying spray drying behavior of ***Lactococcus Lactis*** ssp. *lactis* C2 reported that cellular injury is resulted from dehydration and exposure to high temperatures. The viability of the lactic acid bacteria decreases as the temperature of the drying process increases as depicted with a typical plot in Figure 6.

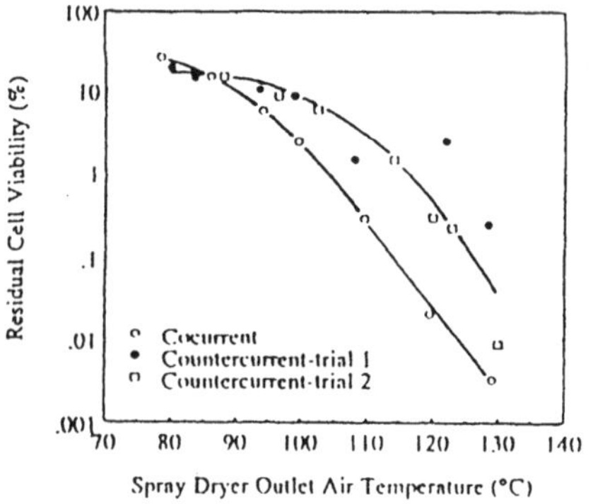

Figure 6. Variation of the residual cell viability with the spray dryer outlet air temperature during spray drying of ***Lactobacillus helveticus*** in 30 % condensed skim milk under various flow conditions (Johnson and Etzel, 1993; © AIChE. Reproduced by permission).

Fu and Etzel (1995) reported that shear in the atomizer did not cause cellular injury. The lag time before lactic acid production from a preparation was used to asses the injury of the cells and it was reported to be essentially independent of the treatments.

Freeze-drying of the microorganisms is a two step process, where the microorganisms are frozen first, then subjected to drying under vacuum at low temperature. The bacterial cells are likely to lose their viability during freeze-drying and the subsequent storage period. The major factors that cause the injury include cold shock, change of the cell wall permeability and the metabolic injury (Fennema *et al.*, 1973; Bozoğlu *et al.*, 1987). Cold shock may be accompanied by damage to the cell membranes or by formation of intracellular ice. Ice is pure water, therefore the concentration of the solute in the remaining portions of the cytoplasm will increase; this may cause a change in the intracellular pH and ionic strength and may also increase the rates of the detrimental chemical reactions. Injury may also cause an increase in the permeability of the cell membrane. Metabolic or enzymatic activities of the injured cells may also change. Metabolically injured cells may be detected by the changes in their growth requirements, i.e. they may grow on complex media, but not on a media formed by simple salts (Ray, 1973). A freeze drying process may be considered successful if a sufficient number, i.e. 0.1-1 % of the original population, remains viable after reconstitution. Cryoprotective additives are employed during freeze drying. Ability of the more effective additives to protect the microorganisms is usually attributed to their capacity to bind water, inhibit either intracellular or extracellular ice formation and contribution to the structure of the micro-environment around the cell wall.

There are large number of studies in the literature as reviewed by Bozoğlu *et al.* (1986) that much of the injury that the microbial cells are subjected to during and after freeze drying is surface related phenomena. Survival kinetics of the lactic acid starter cultures were modeled considering the microorganism and the external medium interfacial area as the critical factor determining the resistance of the microorganisms to freeze drying and the following correlation was suggested (Bozoğlu *et al.*, 1986):

$$\ln\left(\frac{x_f}{x_0}\right) = \alpha + \beta \log x_0 \qquad [23]$$

where x_0 is the initial concentration of the viable microorganisms before freeze-drying, x_f is the concentration of the viable microorganisms immediately after freeze-drying, α and β are constants. Equation [23] is based on the assumption that the interfacial area between each cell and the external medium decreases with cell concentration due to the shielding effect of the microorganisms on each other. Comparison of Equation [25] with experimental data is depicted in Figure 7.

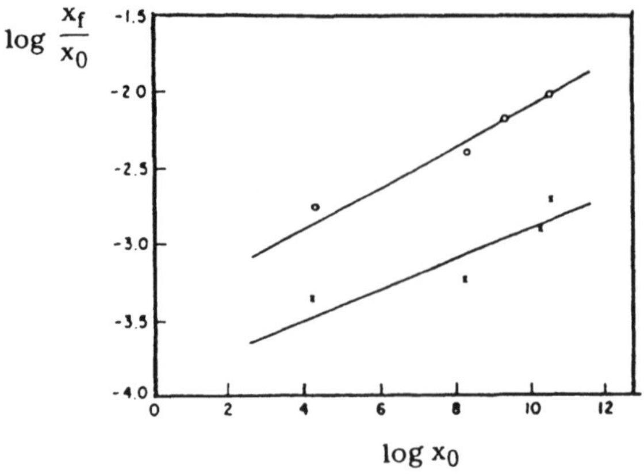

Figure 7. Comparison of Equation [23] (—) with the experimental data (o, x). (–o–) ***S. thermophilus***: $\ln \frac{x_f}{x_0} = -3.4 + 0.13 \log x_0$, (–x–) ***L. bulgaricus***: $\ln \frac{x_f}{x_0} = -3.9 + 0.10 \log x_0$. **Reprinted by permission of the publisher from Survival Kinetics of Lactic Acid Starter Cultures During and After Freeze Drying (Bozoğlu *et al.*, 1987) *Enzyme and Microbial Technology*, 9(9), 531-537.**

The freeze dried microorganisms lose their activity, i.e. amount of lactic produced per unit biomass per time, during the storage. Death of the microorganisms were simulated with equation [4]; activity loss was described as (Alaeddinoğlu *et al.* 1989):

$$\frac{d\pi}{dt} = -k_a \pi \qquad [24]$$

where π is the activity determined at time t (t = 0 indicates the beginning of the storage of the freeze-dried microorganisms) and k_a is the rate constant of the activity loss. Activity and viability loss reported to be parallel during storage, with occasional deviation from this trend caused by presence of the injured microorganisms that did not produce lactic acid (Figure 8).

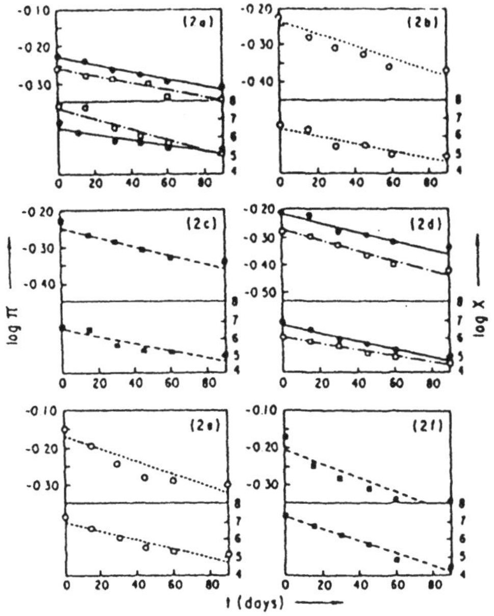

Figure 8. Survival and activity of freeze-dried *Streptococcus thermophilus* cells in sucrose containing media (–•— 0.06, .. o .. 0.15, -■- - 0.29 and - - □ - - 0.44 M sucrose). Non-fat skim milk media were used in (a-c). Whey media were used in (d-f) Reprinted by permission of the publisher from Activity-loss Kinetics of Freeze-dried Lactic Acid Starter Cultures (Alaeddinoğlu *et al.* 1989,) *Enzyme and Microbial Technology*, **11**(11), 765-537. © 1989 Elsevier Science Ltd.

Alpas (1995) studied drying behavior of the baker's yeast in computer controlled tunnel dryer (Özilgen *et al.*, 1995) where percentage of the dry solids obtained after 180 minutes of drying experiments was between 80 to 96.5 %. Commercial baker's yeast consist of *Saccharomyces cerevisiae*, lactic acid bacteria, and low number of contaminating microorganisms (Ponte and Tsen, 1978). During the drying process microbial death was modeled with Equation [4], and weight loss was expressed as (Alpas, 1995):

$$\frac{d\omega}{dt} = k_w(\omega - \omega^*) \qquad [25]$$

where k_w is a weight loss constant, ω is the dimensionless weight calculated by dividing the weight of the slice at time t to its initial weight at the beginning of an experiment, and ω^* is the lowest attainable value of ω when moisture contents of the dry yeast comes in equilibrium with that of ambient air. Values of ω^* were between 0.37 and 0.44 (dimensionless) (Alpas, 1995). Agreement of the model with the experimental data is exemplified with an arbitrarily chosen typical set of data in Figure 9. The fresh compressed yeast pellets were dried for 45, 60, 90, 120, 150 and 180 minutes under constant experimental conditions. The weight loss constant k_w was evaluated in each of these experiments separately and its mean value was used for further analysis. The pooled variance of the weight loss constant k_w was 4.7×10^{-3} min^{-1}. Variance of parameter k_w had the same order of magnitude under all of the experimental conditions and its value was not correlated with the length of the experiments, temperature or the velocity of the air (Alpas, 1995).

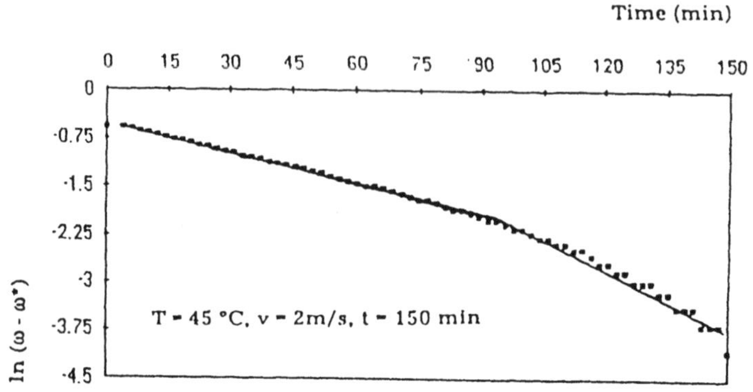

Figure 9. Comparison of the drying model (——) with a typical set of experimental data (■) (Alpas, 1995).

An Arrhenius type of an expression was used to model the temperature effects on parameter k_w (Alpas, 1995):

$$k_w = k_{w0} \exp\{-\frac{E_{aw}}{R\,T}\} \qquad [26]$$

where k_{w0} is pre-exponential constant, E_{aw} is activation energy, R is gas constant, and T is absolute temperature. Constants k_{w0} and E_{aw} were evaluated with the typical Arrhenius plots after taking logarithm of Equation [26] as exemplified with typical sets of data in Figure 10. Values of constants k_{w0} and E_{aw} were found to be dependent on the air flow rates as depicted in Figure 11. During the first falling rate period water is removed from the surface, subsequently the dry zones enlarge in expense of the wet regions. Heat and mass transfer rate constants and consequently parameter k_w increase with the velocity of the air.

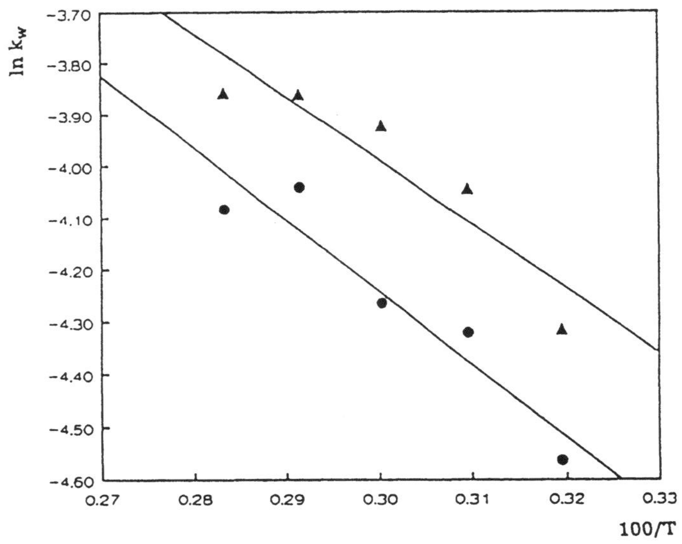

Figure 10. Arrhenius plots for evaluation of constants k_{w0} and E_{aw}. Experimental data were shown in symbols with (▲) 2.0 m/s and (•) 2.5 m/s of air velocity. The solid lines are the best fitting lines (Alpas, 1995).

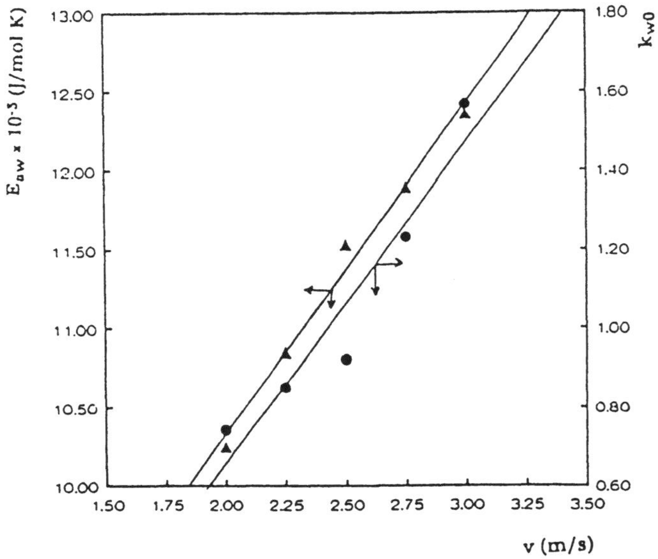

Figure 11. Variation of activation energy (E_{aw}) and pre exponential constant (k_{w0}) with velocity of the air. Experimental data are shown in symbols: (▲) E_{aw}, (•) k_{w0}; (—) best fitting lines. Equations of the lines: $E_{aw} = 2.1x10^3$ v + $6.1x10^3$ (correlation coefficient = 0.99); k_{w0} = 0.81 v - 0.97 (correlation coefficient = 0.96), where v is the velocity of air expressed in m/s (Alpas, 1995)

Constants of the Arrhenius type expressions are generally inter-related. Under slightly different experimental conditions any change in parameter k_{w0} may be compensated by the changes in E_{aw}:

$$\ln k_{w0} = \alpha E_{aw} + \beta \qquad [27]$$

Equation [27] is referred to as a compensation relation. Although parameters k_{wo} and E_{aw} changes with the experimental conditions parameters α and β are constants. Equation [27] and its derivatives are generally observed with various physical, chemical and biological reactions and may be used in process design. Review of the literature and description of use of the compensation relations in process design is described in detail elsewhere (Özilgen and Özilgen, 1995). Comparison of Equation [27] with the experimental data is depicted in Figure 12.

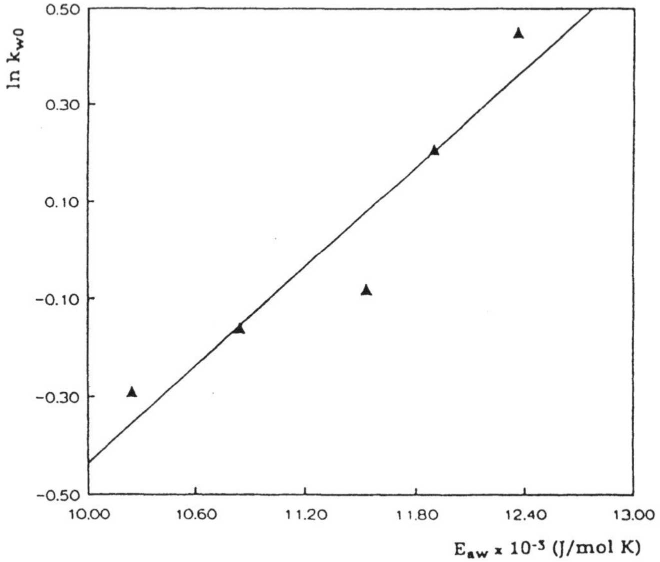

Figure 12. Compensation relation for the Arrhenius expression parameters of the first falling rate drying constant k_w. Experimental data are shown in symbols (▲) Equation of the best fitting line (—) is: $\ln k_{wo} = 3.38 \times 10^{-4} E_{aw} - 3.82$ (correlation coefficient = 0.95) (Alpas, 1995).

A kink was observed about 75-110 minutes after beginning of the drying process depending on the experiment conditions as exemplified in Figure 8. The second region of the plot observed at the later stages of the experiments was attributed to the second falling rate period of the drying process. Drying rate constant referring to the second rate period was denoted as k_{w2} its mean value and variance were 4.3×10^{-2} min^{-1} and 9.8×10^{-3} min^{-1}, respectively. There was no clear cut transition between the falling rate periods, therefore the distinction was made intuitively to obtain the highest correlation coefficients while evaluating the constants k_w and k_{w2}. The second falling rate period is essentially a diffusion rate controlled process, therefore parameter k_{w2} was not affected by the air velocity. Parameter k_{w2} was not affected by the air temperature and the length of the experiments. Having no affect of temperature on k_{w2} may indicate that the diffusion constant of water in the yeast cells did not change much with temperature within the range of the experiments.

Data for the survival kinetics of the microorganisms were obtained during the same experiments with weight loss. Death rate of the microorganisms were described with Equation [4]. Temperature effects on the death rate constant was described with the Arrhenius expression, similarly as that of the first falling rate period weight loss constant k_w. Effects of the air flow rate on the parameters of the Arrhenius expression for the death rate constant k_d were similar to those of the weight loss constant k_w and also kinetic compensation relation similar to Equation [10] was observed. Since the same mathematical models (with different parameter values) described the effects of the air velocity and temperature on the microbial death and first falling rate drying constants, we tried to see if there were any correlation between these two sets of constants. Two linear relations were observed in Figure 13. The first relation was valid at 2.5 and 3.0 m/s of air velocities and the second relation was valid at 2.0 m/s of air velocity. Figure 13 implies that microbial death and drying occurred together, but microbial death was slower at low air velocity.

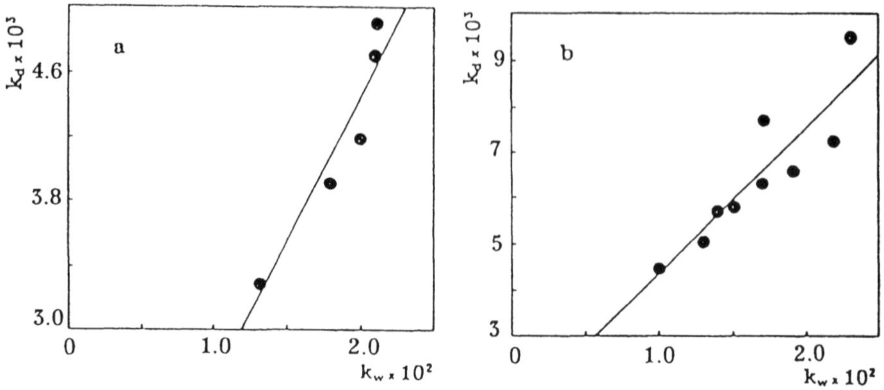

Figure 13. Correlation between the microbial death and first falling rate drying constants: (▲) experimental data, (—) best fitting lines. **(a)** Velocity of air is 2.0 m/s; $k_d = 8.4 \times 10^{-4} + 0.179\ k_w$ (correlation coefficient = 0.94); **(b)** Velocity of air is 2.5 and 3.0 m/s; $k_d = 1.2 \times 10^{-3} + 0.317\ k_w$ (correlation coefficient = 0.90) (Alpas, 1995).

References

Alaeddinoğlu, G., Güven, A. and Özilgen, M. Activity-loss kinetics of freeze-dried lactic acid starter cultures. *Enzyme and Microbial Technology*, **11**, 765-769 (1989)

Alpas, H. Computer Controlled Active Dry Yeast Drying. MS. Thesis. Middle East Technical University, 1995

Bekker, M.J. and Rapoport, A.I. (1987) Conservation of yeasts by dehydration. *Advances in Biochemical Engineering and Biotechnology* **35**, 127-171

Bozoğlu, T.F., Özilgen, M. and Bakır, U. Survival kinetics of lactic acid starter cultures during drying and after freeze drying. *Enzyme and Microbial Technology*, **9**, 531-537 (1987)

Fennema, O.R, Powrie, W.D. and Marth, E.H. Low-Temperature Preservation of Foods and Living Matter. Marcel Dekker, Inc. New York, 1973

Fu, W-Y and Etzel, M.R. Spray drying of *Lactococcus lactis* ssp. *lactis* C2 and cellular injury. *Journal of Food Science*, **60**, 195-200 (1995)

Geankoplis, C.J. *Transport Processes and Unit Operations*, 2nd ed. Allyn and Bacon, Inc. MA, 1983

Johnson, J. A.C. and Etzel, M.R. Inactivation of lactic acid bacteria during spray drying. In *Food Dehydration*, Barbosa-Cánovas, G.V. and Okos, M.R. (editors). AIChE Symposium Series No 297; Volume 89, p. 98-107, 1993.

Labuza, T.P. and Simon, I.B. (1970) Surface tension effects during dehydration. Part 1. Air drying of apple slices. *Food Technol.* **24**(6) 712-715

Luedeking, R. and Piret, E.L. A kinetic study of lactic acid fermentation. Batch process at controlled pH. ***Journal of Biochemical and Microbiological Technology and Engineering*** **1**, 393-412 (1959)

Marinez-Anaya, M.A., Pitarch, B., Bayarri, P. and Beneditto De Barber, C. Microflora of the sourdoughs of wheat flour bread. X. Interactions between yeasts and lactic acid bacteria in wheat doughs and their effects in bread quality. ***Cereal Chemistry,*** **67**, 85-91 (1990)

McCabe, W.L., Smith, J.C. and Harriot, P. ***Unit Operations of Chemical Engineering.*** 4th ed. McGraw-Hill, 1985

Mulchandani, A. and Luong, J.H.T. Microbial growth kinetics revisited. ***Enzyme and Microbial Technology,*** **11**, 66-72 (1989)

Özen, S. and Özilgen, M. Effects of substrate concentration on growth and lactic acid production by mixed cultures of ***Lactobacillus bulgaricus*** and ***Streptococcus thermophilus. Journal of Chemical Technology and Biotechnology,*** **54**, 57-61 (1992)

Özilgen, M. Ollis, D.F. and Ogrydziak, D. Kinetics of batch fermentations with ***Kluyveromyces fragilis. Enzyme and Microbial Technology,*** **10**, 165-172 and 640 (1988)

Özilgen, M., Çelik, M. and Bozoğlu, T.F. Kinetics of spontaneous wine production. ***Enzyme and Microbial Technology,*** **13**, 252-256 (1991)

Özilgen, M., Güvenç, G., Makaracı, M. and Tümer, İ. Colour change and weight loss of apple slices during drying. ***Zeitschrift für Lebensmittel-Untersuchung und-Forschung,*** **201**, 40-45 (1995)

Özilgen, S. and Özilgen, M. (1995) Kinetic compensation relations: Tools for design in desperation. ***J. Food Eng.*** (in press)

Özilgen, M. ***Chemical Engineering Applications For Food Process Modeling & Control,*** Gordon and Breach Publishing Associates Ltd. Reading, UK (in press), 1996

Peppler, H.J. (1979) Production of yeast and yeast products, in ***Microbial Technology,*** Volume 1. Peppler, H.J. and Perlman, D. editors, Academic Press, New York

Ponte, J.G. and Tsen, C.C. (1978) Bakery products, in ***Food and Beverage Mycology,*** Beuchat, L.R. editor, Avi Pub. Co. Westport, Connecticut

Porter, H.F., McCormick, P.Y., Lucas, R.L. and Wells, D.F. Gas-solid systems, in ***Chemical Engineer's Handbook,*** McGraw-Hill Kogakusha, Japan, 1973

Ray, B. and Speck, M.L. Freeze-injury in Bacteria. ***CRC Reviews in clinical and Laboratory Sciences,*** **4**, 161-213 (1973)

Reid, G. (1982) Production of baker's yeast, in ***Industrial Microbiology,*** 4th ed. Reed, G. editor, Avi Pub. Co. Westport, Connecticut

Reed, G. and Nagodawithana, T.W. (1991) ***Yeast Technology,*** 2nd ed. Van Nostrand Reinhold, New York

Treybal, R.E. ***Mass Transfer Operations,*** 3rd ed. McGraw-Hill Kogakusha Ltd, Tokyo, 1980

Yöndem, F., Özilgen, M. and Bozoğlu, T.F. Kinetic aspects of leavening with mixed cultures of ***Lactobacillus plantarum*** and ***Saccharomyces cerevisiae. Lebensmittel Wissenschaft und Technologie,*** **25**, 162-167 (1992)

THE STABILISATION OF BIOCATALYTIC PROTEINS AGAINST INACTIVATION

Altan Erarslan[1,2]

[1]Kocaeli University, Faculty of Arts and Sciences, Department of Chemistry, Section of Biochemisty, 41300 İzmit - Kocaeli, Turkey
and
[2]The Scientific and Technical Research Council of Turkey, Marmara Research Centre, Institute of Genetic Engineering and Biotechnology, PO. Box 21, 41470 Gebze - Kocaeli, Turkey

ABSTRACT: The methods and Stabilisation methods for thermal inactivation of enzymes are classified as 1. Screening of micro-organisms for enzymes with enhanced intrinsic stability, 2. Addition of stabilising agents, 3. Chemical modification of enzyme molecules, 4. Immobilisation of enzyme molecules. The criteria for the stabilization of biocatalytic proteins against irreversible inactivation are described with numerous applications.

Introduction

Many suggestions have been made for the biotecnological applications of enzymes and several of them have been commercialised. Today enzyme technology widely is applied in such industries as detergent, dairy, pharmaceutics, distilling, brewing, fruit, wine, milling, baking and textile. In addition, enzymes are used in chemical and clinical analyses in medicine. Although in most of the technological processes immobilized enzymes are used, the majority of the enzymes on the market are in their native form. Enzymes should be stable for very long times under operational conditions in order to be suitable for technological applications. The lengthening of operational stability will reduce the number of enzyme replacements and thus decrease the cost of enzyme preparations, which can be rather expensive. The cost of enzymatic processes will also be reduced. The main reason for the decrease of operational stability is the inactivation of the enzymes. Enzyme inactivation can take place by the action of several factors such as heat, medium pH, denaturing agents,

NATO ASI Series, Vol. H 98
Lactic Acid Bacteria:
Current Advances in Metabolism, Genetics and Applications
Edited by T. Faruk Bozoğlu and Bibek Ray

oxygen and proteases. Therefore, the stabilization of enzymes against inactivation is generally indispensable for their industrial applications.

Stabilization Methods

Four major methods can be applied in order to stabilise soluble enzymes. These are: 1. Screening of micro-organisms for enzymes with enhanced intrinsic stability, 2. Addition of stabilising agents, 3. Chemical modification of enzyme molecules, 4. Immobilisation of enzyme molecules.

1. Screening of Micro-organisms: A number of micro-organisms are able to survive in extreme conditions such as very hot or cold temperatures, strong acidic and alkaline pH values, saline media of very high ionic strength. Marine micro-organisms have been observed at depths that correspond to a pressure of hundreds of bars. In many of these organisms specifically "stabilised" proteins have been identified and characterised. However, to find a few microbial source involves a laborious and time consuming screening of microbial strains in the hope of finding one that possesses an enzyme with the desired properties such as stability, catalytic activity and specificity.

2. Enzyme Stabilisation by Additives: The addition of various compounds into enzyme solutions for their stabilisation has been known in a very long time. The effect of these stabilising agents will be discussed in several sections.

2.1. Addition of Substrates and Other Ligands: In general the conformation of proteins at binding sites is less stable than at other areas of the molecule. In other words the point of instability is closer to the catalytic site of enzyme molecule. Binding of substrates and other ligands may lead to stabilisation or labilization of an enzyme or to no effect at all. For the case of reversible thermal inactivation of enzymes, the stabilising effect of substrates and ligands can be explained as follows: Assuming that the active sites exist only in the native form of enzyme, it is obvious that since most ligands bind only to the native enzyme and will cause the shifting of the equilibrium of "*native enzyme* $\leftrightarrow$ *thermo unfolded enzyme*" to the left, consequently enzymes will be stabilised against thermal inactivation. Hence if unfolding is a critical step in the irreversible thermal inactivation of enzymes, additions of ligands and substrates should stabilise enzymes (1,2).

A few specific examples have been reported in the literature relating the enzyme stabilisation by substrate addition. Lactate dehydrogenase is stabilised towards thermal inactivation by its substrate, NAD, NADH, lactate and effectors such as fructose diphosphate. The thermal stability of glucoamylase is significantly enhanced in the presence of its substrate analogues glucose, δ-glucono-lactone and lactose. β-glucosidase are stabilised by glucose and fructose against to denaturation by urea (2). Investigations of the kinetic effect of the substrate (maltodextrins) and the product (maltose) on the rate constant of irreversible thermal inactivation of *Aspergillus niger* glucoamylase at 75°C showed that both ligands substantially stabilise the enzyme, reducing the rate constant of thermal inactivation by 15 and 40 fold for the immobilised and free glucoamylase respectively (2). The invitro stabilisation of δ-(L-α-aminoadipyl)-L-cysteinyl-D-valine (ACV) synthetase (ACVS) from *Streptomyces clavuligerus* by its cofactor and substrate has been recently reported (3). The stability of crude enzyme is substantially increased by dithiothretiol (DTT) and the cofactor Mg^{+2}. Addition of the substrate L-valine along with the cofactors (ATP and Mg^{+2}) raises the thermal inactivation temperature, and increases the stability of enzyme at low temperatures. Amino acids capable of replacing L-valine as ACVS substrate generally stabilise the enzyme.

2.2. Addition of Neutral Salts: Many investigations have been made concerning the stabilising effect of salts on enzyme molecules. All studies in this area can be divided into two groups (1): 1. The reaction of relatively low concentrations of divalent ions such as Ca^{+2}, Mg^{+2}, 2. The effect of relatively high concentrations of various neutral salts.

2.2.1. Low Concentrations: The stabilising effect of salt cations at low concentrations can be defined as their interactions with metaloenzymes that are the proteins having strong bindings with metals. Divalent metal ions such as Ca^{+2}, Zn^{+2}, Cu^{+2}, Mg^{+2} peform important functions in the activities of many proteins. Some examples are as follows: Cu on hemocyanins in oxygen transport; Ca on trypsinogen in zymogen activation; Zn on carboxy peptidase in catalytic function; Ca on α-amylase in maintenance of tertiary structure (1,2). The participation of metal ions in the catalytic function of enzymes must be considered as either their incorporation into the active site of enzyme, or participation in the formation of enzyme substrate complexes as well as in the removal of substrates or participation of allosteric effects. Stabilisation by added salts may be due to interaction using either of these mechanisms. In maintaining the tertiary structures of some enzymes and proteins, Ca^{+2} plays a most important role. As an example, α-amylase from *Bacillus caldolyticus* is greatly stabilised by bound Ca^{+2} ions. After removal of Ca^{+2} ions with complex forming agents, thermal stability is lost. However Sr cannot substitute for Ca^{+2} ions. Ca ions^{+2} show a critical specificity for this stability (1).

2.2.2. High Concentrations: At high concentrations, neutral salts may bind to protein-charged groups or dipoles (*salting in*) or reduce the solubility of protein-hydrophobic groups by increasing the ionic strength of solution (*salting out*). The salting out of hydrophobic residues from the surface into the interior of an enzyme molecule will conformationally "compress" the enzyme. Such enzyme will be more resistant to thermal unfolding and thus could demonstrate a higher thermal stability (2). The effect of salts on protein solubility generally corrected with the Hofmeister lyotropic series. In accordance with the Hofmeister lyotropic series, the stabilising effect of cations and anions should decrease in the following order (4): $(CH_3)_4N^+ > (NH_4)^+ > K^+ > Na^+ > Mg^{+2} > Ca^{+2} > Ba^{+2}$ and $SO^{-2} > Cl^-$ $> Br^- > NO_3 > ClO_4 > SCN^-$. Therefore, $(NH_4)_2SO_4$ should be a strong stabiliser, whereas NaSCN destabilises enzymes.

2.3. Addition of Polyols and Sugars: Another possible approach to increase enzyme stability is to use additives such as sugars and polyols that preserve the biocatalyst activity by modulating its micro environment. The addition of these solutes has been reported to prevent the enzyme from unfolding by strengthening hydrogen bonds or by strengthening hydrophobic interactions or both. The efficiency of an additive to preserve enzymatic activity is dependent both on its chemical character and on the enzyme itself. With this method, enzymes are used in conditions similar to those of industrial processes (heterogeneous, with low water content, non-Newtonian media). Enzyme stability may often be correlated to water activity (4,5).

The effect of sorbitol, mannitol and glycerol on storage stability of invertase at temperature intervals among 55-71°C has been investigated (4). The nature of polyol appeared to be very important for enzyme stability. The decrease of water activity in the storage medium lead to a thermostabilizing effect, but the different polyols showed different effects. There was no residual activity, whatever the temperature when the water activity was lower than 0.8. Only sorbitol showed a good protective effect. The effects of ethanol, glycerol, sucrose, sorbitol and polyethylene glycol on the stability of lysozyme as a function of storage medium temperature and additive concentration has also been investigated. The highest stabilisation effect was obtained with sorbitol and sucrose, mainly for high additive concentrations at high temperatures. Enzyme was also stabilised by glycerol except when concentrated to 80%. The other tested additives did not show enzyme thermostability. These results were shown that enzyme stability cannot be directly related to water activity (4). An increased stability of yeast alcoholdehydrogenase at different temperatures was observed by the addition of sorbitol to the storage medium. This result was obtained with the enzyme in its native and immobilised form (4). The stability of yeast alcoholdehydrogenase was also studied for the enzyme stored in the presence of different polyols having an OH/C ratio equal to one. The effects of arabitol, adonitol, xylitol, sorbitol and mannitol were compared (5). The half life of enzyme was increased with the addition of all these polyols except adonitol. The influence of two pairs of epimer polyols on enzyme stability (*xylitol-adonitol* and *sorbitol-mannitol*) was also compared. Sorbitol and mannitol were influenced in the same order of range. Xylitol exhibited a good protective effect on the enzyme, whereas adonitol destabilised the protein activity. On the other hand, an enhancement was reported on the storage stability of enzyme as a function of temperature with increase of sorbitol concentration in the storage medium (5).

The influence of various additives such as polyhydric alcohols, sugars, polyethylene glycol and substrates on the thermostability of *Bacillus licheniformis* α-amylase has recently been studied (6). Results obtained were shown a thermostabilising effect of most of the additives tested (ethylene glycol, glycerol, sorbitol, glucose, maltose, maltotriose, and polyethylene glycol). The stabilising effect appears to be strongly dependent on the additive concentration. The thermostability of α-amylase at 75°C was also studied in organic solvents such as octane, dioxane, dimethylsulphoxide (DMSO) and butanol. The stability was found to depend on the nature of solvent. An increase was observed in thermostability in the case of butanol and dioxane, however it was not as much as octane. Complete inactivation was observed in the presence of DMSO.

2.4. Addition of Polymers: The addition of proteins and stabilising agents for enzyme solutions is a well-known procedure. Ovalbumin, bovine albumin or gelatin have been used for the stabilisation of α-amylase, β-amylase and lactate dehydrogenase respectively. Enhanced activity of α-amylase in the presence of 2% dextrane 250 was also observed (1). On the other hand, an improved thermal stability of glucoseoxidase was reported after addition of 1% polyvinyl alcohol (PVA). The stabilising effect of polymers on glucoseoxidase was increased with increasing hydrophobicity (polyvinyl acetate > polyvinyl pyrrolidone-polyvinyl acetate > poly-vinyl pyrrolidone > polyvinyl pyrrolidone-polyvinyl alcohol) and was independent of the molecular weight of polymer. A considerable increase on the thermostability of glucosedehydrogenase from *Bacillus megaterium* IWG3 has been recently reported when polyethylene imine (PEI), a water soluble cationic polymer, was added to solutions of enzyme at a molar ratio of PEI to enzyme greater than 10 (7). The half life of enzyme was increased by about 8000 times that of the free enzyme with an increase in the molar ratio of PEI to glucosedehydrogenase.

3. Stabilisation of Soluble Enzymes by Chemical Modification: Chemically modified soluble enzymes show a number of potential advantages. They often exhibit higher catalytic activity and they may be more stable in liquid formulations, such as detergents. Chemical modifications of proteins have been extensively studied for many years but much less work has been devoted to the influence of chemical modification on enzyme stability. The stabilisation of soluble enzymes by chemical modifications will be discussed in the following three different sections (1): 1. Stabilisation by the modification of *key* functional groups in the tertiary structure of proteins, 2. Stabilisation by intramolecular cross-linking with bifunctional reagents, 3. Stabilisation by soluble conjugates of proteins with polysaccharides.

3.1. Stabilisation by the Modification of *Key* Functional Groups in the Tertiary Structure of Proteins: Modifications can be made either by acylation or alkylation of protein molecules (1).

3.1.1. Acylation of Protein Molecules: There is only a few reports in the literature concerned with the stabilisation of enzyme by acylation of protein molecules. In most cases, enhanced thermal stability was observed which was dependent on the amount of substitution and the type of acylating agent used. In a systematic investigation on the effects of various acylating agents on the thermostability of α-amylase, it has been reported that thermostability is increased in the following sequence: [*no substituent* ($-NH_2$) ≤ *dimethyl carbamoyl, ethoxy formyl* < *formyl* < *acetyl* < *dimethyl propanoyl*]. In addition, hydrophobicity of substituents is also of critical importance, a decrease in the thermal stability of acylated α-amylase has been reported when acyl groups have very high chain lengths such as palmitylation (1).

The thermostabilisation of α-chymotrypsin by acylation with succinic and acetic anhydrides has been reported (8). It has been determined that modification of the majority of titrated amino groups (approximately 80% or higher) caused a sharp increase in thermostability (by 120 times), but nevertheless, the complete substitution of all the titrated amino groups again leads to enzyme destabilisation.

3.1.2. Reductive Alkylation of Protein Molecules: Many simple aldehydes and ketones give rapid and reversible reactions with the amino groups of proteins, and as a result of these reactions the Schiff bases are formed. They are very stable compounds in dilute aqueous solutions, but extensive modification of protein amino groups can be obtained by reduction of the Schiff base to a stable secondary amine. Thus the reductive alkylation of protein molecules with many alkylating agents such as aldehydes and ketones under very mild conditions using sodium borohydride as reductant can be achieved. The α- and ε-amino groups of protein molecules are easily modified, but other common protein groups are not affected. Using a large variety of aldehydes and ketones, the size, the shape and the hydrophobicity of added substituents can easily be varied. By using formaldehyde and acetaldehyde, the added substitu-ents are small and generally have the least effect on physical and chemical properties of protein molecules. With larger and generally more hydrophobic carbonyl compounds greater changes can be expected. In many cases catalytic activities or other biological properties of proteins are unaffected or even enhanced by reductive alkylation. Methods of reductive alkylation has been described elsewhere (8,9,10).

A good example to the stabilisation of enzymes by reductive alkylation is the stabilisation of chmotrypsin (8,10). The enzyme was modified by treatment with acrolein followed by reduction of Schiff bases by $NaBH_4$. As a result, enzyme is retained almost all of its activity. The degree of stabilisation can be defined as the ratio of the first order rate constants of inactivation for the native enzyme as compared to the modified enzyme. With stabilisation this ratio should be increased. The alkylation of NH_2 groups that were the first to undergo modification did not affect the thermostability of the enzyme. However, alkylation of the less reactive or less accessible NH_2 groups caused a drastic increase in the stability (11). Then with complete modification of all NH_2 groups titratable with picrylsulfonic acid, thermostability decreased sharply.

3.2. Stabilisation of Enzymes by Intramolecular Cross-linking Using Bifunctional Reagents: The reinforcement of the compact tertiary structure of protein and enzyme molecules with cross-linking agents is another approach for their stabilisation against inactivation. Chemical cross links will reticulate the molecule and decrease the rate of inactivation. The fixation of enzyme globule can be achieved by both intra- and intermolecular cross linkings (12). Chemical cross linking can be most easily achieved with bifunctional reagents. These chemicals react with nucleophilic side chain of amino acids, such as the sulphydryl group of cysteine, the amino groups of lysine and N-terminal amino acids, the carboxyl groups of aspartic and glutamic acids and the C-terminal amino acids, the imidazolyl group of histidine, and the thio ether group of methionine (12). The specificity of cross linkers for a specific amino acid side chain depends on the relative reactivity of the nucleophile. In general, the thiolate ion is the most nucleophilic. Thus sulphydryl groups of proteins will react with most of the cross linkers at alkaline pH (12). If the free sulphydryl groups are not available in protein molecule, the amino group becomes the major target of reaction (12). The bifunctional cross linking reagents can be classified into three groups. as follows: 1. Zero-length bifunctional cross linkers, 2. Homobifunctional cross linkers, 3. Heterobifunctional cross linkers.

3.2.1. The Zero-length Cross linkers: These are the cross linkers that provide the direct cross linking of two protein molecules without the introduction of any spacer group. The zero-length cross linkers such as carbodiimides, isoxazolium derivatives, chloroformates and carbonyldiimidazole links carboxyl and amino groups to each other. The cross linkers such as cupric di (1,10-phenanthroline) and 2,2'-dipyridyldisulphide makes disulphide bonds (12). These reagents react by activating one of the molecules to form an active intermediate, and then react with a second component.

3.2.2. Homobifunctional Cross linkers: These cross linkers contain two reactive functionalities that will react with amino acid side chains and bringing two components together. The reactive groups are located at the two ends of molecule. When a cross linker contains two identical reactive groups it is referred to as a homofunctional reagent. Homofunctional cross linkers can be considered under two different sections as diacylating and dialkylating cross linkers (12).

3.2.2.1. Diacylating Homobifunctional Cross linkers: These cross linkers include bis-imidoesters, bis-succinimidyl derivatives, di-isocyanates, di-isothiocyanates, di-sulphonyl halides, bis-nitrophenylesters, dialdehydes and diacylazides. Some examples of diacylating homobifunctional reagents are disuccinimidyl suberate, dimethyl maloimidate, 1,4-dicyanato-benzene and p-phenylene-diisothiocyanate. Glutaraldehyde is the most extensively used dialdehyde for cross linking. Diacylating agents are generally amino group selective cross-linkers (12).

3.2.2.2. Dialkylating Homobifunctional Cross linkers: These cross linkers include bis-maleimides, bis-haloacetyl derivatives, di-alkyl halides and bis-oxiranes. Some examples of dialkylating homobifunctional reagents are N,N'-methylenebismaleimide, α,α'-diiodo-p-xylene sulfonic acid, di(2-chloroethyl)sulfone and bis(3-nitro-4-fluorophenylsulfone. Dialkyl-ating agents are generally sulphydryl-specific cross linkers (12).

3.2.3. Heterobifunctional Cross linkers: These cross linkers have two different reactive groups that may act as a combination of acylating and alkylating agents (12). For instance, one end of the cross linker may be a sulphydryl group-specific and other end may be an acylating agent for amino group selectivity.

3.2.3.1. Photoactivable Heterobifunctional Cross linkers: These cross linkers are completely inert in the dark. Photoactivable groups can be classified into the categories of precursors of either *nitrenes* or *carbenes*. Nitrenes are generated upon heating of photolysis of azides. Electron-deficient nitrenes are very reactive compounds and therefore, they are considered to be nonspecific reagents. They can potentially react with a variety of chemical bonds including N-H, O-H, C-H and C=C (13). Some photoactivable cross-linking reagents consists of diazocompounds that form an electron-deficient carbene upon photolysis. These cabenes undergo a variety of reactions including insertion into C-H bonds, addition of double bonds, hydrogen abstraction and coordination to nucleophilic centres to give carbon ions (13). These compounds can be used for reactions with relatively inert amino acid side chains. Some examples to photoactivable bifunctional reagents are N-succinimidyl-5-azido-2-nitro-benzoate and 1-azido-5-naphthaleneisothiocyanate.

3.2.3.2. Chemically Activable Reagents: These reagents convert a non reactive group to a highly reactive intermediate like the zero-length cross linker (12). The activation of carboxyl groups with carbodiimide and N-hydroxysuccinimide to form reactive ester can be given as an example to these reagents. Reactive ester can easily react with the amino groups of protein molecules. *Cis*-vicinal diols can also be oxidized with periodate to dialdehydes which form Schiff bases with amino groups (12). The Schiff bases may be stabilized with reducing agents in the process referred to as reductive alkylation. This method is particularly useful for cross linking glycoproteins.

Spaces may be introduced between the cross linked components during the activation process. Diamines and dicarboxylates of different chain lengths can be linked to carboxyl and amino groups, respectively with carbodiimides (12). The success or failure of the enzyme stabilisation procedure by using intramolecular cross linking technique may depend on the length of bifunctional molecule corresponds to the distance between the possible centres of attachment on the protein globule. Consequently, for any protein, there is an optimal size of cross linker and this size must be determined (11). The effect of intramolecular cross-linkages of different length on the thermostability of α-chymotrypsin has been reported (11,14). The carboxy groups of the enzyme were first activated by 1-ethyl-3-(3-di-methylaminopropyl) carbodiimide; then the activated protein was treated by diamines of $H_2N(CH_2)_nNH_2$ type with n ranging from 0 to 12. Fig. The dependence of the first order rate constant for thermoinactivation of chymotrypsin cross linked by various diamines have shown a sharp minimum at $n = 4$, although at $n = 2,5$ and 6 stabilisation of enzyme was also occurred. The reason for obtaining of greatest stabilisation by tetramethylenediamine treatment may be due to the fact that this reagent has the best length fitting to the distance between carboxy groups of the protein molecule (11,14).

It may be expected that the existence of more cross links in a protein molecule tends it to be more stable against unfolding and hence against inactivation. The number of cross-linkages is determined by the quantity and relative position of functional groups on the surface of the protein molecule that interact with the cross linking agent. This quantity may be increased by premodifying the protein in a certain way. The modification of α-chymotrypsin by succinic anhydride to increase the number of carboxyl groups in its surface layers has been reported (11,14). The succinylated chymotrypsin was treated with diamines of different length. Maximum stabilisation effect was 7 higher than for non succinylated enzyme, and maximum stabilizing effect was produced not by tetramethylenediamine but rather by a shorter bifunctional reagent, ethylenediamine. This suggests a greater surface concentration of carboxyl groups in the succinylated chymotrypsin molecule as compared to native chymotrypsin. Thus, the premodifica-tion of the enzyme makes possible regulation of stabilization effect both with respect to the degree and the optimal length of the cross-linkage (11,14).

3.2.4. Some Applications of Bifunctional Reagents to Protein Stability: Thermal inactivation kinetics of native and glutaraldehyde cross-linked forms of penicillin G acylase obtained from a mutant of *Escherichia coli* ATCC 11105 were recently reported (15). Apparent activation energies of thermal inactivation of both native and cross-linked forms of enzyme were calculated to be [57.71 ± 8.46] and [67.11 ± 13.83] kcal mol^{-1} respectively. This slight increase in activation energy suggested that glutaraldehyde cross linking did not markedly protect against thermal inactivation. Cross linked enzyme did, however, have a significant improved half life at temperatures between 40°C and 50°C.

Another example is the stabilisation of restriction endonuclease *Bam* HI from *Bacillus amyloliquefaciens* H several cross-linking agents such as glutaraldehyde (GA), dimethyl-adipimate (DMA), dimethylsuberimidate (DMS), and dimethyl3,3'-dithio-bis-propionimidate (DTBP) (16). Reaction with GA did not show any enhanced thermostability. However, the DMA-, DMS- and DTBP-cross linked preparations of *Bam* HI showed significant improvement in thermal stability. The half lives of these preparations at 35°C were 4.0, 5.25, and 5.5 in respectively, whereas the native enzyme showed a half life of 1.2 h only. The apparent values of inactivation energies for denaturation of native, DMA-, DMS- and DTBP-cross linked *Bam* HI was 2.63, 5.24, 6.55 and 9.2 kcal mol^{-1}, respectively. The DTBP- cross linked *Bam* HI was, therefore, the best themostable preparation among those tested.

3.3. Covalent Attachment of Enzymes to Polysaccharides: Some proteins contain carbohydrates linked to *surface* amino acids.Glycoproteins often exhibit increased stability towards proteolysis, heat, storage or denaturants which in many cases seems to be due to the carbohydrate part of the molecule. In addition, most glycoproteins exhibit high water solubility. Thus it may be considered to obtain stabilized water-soluble enzyme-polymer conjugates by covalent attachment of enzymes to activated polysaccharides or carbohydrates (3). Various methods and procedures have been reviewed related with the binding of enzymes to activated polysaccharides (17,18). Wykes et al (19) are prepared water-soluble enzyme polymer derivatives of α-amylase from soluble dextrane, DEAE-dextrane and CM-cellulose. The water-soluble CM-cellulose derivative is showed enhanced thermal stability.

Continuous hydrolysis of casein in an ultra filtration membrane reactor by means of dextrane-conjugated tyrpsin was carried out by O'neill et al (20). 42% residual activity of poly-aldehyde-dextrane trypsin conjugate after 34 days at 30°C has been observed, while native trypsin had lost all its activity in three days. 99% residual activity of a lysozyme conjugate after 30 min at 100°C has been reported, whereas native lysozyme had lost 80% of its activity under these conditions (1).

In natural glycoproteins each carbohydrate chain is attached to the polypeptide backbone through a single linkage, whereas synthetic polysaccharide-protein conjugates may contain additional intramolecular cross-links. These additional linkages provides additional rigidity and consequently additional stability to the native conformation. Polysaccharide-protein conjugates are generally more resistant to the action of proteolytic enzymes. One reason for this may be the reduction of the number of lysine residues on protein molecules by glycosidation procedures. Thus, proteolysis by proteases with lysine specificity such as trypsin is strongly reduced (1).

4. Stabilisation of Enzymes by Multipoint Attachment to Solid Supports

4.1. Multipoint Covalent Attachment of the Enzyme to the Complementary Surface of the Support: In generals, the multipoint binding of enzyme molecules to the solid supports is make the conformation of the molecule more rigid and hence more stable against unfolding and inactivation. Even when a multipoint binding is realized, it is only a small part of the enzyme molecule surface that is bound the support, hence one should not expect rigidity over the whole macromolecule. The problem is to provide a support with a surface complementary to that of the enzyme molecules. Only then can one expect a multipoint binding resulting in thermostabilisation of the enzymes (21). One approach to the preparation of a complementary polymeric support is based on modifying the enzyme by an analogue of the monomer and then copolymerizing the resulting preparation with the monomer (11,21).

This gives an enzyme chemically incorporated into the three dimensional lattices of the polymer gel, the points of the enzyme-support binding being the centres of pre modification of the enzyme molecule. As an Analogue of the monomer amino group modifiers can be used. This will provide a multipoint binding of the monomer analogue with the enzyme molecule. Some examples include acryloyl chloride and methacryloyl chloride that acylates NH_2 and OH groups of proteins, and acrolein that reacts with NH_2 groups with subsequent reaction of Schiff's bases by sodium borohydride to give alkylated proteins. Acrylamide, sodium-methacrylate, and 2-hydroxyethyl-methacrylate can be used as comonomers and N,N'-methylenebisacrylamide as cross linking agents (11,21).

The scale of the thermostability of immobilized chymotrypsin was defined by the comparison of free and immobilized enzymes inactivation rates in Arrhenius coordinates (21). At 60°C acryloylated chymotrypsin entrapped in polymethacrylate and polyacrylamide gels in 1000 and 200 times as stable as the free native enzyme, respectively. It was found that as the number of acryloylated amino groups of the immobilized enzyme increases or in other words the number of covalent linkage between protein molecule and the polymer support increases enzyme becomes more and more thermostable (21). For instance, the effective inactivation rate constant of α-chymotrypsin at 60°C decreases more than a 1000 fold if the number of cross linkages between the enzyme and polymethacrylate is increased by about 75% (21). The increased thermostability of chymotrypsin cannot be explained by the chemically modification of enzyme. Because as it was shown that acryloyl cchymotrypsin is even less stable than the native enzyme regardless of the degree of modification. The nature of the gel is also not the reason of the enzyme stabilisation. Increased in the thermostability was observed if the enzyme chemically entrapped in polylectolyte polymethacrylate or electroneutral polyacrylamide gels (21). The critical factor for enzyme stabilisation observed is the chemical entrapment of the enzyme molecule in the three dimensional structures of the support. When the protein is fixed, the unfolding of the protein molecule on heating is hindered. Thereby, native, catalytically active conformation of the active centre may only be maintained if the enzyme-support interaction is multipoint (21).

4.2. Multipoint Non covalent Interaction of Enzyme With Support: Multipoint interaction of an enzyme with support can also provide the stabilization of enzyme when it has been immobilized by physical methods such as entrapment in gels, adsorption on supports ect. (22,23,24). The thermal stability of chymotypsin and tyrpsine mechanically entrapped in methacrylate and polyacrylamide gels have been studied by Martinek et al (25). Thermostability of entrapped enzyme is higher than enzyme in aqueous solution. At the gel concentration of 0 - 30% (w/w), the rate of thermo-inactivation of chymotrypsin at 60°C is almost constant, but drops sharply as the concentration of gel is raised. In 50% (w/w) gel concentration, 10^5 times stabilisation was observed (25).

The possible mechanism of stabilisation of enzymes mechanically entrapped into polymer matrices may be due to the hindrance of rotational and transitional movements of enzyme molecule within matrix. Enzyme-support interactions in concentrated gels should be multipoint and because of the steric reasons polymer chains can be adhered to the enzyme molecule from all sides. These bonds are the electrostatic and hydrogen bonds and they are relatively weak, however, such multipoint interaction may lead to sharp increase in the thermostability of immobilized enzyme (11,25).

5. Stabilisation of Enzymes Against Inactivation Under the Action of Extreme pH Values: Many enzymes rapidly and irreversibly loose their catalytic activity at pH values that far from the optimal pH range of enzymatic reaction. However, when the enzyme is immobilized its pH stability can be changed due to the alteration of the micro environment of enzyme or the alteration of the conformation of enzyme as result of immobilisation.

5.1. Ridification of Native Enzyme Conformation: pH inactivation of enzymes takes place by unfolding of protein molecule as result of the alteration of the balance of electrostatic and hydrogen bonds. The change on hydrogen bonds occurs with the alteration of the ionization state of ionogenic groups on proteins induced by pH. Consequently all the methods for thermostabilisation of enzymes discussed above should be valid for the stabilisation of enzymes against pH.

The pH dependence of the rate of thermoinactivation of free and mechanically entrapped glucose oxidase (11) has been investigated. The inactivation rate of free enzyme at 56°C has increased sharply with the increase of pH. Entrapment by copolymerization of ethylmethacrylate and methacrylic acid caused to considerable increase of the thermostability of glucose oxidase. The rate of thermoinactivation of entrapped enzyme at 65°C was nearly insensitive to pH changes as result of immobilisation (11). this may be explained by the formation of hydrogen and electrostatic bonds between enzyme molecule and polymeric support.

5.2. Shift of the pH Profile of Enzyme Inactivation Rate: On immobilisation of enzymes on polyanions, the pH profile of the enzyme activity shifts towards the alkaline region, whereas on immobilization on polycations the shift is toward the acidic region. The reason of this may be due to the higher and lower concentrations of hydroxonium ions in the support phase as compared to the solution phase for the former and latter immobilizations respectively. The shift of the pH profile on the catalytic activity of enzymes immobilized on electroneutral support should also take place because protons are distributed between the support and the medium in a certain specific way (11). The similar effects should take place during enzyme inactivation. Any difference in the pH of the solution and the support phases should induce a shift in the apparent pH profile of the inactivation (11). It was demonstrated that enzymes immobilized in a polyanionic support have a higher stability in alkaline pH; whereas enzymes immobilized in polycationic supports become more stable in acidic solutions. The pH profile of the inactivation rate constants of native and immobilized penicillin acylase shows that enzyme inactivation shift depending on the nature of the support, in complete agreement with the theory (11).

6. Enzymatic Reactions in Organic Solvents With a Low Content of Water: Many chemical reactions can thermodynamically favour the synthesis of a desired end product only in certain organic solvents. This is related with the specific solvation effects and the solubility of the individual components of the reaction. Sometimes in addition to the required product, water is formed, and in aqueous solutions the equilibrium is shifted towards the starting substances (11,26). To overcome this difficulty, water should be replaced by a non aqueous solvent as a reaction medium. However, when water is replaced by an organic solvent as a reaction medium, the catalytic activity of enzymes may dramatically decreases or their substrate specificity may lose or complete denaturation of enzyme may take place. A practical solution of this problem is to use an organic component that produces the lowest denaturing effect on enzyme (11,26). However, two other approaches may also make for to overcome enzyme inactivation. These are: 1. The use of biphasic aqueous organic systems, such as water:water-immiscible organic solvent, 2. Entrapment of enzymes in *reversed* surfactant micelles.

6.1. Enzymatic reactions in biphasic systems: In this systems enzyme will stay only in aqueous phase. Most of the enzymes are easily dissolve in water but they are almost insoluble in hydrophobic solvents. The substrates will stay in organic phase that consisted by solvents immiscible with water such as chloroform, ether, hydrocarbons. Thus substrates can freely diffuse to water phase, where they will undergo a chemical conversion; and the products will diffuse back to the organic phase (11). Due to the absence of contact between enzyme and non aqueous component of reaction mixture the problem of enzyme stabilization is eliminated in biphasic systems. In addition the existence of non aqueous component in the medium at high content may provide solubility of hydrophobic reagents such as steroids that are sufficient for preparative synthesis (11). The enzymatic system water:water-immiscible organic solvent may be prepared as an emulsion of an aqueous solution of the enzyme in an organic medium. It can also be prepared by suspending the porous particles impregnated by an aqueous solution of the enzyme in an organic medium (11).

6.2. Catalytic by water-soluble enzymes entrapped in *reversed* micelles of surfactants in organic solvents: The possibility of stabilizing water-soluble enzymes against the inactivating action of organic solvents by means of surfactants has been reported by Martinek et al (26). Several enzymes such as α-chymotrypsin, trypsin, pyrophosphatase, peroxidase, lactate dehydrogenase, and pyruvate kinase were used to demonstrate that enzymes can be entrapped into reversed micelles formed by surfactants such as di(2-ethyl)ester of sulfosuccinic acid sodium salt (Aerosol OT), cetyltrimethylammonium bromide in an organic solvent like benzene, chloroform, octane, cyclohexane. The enzymes solubilized in this way retain their catalytic activity and substrate specificity (26).

When enzyme molecule is entrapped in a reversed micelle, it is protected against denaturation because of the formation of an interface between protein molecule and the organic solvent phase that is stabilized by the surfactant molecules. As a result, enzyme molecule cannot come into direct contact with the unfavourable organic medium (11). The research work on the oxidation of pryogallol by peroxidase in reversed micelles was showed that, when water is replaced by a medium consisting a reversed micelle in octane, the kinetic mechanism of catalysis changes, because substrate inhibition that appear during the reactions in water is disappear in reaction medium consisted of reversed micelles in octane (11).

7. Stabilisation of Enzymes by Antibodies: Recent development in molecular biology is made possible the modification of the surface of biologically active proteins by speci-fic antibodies. Some antibodies may lead to covalent and non covalent interactions at sites whe-re protein unfolding is initiated, or where proteolysis occurs, thereby stabilising the protein. Shami et al (27) were reported the stabilisation of α-amylase, glucoamylase, and subtilisin by polyclonal antibodies and L-asparaginase by selected monoclonal antibodies. The principle of work to prepare the enzyme-antibody complexes with increasing concentrations of specific antibodies. Then the complexed and the free enzyme was pre incubated for different times, exposed to various inactivating conditions. Stability was assessed by determining the residual activity of the enzyme. The reason of stabilisation could be explained as follows (27): a) The active site of the enzyme used may be located in a *blind* spot for the immune system, b) The inhibitory antibodies may have lower affinity and quantity and their binding to the active site of enzyme may be hindered by high affinity antibodies. The thermostability of enzyme-antibody complexes was much higher than that of the free enzymes. Activity half-lives were as much as 200 times as long. L-Asparaginase is lost 98% of its activity in 50 min when exposed to pH:3.0, while antibody-protected enzyme retained close to 40% activity at 50 min 25% for the next 17 hours. Subtilisin is retained only 25% of its original activity when exposed to 0.05% NaOCl for 30 min, while the antibody-protected enzyme retained close to 80% of its activity. Glucoamylase is retained only 10% of its activity when pre incubated with 2.5% alcohol for 25 hours, while antibody protected enzyme retained 98% of its activity (27). Monoclonal antibody-L-asparaginase complex was shown excellent stability against biological inactivation by proteolysis. The protected trypsin-treated enzyme was virtually as active as the trypsin-untreated control, whereas trypsin-treated unprotected enzyme lost most of its activity (27).

These examples show that stabilisation by antibodies might be afforded to any protein. However, the production of monoclonal antibodies via mammalian cell culture is more costly by at least two orders of magnitude than *target* protein production in microorganisms. The production of monoclonal antibodies in microorganisms could be the solution of this problem.

References

1. Schmid, R.D., Stabilized soluble enzymes, Adv. Biochem. Eng., **12**, 41, (1979).

2. Klibanov, A.M., Stabilization of enzymes against thermal inactivation., Adv. Appl. Microbiol., **29**, 1, (1983).

3. Zhang, J., Demain, A.L., In vitro stabilization of ACV synthetase activity from *Streptomyces clavuligerus*, **37**, 97, (1992).

4. Larreta-Garde, V., Xu, Z.F., Biton, J., Thomas, D., Stability of enzymes in low water activity media., "Biocatalysis in Organic Media", (J. Tramper & M.D. Lilly Eds), Proc. Int. Symp., Wageningen, The Netherlands, pp.247-252, (1986).

5. Larreta-Garde, V., Xu, Z.F., Thomas, D., Behavior of enzymes in the presence of additives, Enzyme Engineering, (H.W. Blanch & A.M. Klibanov Eds), **9**, pp.294-298, (1988).

6. Ashter, M., Meunier, J.C., Increased thermal stability of *Bacillus licheniformis* α-amylase in the presence of various additives, Enzyme Microb. Technol., **12**, 902, (1990).

7. Teramoto, M., Nishibue, H., Okuhara, K., Ogawa, H., Kozono, H., Matsuyama, H., Kajiwara, K., Effect of addition of polyethyleneimine of thermal stability and activity of glucose dehydrogenase., Appl. Microbiol. Biotechnol., **38**, 203, (1992).

8. Torchillin, V.P., Maksimenko, A.V., Smirnov, V.N., Berezin, I.V., Klibanov, A.M., Martinek, K., The principles of enzyme stabilization, IV. Modification of 'key' functional groups in the tertiary structure of proteins., Biochim. Biophys. Acta, **567**, 1, (1979).

9. Means, G.E., Reductive alkylations of amino groups., in "Methods in Enzymology", (C.H.W. Hirs & S.N. Timasheff Eds), Academic Press, New York, **Vol:57**, pp.469-479, (1977).

10. Jenfont, N., Dearborn, D.G., Protein labelling by reductive alkylation., in "Methods in Enzymology", (C.H.W. Hirs & S.N. Timasheff Eds), Academic Press, New York, **Vol:91**, pp.570-579, (1983).

11. Martinek, K., Mozhaev, V.V., Berezin, I.V., Stabilization and reactivation of enzymes, in "Enzyme Engineering - Future Directions", (L.B. Wingard, I.V. Berezin & A.A. Klyosov Eds), Plenum Press, New York, pp.3-54, (1980).

12. Wong, S.S., Wong, L.C., Chemical cross linking and the stabilization of proteins and enzymes., Enzyme Microb. Technol., **14**, 866, (1992).

13. Ji, T.H., Bifunctional reagents., in "Methods in Enzymology", (C.H.W. Hirs & S.N. Timasheff Eds), Academic Press, New York, **Vol:91**, pp.580-609, (1983).

14. Torchillin, V.P., Maksimenko, A.V., Smirnov, V.N., Berezin, I.V., Klibanov, A.M., Martinek, K., The principles of enzyme stabilization, III. The effect of the length of intra-molecular cross-linkages on thermostability of enzymes, Biochim. Biophys. Acta, **522**, 277, (1983).

15. Erarslan, A., Koçer, H., Thermal inactivation kinetics of penicillin G acylase obtain-ed from a mutant derivative of *Escherichia coli* ATCC 11105., J. Chem. Tech. Biotechnol., **55**, 79, (1992).

16. Dubey, A.K., Bisaria, V.S., Mukhopadyay, S.N., Ghose, T.K., Stabilization of restriction endonuclease *Bam* HI by cross-linking reagents., Biotechnol. Bioeng., **33**, 1311, (1989).

17. Porath, J., Axén, R., Immobilization of enzymes to agar, agarose and sephadex supports, in "Methods in Enzymology", (Ed. K. Mosbach), Academic Press, New York, **Vol:44**, pp.19-46, (1976).

18. Zaborsky, O.R., Immobilized Enzymes, CRC Press, (1973).

19. Wykes, J.R., Dunnill, P., Lilly, M.D., Immobilization of α-amylase by attachment to soluble support materials, Biochim. Biophys. Acta, **250**, 522, (1971).

20. O'Neill, S.P., Wykes, J.R., Dunnill, P., Lilly, M.D., An ultra filtration-reactor system using a soluble immobilized enzyme, Biotechnol. Bioeng., **13**, 319, (1971).

21. Martinek, K., Klibanov, A.M., Goldmacher, V.S., Berezin, I.V., The principles of enzyme stabilization, I. Increase in thermostability of enzymes covalently bound to a complementary surface of a polymer support in a multipoint fashion, Biochim. Biophys. Acta, **485**, 1, (1977).

22. Mosbach, K., Methods in Enzmology, **Vol:44**, Academic Press, New York, (1976).

23. O'driscoll, K.F., Preparation and properties of gel entrapped enzymes., Adv. Biochem. Eng., **4**, 155, (1976).

24. Woodward, J., Immobilized Cells and Enzymes - A Practical Approach, IRL Press, Oxford, (1985).

25. Martinek, K., Klibanov, A.M., Goldmacher, V.S., Tchernsheva, A.V., Mozhaev, V.V., Berezin, I.V., Glotov, B.O., The principles of enzyme stabilization, II. Increase in the thermostability of enzymes as a result of multipoint non covalent interaction with a polymeric support, Biochim. Biophys. Acta, **485**, 13, (1977).

26. Martinek, K., Levashov, A.V., Klyachko, N.L., Pantin, V.I., Berezin, I.V., The principles of enzyme stabilization, VI. Catalysis by water-soluble enzymes entrapped into reversed micelles of surfactants in organic solvents, Biochim. Biophys. Acta, **657**, 277, (1981).

27. Shami, E.Y., Rothstein, A., Ramjeesingh, M., Stabilization of biologically active proteins, Tib Tech, 7,186, (1989).

INDEX

NATO ASI Series H

Vol. 21: **Cellular and Molecular Basis of Synaptic Transmission.** Edited by H. Zimmermann. 547 pages. 1988.

Vol. 22: **Neural Development and Regeneration. Cellular and Molecular Aspects.** Edited by A. Gorio, J. R. Perez-Polo, J. de Vellis, and B. Haber. 711 pages. 1988.

Vol. 23: **The Semiotics of Cellular Communication in the Immune System.** Edited by E.E. Sercarz, F. Celada, N.A. Mitchison, and T. Tada. 326 pages. 1988.

Vol. 24: **Bacteria, Complement and the Phagocytic Cell.** Edited by F. C. Cabello und C. Pruzzo. 372 pages. 1988.

Vol. 25: **Nicotinic Acetylcholine Receptors in the Nervous System.** Edited by F. Clementi, C. Gotti, and E. Sher. 424 pages. 1988.

Vol. 26: **Cell to Cell Signals in Mammalian Development.** Edited by S.W. de Laat, J.G. Bluemink, and C.L. Mummery. 322 pages. 1989.

Vol. 27: **Phytotoxins and Plant Pathogenesis.** Edited by A. Graniti, R. D. Durbin, and A. Ballio. 508 pages. 1989.

Vol. 28: **Vascular Wilt Diseases of Plants. Basic Studies and Control.** Edited by E. C. Tjamos and C. H. Beckman. 590 pages. 1989.

Vol. 29: **Receptors, Membrane Transport and Signal Transduction.** Edited by A. E. Evangelopoulos, J. P. Changeux, L. Packer, T. G. Sotiroudis, and K.W.A. Wirtz. 387 pages. 1989.

Vol. 30: **Effects of Mineral Dusts on Cells.** Edited by B.T. Mossman and R.O. Begin. 470 pages. 1989.

Vol. 31: **Neurobiology of the Inner Retina.** Edited by R. Weiler and N.N. Osborne. 529 pages. 1989.

Vol. 32: **Molecular Biology of Neuroreceptors and Ion Channels.** Edited by A. Maelicke. 675 pages. 1989.

Vol. 33: **Regulatory Mechanisms of Neuron to Vessel Communication in Brain.** Edited by F. Battaini, S. Govoni, M.S. Magnoni, and M. Trabucchi. 416 pages. 1989.

Vol. 34: **Vectors asTools for the Study of Normal and Abnormal Growth and Differentiation.** Edited by H. Lother, R. Dernick, and W. Ostertag. 477 pages. 1989.

Vol. 35: **Cell Separation in Plants: Physiology, Biochemistry and Molecular Biology.** Edited by D. J. Osborne and M. B. Jackson. 449 pages. 1989.

Vol. 36: **Signal Molecules in Plants and Plant-Microbe Interactions.** Edited by B.J.J. Lugtenberg. 425 pages. 1989.

Vol. 37: **Tin-Based Antitumour Drugs.** Edited by M. Gielen. 226 pages. 1990.

Vol. 38: **The Molecular Biology of Autoimmune Disease.** Edited by A.G. Demaine, J-P. Banga, and A.M. McGregor. 404 pages. 1990.

NATO ASI Series H

Vol. 39: **Chemosensory Information Processing.**
Edited by D. Schild. 403 pages. 1990.

Vol. 40: **Dynamics and Biogenesis of Membranes.**
Edited by J. A. F. Op den Kamp. 367 pages. 1990.

Vol. 41: **Recognition and Response in Plant-Virus Interactions.**
Edited by R. S. S. Fraser. 467 pages. 1990.

Vol. 42: **Biomechanics of Active Movement and Deformation of Cells.**
Edited by N. Akkas. 524 pages. 1990.

Vol. 43: **Cellular and Molecular Biology of Myelination.**
Edited by G. Jeserich, H. H. Althaus, and T. V. Waehneldt. 565 pages. 1990.

Vol. 44: **Activation and Desensitization of Transducing Pathways.**
Edited by T. M. Konijn, M. D. Houslay, and P. J. M. Van Haastert. 336 pages. 1990.

Vol. 45: **Mechanism of Fertilization: Plants to Humans.**
Edited by B. Dale. 710 pages. 1990.

Vol .46: **Parallels in Cell to Cell Junctions in Plants and Animals.**
Edited by A. W Robards, W. J . Lucas, J . D. Pitts, H . J . Jongsma, and D. C. Spray. 296 pages. 1990.

Vol. 47: **Signal Perception and Transduction in Higher Plants.**
Edited by R. Ranjeva and A. M. Boudet. 357 pages. 1990.

Vol. 48: **Calcium Transport and Intracellular Calcium Homeostasis.**
Edited by D. Pansu and F. Bronner. 456 pages. 1990.

Vol. 49: **Post-Transcriptional Control of Gene Expression.**
Edited by J. E. G. McCarthy and M. F. Tuite. 671 pages. 1990.

Vol. 50: **Phytochrome Properties and Biological Action.**
Edited by B. Thomas and C. B. Johnson. 337 pages. 1991.

Vol. 51: **Cell to Cell Signals in Plants and Animals.**
Edited by V. Neuhoff and J. Friend. 404 pages. 1991.

Vol. 52: **Biological Signal Transduction.**
Edited by E. M . Ross and K . W. A. Wirtz. 560 pages. 1991.

Vol. 53: **Fungal Cell Wall and Immune Response.**
Edited by J. P. Latge and D. Boucias. 472 pages. 1991.

Vol. 54: **The Early Effects of Radiation on DNA.**
Edited by E. M. Fielden and P. O'Neill. 448 pages. 1991.

Vol. 55: **The Translational Apparatus of Photosynthetic Organelles.**
Edited by R. Mache, E. Stutz, and A. R. Subramanian. 260 pages. 1991.

Vol. 56: **Cellular Regulation by Protein Phosphorylation.**
Edited by L. M. G. Heilmeyer, Jr. 520 pages. 1991.

NATO ASI Series H

Vol. 57: **Molecular Techniques in Taxonomy.**
Edited by G . M . Hewitt, A. W. B. Johnston, and J. P. W. Young .
420 pages. 1991.

Vol. 58: **Neurocytochemical Methods.**
Edited by A. Calas and D. Eugene. 352 pages. 1991.

Vol. 59: **Molecular Evolution of the Major Histocompatibility Complex.**
Edited by J. Klein and D. Klein. 522 pages. 1991.

Vol. 60: **Intracellular Regulation of Ion Channels.**
Edited by M. Morad and Z. Agus. 261 pages. 1992.

Vol. 61: **Prader-Willi Syndrome and Other Chromosome 15q Deletion Disorders.**
Edited by S. B. Cassidy. 277 pages. 1992.

Vol. 62: **Endocytosis. From Cell Biology to Health, Disease and Therapie.**
Edited by P. J. Courtoy. 547 pages. 1992.

Vol. 63: **Dynamics of Membrane Assembly.**
Edited by J. A. F. Op den Kamp. 402 pages. 1992.

Vol. 64: **Mechanics of Swelling. From Clays to Living Cells and Tissues.**
Edited by T. K. Karalis. 802 pages. 1992.

Vol. 65: **Bacteriocins, Microcins and Lantibiotics.**
Edited by R. James, C. Lazdunski, and F. Pattus. 530 pages. 1992.

Vol. 66: **Theoretical and Experimental Insights into Immunology.**
Edited by A. S. Perelson and G. Weisbuch. 497 pages. 1992.

Vol. 67: **Flow Cytometry. New Developments.**
Edited by A. Jacquemin-Sablon. 1993.

Vol. 68: **Biomarkers. Research and Application in the Assessment of Environmental Health.** Edited by D. B. Peakall and L. R. Shugart.
138 pages. 1993.

Vol. 69: **Molecular Biology and its Application to Medical Mycology.**
Edited by B. Maresca, G. S. Kobayashi, and H. Yamaguchi. 271 pages.
1993.

Vol. 70: **Phospholipids and Signal Transmission.**
Edited by R. Massarelli, L. A. Horrocks, J. N. Kanfer, and K. Löffelholz.
448 pages. 1993.

Vol. 71: **Protein Synthesis and Targeting in Yeast.**
Edited by A. J. P. Brown, M. F. Tuite, and J. E. G. McCarthy. 425 pages.
1993.

Vol. 72: **Chromosome Segregation and Aneuploidy.**
Edited by B. K. Vig. 425 pages. 1993.

Vol. 73: **Human Apolipoprotein Mutants III. In Diagnosis and Treatment.**
Edited by C. R. Sirtori, G. Franceschini, B. H. Brewer Jr. 302 pages. 1993.

NATO ASI Series H

NATO ASI Series H

Vol. 91: Trafficking of Intracellular Membranes:
From Molecular Sorting to Membrane Fusion.
Edited by M. C. Pedroso de Lima, N. Düzgüneş, D. Hoekstra.
371 pages. 1995.

Vol. 92: Signalling Mechanisms – from Transcription Factors to Oxidative Stress.
Edited by L. Packer, K. Wirtz. 466 pages. 1995.

Vol. 93: Modulation of Cellular Responses in Toxicity.
Edited by C. L. Galli, A. M. Goldberg, M. Marinovich.
379 pages. 1995.

Vol. 94: Gene Technology.
Edited by A. R. Zander, W. Ostertag, B. V. Afanasiev, F. Grosveld.
556 pages. 1996.

Vol. 95: Flow and Image Cytometry.
Edited by A. Jacquemin-Sablon. 241 pages. 1996.

Vol. 96: Molecular Dynamics of Biomembranes.
Edited by J. A. F. Op den Kamp. 424 pages. 1996.

Vol. 97: Post-transcriptional Control of Gene Expression.
Edited by O. Resnekov, A. von Gabain. 283 pages. 1996.

Vol. 98: Lactic Acid Bacteria: Current Advances in Metabolism, Genetics and Applications.
Edited by T. F. Bozoğlu, B. Ray. 412 pages. 1996.